D1000661

96 Springer Series in Solid-State Sciences

Edited by Manuel Cardona

Springer
Berlin
Heidelberg
New York
Barcelona
Budapest
Hong Kong
London
Milan
Paris
Santa Clara
Singapore
Tokyo

Springer Series in Solid-State Sciences

Editors: M. Cardona P. Fulde K. von Klitzing H.-J. Queisser

Managing Editor: H. K.V. Lotsch   Volumes 1–89 are listed at the end of the book

Akira Isihara

Electron Liquids

With 116 Figures

Second Edition

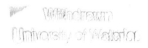

Springer

Professor Akira Isihara, Ph. D.

JIDECO of Bardstone, Inc., 901 Withrow Court, Bardstown, KY 40004, USA

Series Editors:

Professor Dr., Dres. h. c. Manuel Cardona
Professor Dr., Dres. h. c. Peter Fulde*
Professor Dr., Dres. h. c. Klaus von Klitzing
Professor Dr., Dres. h. c. Hans-Joachim Queisser

Max-Planck-Institut für Festkörperforschung, Heisenbergstrasse 1, D-70569 Stuttgart, Germany
* Max-Planck-Institut für Physik komplexer Systeme, Bayreuther Strasse 40, Haus 16
 D-01187 Dresden, Germany

Managing Editor:

Dr.-Ing. Helmut K.V. Lotsch

Springer-Verlag, Tiergartenstrasse 17, D-69121 Heidelberg, Germany

Library of Congress Cataloging-in-Publication Data

Isihara, A. (Akira) Electron liquids / Akira Isihara. – 2nd ed. p. cm. – (Springer series in solid-state sciences; 96)
Includes bibliographical references and index. ISBN 3-540-62789-8 (Berlin: alk. paper) 1. Solid state physics.
2. Electronic structure. 3. Electro-holw droplets. 4. Plasma (Ionized gases) I. Title. II. Series. QC176.I75 1997
530.4'12–dc21 97-25655 CIP

ISSN 0171-1873
ISBN 3-540-62789-8 2nd Edition Springer-Verlag Berlin Heidelberg New York
ISBN 3-540-53598-5 1st Edition Springer-Verlag Berlin Heidelberg New York

© Springer-Verlag Berlin Heidelberg 1993, 1998
Printed in Germany

Typesetting: M. Hillen, Heidelberg
Cover design: *design & production* GmbH, Heidelberg

SPIN: 10569064   54/3144 – 5 4 3 2 1 0 – Printed on acid-free paper

Preface

Several years have passed since the first edition of this book was published. During this period, significant developments in the study of electron systems have taken place, especially in the areas of high-T_c superconductivity and the quantized Hall effect. These developments, and such fascinating subjects as crystallization and the stability of matter are included in the second edition.

Bardstown, KY   *A. Isihara*
June 1997

Preface to the First Edition

The study of electronic properties reveals a common basis for a variety of systems, including gaseous plasmas, ionic solutions, metals, and semiconductors. This study started with one-electron properties in free space, as discussed in solid-state books. However, significant progress has been made recently in more realistic and complicated cases with interactions, confinements, impurities, and fields. Moreover, the recent discoveries of the quantum Hall effect, high-T_c superconductors, and localization phenomena, along with the introduction of low-dimensional materials have opened new areas and have led to a tremendous number of articles in existing journals and even new specialized journals. This book has been written to provide a new, comprehensive review on electronic properties in such diverse areas and materials.

The title indicates emphasis on electron correlations. Chapter I starts with an introductory description of electron systems, including classification, characterization, and models. It provides the reader with a general account of the amazingly diverse electron systems. It is followed by discussions on strongly coupled gaseous plasmas, electron-hole liquids, magnetic response, lowdimensional systems, heavy Fermions, high-T_c superconductivity, localization, the quantum Hall effect, phase transition, and the stability of matter.

For simplicity the natural unit in which $\hbar = 1$ and $2m = 1$ is used, particularly for complicated expressions, m being the electron mass. However, ordinary units are restored whenever desirable. This restoration can easily be made by considering dimensions.

Each chapter has been written so as to be self-contained, so that the reader can start at any chapter without going through previous chapters. Nevertheless, an effort has been made to maintain uniformity and coherence as a single book.

The author had an opportunity to meet with Professor Cardona who visited Buffalo in the spring of 1989. The present book has emerged as a result of his encouragement and stimulation. The patience and efforts of Dr. Lotsch have also contributed to its completion. The author is especially grateful to Atsushi and Hikaru who gave him many constructive suggestions. He is indebted to various investigators, physical societies, and publishers for permission to reproduce figures. These include the American Physical Society, the Physical Society of Japan, Progress of Theoretical Physics, Nature, Physica Status Solidi, Pergamon Press, Inc., North-Holland Publishing Company, Plenum Press, Gordon & Breach Science Publishers, Inc., and IOP Publishing Ltd. The author's original work in this book was supported by the National Science Foundation and the Office of Naval Research.

Buffalo, NY   *A. Isihara*
July 1992

Contents

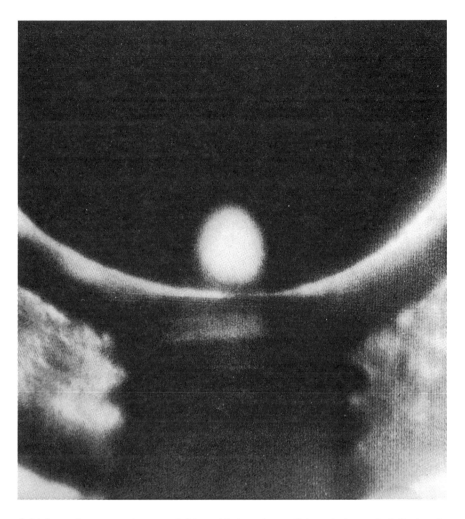

A high-purity germanium crystal is cut into a 4 mm disk and stressed at the top by a nylon screw. The sample is immersed in liquid helium at 2 K and excited from the far side by an argon-ion laser. The electrons and holes produced by photoexcitation condense into electron-hole droplets and are drawn into the region beneath the screw, where the energy gap is lowered by a local strain. The droplets coalesce into a single large drop at the strain maximum, or energy maximum. Bright emission from the drop is produced by the radiative recombination of electrons and holes at about $1.75\,\mu$m wavelength, which is detected with an infrared vidicon tube. [J.P. Wolfe, W.L. Hansen, E.E. Haller, R.S. Markiewicz, C. Kittel, C.D. Jeffries: Phys. Rev. Lett. *34*, 1292 (1975)]

1. General Description of Electron Systems

In this chapter typical electron systems in various dimensions are introduced, together with some basic theoretical parameters and models.

1.1 Classification of Electron Systems

The theory of electron correlations has been one of the most fundamental subjects in the studies of ionic solutions, plasmas, metals, and other systems in which charged particles can move. The properties of metals and gaseous plasmas depend mainly on the electrons, which are lighter and respond to an external probe more effectively than the ions. A one-component plasma, often abbreviated as OCP, has been adopted as a model in which the electrons are imbedded in a uniform cloud of neutralizing positive charges formed by the ions. That is, the ions play only an indirect role of neutralizing the system. A great deal of theoretical progress has been based on this model. Moreover, remarkable experimental discoveries have triggered developments of new theories and experiments in the past two decades, especially on low-dimensional electron systems.

Statistical properties of an electron system depend on temperature and density, which also determine the appropriate theoretical approaches. Therefore, it is convenient to present a general classification of the phase of an electron system. This can be achieved by using dimensionless parameters. First, an electron system is more or less classical at high temperatures and low densities. This classical region is determined by the smallness of the ratio of the average Coulomb energy, which is $e^2/(3/4\pi n)^{1/3}$ in 3D and the thermal energy $k_B T$. Here, n is the number density of electrons. This ratio is commonly expressed as

$$\Gamma = (4\pi n/3)^{1/3} e^2/k_B T, \qquad \text{(3D)}$$
$$\Gamma = \pi^{1/2} e^2 n^{1/2}/k_B T. \qquad \text{(2D)}$$

$$(1.1.1)$$

The system is classical only if Γ is less than 1. The corresponding region for 3D is above the line $\Gamma = 1$ in Fig. 1.1.

While the above parameter is used widely, there is yet another slightly modified parameter. The so-called appears naturally in treating classical plas-

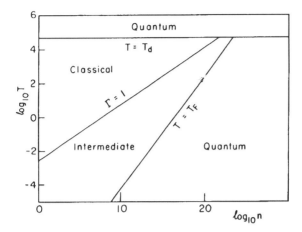

Fig. 1.1. Phase diagram of a 3D electron system

mas. It is defined by the ratio of the Coulomb energy at the Debye screening length l/κ_D and the thermal energy:

$$\varepsilon = e^2 \kappa_D / k_B T. \tag{1.1.2}$$

Here

$$\kappa_D = (4\pi n e^2 / k_B T)^{1/2}, \qquad \text{(3D)}$$
$$\kappa_D = 2\pi n e^2 / k_B T. \qquad \text{(2D)} \tag{1.1.3}$$

Note that Γ and ε are not independent of each other but rather are related as follows:

$$\Gamma^3 = \varepsilon^2/3, \qquad \text{(3D)}$$
$$\Gamma^2 = \varepsilon/2. \qquad \text{(2D)} \tag{1.1.4}$$

Second, the system is quantum mechanical at low temperatures and high densities. The criterion for this region is given by the relation which equates the thermal energy and the Fermi energy:

$$k_B T = \hbar^2 k_F^2 / 2m = k_B T_F.$$

Here T_F is the Fermi temperature. The bottom straight line in Fig. 1.1 represents this relation, below which quantum treatments are necessary. Numerically, the ratio of the thermal energy $k_B T$ and the Fermi energy ε_F is given by

$$k_B T / \varepsilon_F = 0.8 \times 10^{11} T n^{-2/3}. \qquad \text{(3D)}$$

with T in K and n in cm^{-3}. Figure 1.1 corresponds to 3D, but a similar graph is obtained for 2D (Fig. 1.2). In this figure the dashed line indicates how the quantum region determined by $T = T_F$ expands when a small effective mass is used, and the dotted area indicates the depletion of the quantum region

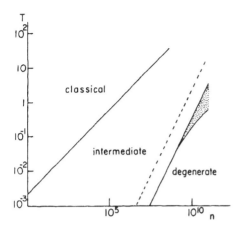

Fig. 1.2. Phase diagram of a 2D electron system. The dashed line indicates use of a small effective mass. and the shaded area represents depletion of the quantum area due to Coulomb interaction

due to Coulomb interaction. Note that the Fermi wave number k_F of an ideal electron gas depends only on density n such that

$$k_F^3 = 3\pi^2 n, \qquad \text{(3D)}$$

$$k_F^2 = 2\pi n, \qquad \text{(2D)} \qquad\qquad\qquad (1.1.5)$$

$$k_F = \pi n/2. \qquad \text{(1D)}$$

However, in the presence of Coulomb interaction, the actual Fermi energy which is the chemical potential depends on e^2. This dependence causes a depletion of the quantum region, as indicated by the dotted region in Fig. 1.2.

In between the above two limits there is a vast intermediate region which is difficult to treat theoretically. This region may be characterized by several combinations of parameters, including the de Broglie thermal wavelength $\lambda = h/(2\pi m k_B T)^{1/2}$, the Debye screening constant κ_D, and the Landau length defined by

$$l_L = e^2/k_B T.$$

For instance, if

$$\lambda/l_L = \eta$$

is small, the system is semiclassical. Note that

$$\eta = 2\pi a_0/\lambda$$

where a_0 is the Bohr radius. If the ratio

$$\lambda/l_L = 1$$

at T_d, we find $T_d = 5.02 \times 10^4$ K for electrons. For temperatures higher than T_d, i.e., if $\lambda > l_L$ the wave packets of electrons overlap and the quantum mechanical diffraction becomes important. For $T < T_d$, this diffraction effect is small, and for $T > T_d$, it depends on $\lambda/n^{-1/3}$ in the 3D case. The top line in Fig. 1.1 represents the relation $T = T_d$. Note that the parameter $\lambda n^{1/3}$, which is of order $1 \times 10^{-7} n^{1/3} T^{-1/2}$, determines the temperature range in which the diffraction effects are important for a given n.

The lengths $l_{\rm L}$, λ and their ratio η depend on temperature. For an electron gas they are listed in Table 1.1.

Table 1.1. Landau length and de Broglie wavelength for an electron gas

T [K]	10^2	10^3	10^4	10^5	10^6
λ [cm]	7.45×10^{-7}	2.35×10^{-7}	7.45×10^{-8}	2.35×10^{-8}	7.45×10^{-9}
l_L [cm]	1.67×10^{-5}	1.67×10^{-6}	1.67×10^{-7}	1.67×10^{-8}	1.67×10^{-9}
η	4.46×10^{-2}	1.41×10^{-1}	4.46×10^{-1}	1.41	4.46

Since different regions in the phase space determined by temperature and density require specific approaches and because the intermediate region has not been well explored theoretically, it is by no means a simple problem to discuss electron correlations in a very wide range. Nevertheless, we shall try to cover many aspects of electron correlations ranging from the high temperature classical region to the low temperature quantum mechanical one. For low temperatures, the density parameter r_s is frequently used. It is defined by an average length available to each electron measured by the Bohr radius such that

$$r_s = \left(\frac{3}{4\pi n}\right)^{1/3} \frac{1}{a_0}, \qquad (3D)$$

$$r_s = \left(\frac{1}{\pi n}\right)^{1/2} \frac{1}{a_0}. \qquad (2D)$$

$$(1.1.6)$$

The parameter r_s is dimensionless and decreases as the density n increases. We remark that the Rydberg unit of energy is $e^2/2a_0 = e^4 m/2\hbar^2 = 13.61\,{\rm eV}$. Hence, in the *natural unit* in which a_0 is set to 1 and e^2 to 2, the Rydberg is the unit of energy. The Rydberg unit appears naturally in electron systems because the Hamiltonian consisting of the kinetic energy and Coulomb potentials can be expressed as

$$\mathcal{H} = \frac{me^4}{\hbar^2} \frac{1}{r_s^2} \left(\sum_i \frac{p_i'^2}{2} + r_s \sum_{i<j} \frac{1}{r_{ij}'} \right). \qquad (1.1.7)$$

Here, the operators in the round brackets are dimensionless, and the Hamiltonian as energy is characterized by the Rydberg unit. When this unit is set to 1, r_s is the key parameter to express the ground state energy. From the above form we find also that for small r_s and at low temperatures, the Coulomb potential can be considered as a perturbation to the first term which represents an ideal electron gas. However, r_s is not necessarily small in actual metals. For example, in alkali metals, r_s is of the order of 3 to 5.

In the so-called *atomic unit*, $\hbar$ and e are set to 1. Hence, the Bohr radius will be 1 and will be the unit for length. On the other hand, the unit for energy will be the *Hartree unit* which is 2 Ryd.

For electrons in solids, an effective mass, which depends on the band structure, should be used. We shall denote such an effective mass by m. When the bare mass must be distinctively shown, the notation m_0 will be adopted. As we shall discuss in the next section, the effective mass of certain 2D systems is small. In such a case, the quantum region in Fig. 1.2 is enlarged, as indicated by the dashed line. In a medium, the Coulomb potential of electrons is screened by its dielectric constant κ. Hence, e^2 must be replaced by e^2/κ, and accordingly an effective Bohr radius must be used. Additional specifications might be necessary for a given electron system, even though the simplest case without medium serves as a model system.

1.2 Low-Dimensional Electron Systems

The one component plasma (OCP) model has been used extensively for gaseous plasmas, metals and ionic solutions. These traditional systems are three-dimensional but in the past two decades one- and two-dimensional electron systems have also attracted considerable attention, not only because they have been realized experimentally, but also because they show unexpected and interesting properties. Some of these properties are shared by the 3D systems while others are unique. Moreover, low-dimensional electron systems are being used for modern electronic devices.

1.2.1 Two-Dimensional Electron Systems

There are several different types of 2D electron systems but the following cases are most important: (a) Interfacial electrons in Si/SiO_2 or (b) GaAs/GaAlAs and (c) surface electrons on liquid He. The densities of these systems in general suggest that the interfacial electrons are more or less quantum mechanical while the electrons on He are classical. Let us discuss these three cases briefly.

(a) Inversion and Accumulation Layers in Si/SiO_2

In the early 1930s, *Lilienfeld* and *Heil* [1.1 a, b] proposed a new device which has eventually been developed into what is now called the MOSFET (metal-oxide-semiconductor-field-effect transistor). Their basic idea was to make an active circuit element by replacing one of the metallic plates of a parallel condenser by a semiconductor plate. Since induced charges on such a plate are more appreciable than in the case of a metallic plate due to the relatively

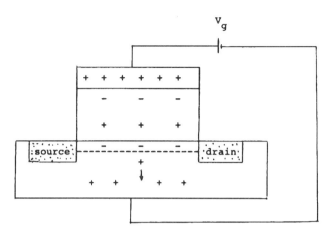

Fig. 1.3. Schematic structure of a Si-MOSFET. The top plate represents the gate, the middle part consists of SiO_2, and the bottom plate represents p-Si

low carrier density in semiconductors and because the induced charges can be controlled by an external bias field, an active circuit element will be obtained. Their idea was then taken up by several authors including *Shockley* and *Pearson* [1.1 c] in the late 1940s but its realization was difficult for the following reason. During their time when germanium was used extensively as a major semiconductor for transistors, its use for a MOSFET was attempted. However, induced charges in germanium did not move well in response to an external field and germanium oxide used as the dielectric in such a condenser was chemically unstable.

In 1960 *Kahng* and *Atalla* [1.2] overcame these difficulties by fabricating the first MOSFET using a thermally oxidized Si. This contributed to the development of the current Si era. Induced charges in the interface of Si/SiO_2, contrary to the case of germanium, move freely in the plane parallel to the condenser plates if an electric field is applied. Since their motion perpendicular to the plates is restricted, causing quantization of energy, they form a 2D electron system if no excitation takes place.

Let us consider the case in which p-type Si with positive carriers and SiO_2 are used respectively as a condenser plate and a dielectric. It is easy to oxidize Si, and SiO_2 is a very good insulator which is chemically stable. The width of the oxide layer is typically of the order of 2000 Å.

The other plate, called the *gate*, is a metal. By applying a plus voltage (*gate voltage*) to the gate one can induce negative charges in the dielectric surface adjacent to it and positive charges on the opposite side, close to the Si plate. The positive carriers near the surface of the p-Si plate will be pushed into the interior by these induced positive charges, leaving negative space charges behind (Fig. 1.3). It turns out that these negative charges are very mobile in the plane parallel to the condenser plates in response to an

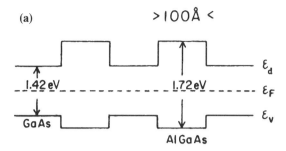

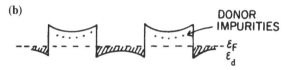

Fig. 1.4. Energy diagram of a quantum well GaAs/GaAlAs. (a) Undoped; (b) doped

external electric field. These negative charges appear in a narrow layer called an *inversion layer*, so called because the sign of these charges is opposite to that of the original carriers. Typically, the width of an inversion layer is of order 100 Å. The density of induced charges can be varied easily in the range $10^{11} - 10^{12}\,\mathrm{cm}^{-2}$. Such a large variation is very advantageous for the study of electron correlations. In metals, it is very difficult to change the density of electrons.

If similar condensers are connected with each other electrically, it is possible to move the negative charges from one condenser to another. The presence and absence of these charges in one particular condenser can be used as a binary code. This is one reason why a Si MOSFET has become important for modern electronic devices. If n-type Si is used as a condenser plate instead of p-type, space charges will be induced in an *accumulation layer*, so called because the induced charges have the same sign as in the bulk carriers.

(b) Superlattices

Another even more important 2D electron system has been realized at the interface between two semiconductors with similar lattice constants [1.3]. By means of computerized molecular beam epitaxy, thin layers of two such semiconductors are alternated to form a *superlattice* which has a periodicity larger than that of the host crystals. The most familiar case is GaAs/ $\mathrm{Ga}_x\mathrm{Al}_{l-x}\mathrm{As}$, in which x is a fraction. The Al atoms do not change the lattice structure of the GaAs lattice but modify the band structure. The alternating layers form a 1D array of potentials, similar to the *Kronig* and *Penney* model [1.4] and

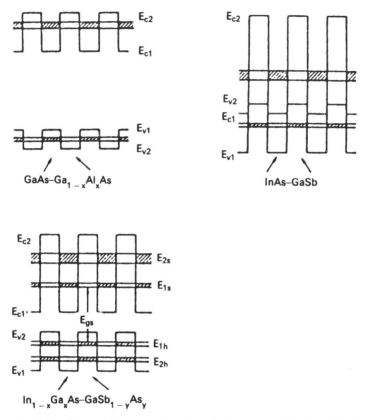

Fig. 1.5. Energy diagrams for three different superlattices. Shaded areas in the subbands represent the regions in which carriers are confined

can be represented by alternating conduction and valence bands as illustrated in Fig. 1.4 a.

One can dope GaAlAs layers with Si atoms as donors and modulate the energy bands such that the Fermi energy moves above the edge of the conduction band of GaAs. The electrons from the donors in GaAlAs will then fall into the conduction band in the narrow valley in the GaAs layers as in Fig. 1.4 b. These electrons can move very freely in the direction perpendicular to the 1D array of semiconductors. Their mobility is very high as reflected in their small effective mass of $0.068\,m_0$ in contrast to $0.19\,m_0$ in Si MOSFETs. Typically their density is of order 10^{12} cm^{-2}. Although this density cannot be changed freely, their properties in a magnetic field depend generally on $\nu = n/(eH/ch)$ so that the magnetic field H can be used instead of n to change ν. The parameter ν is the *filling factor* of the ground Landau level if $\nu < 1$.

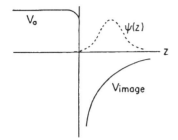

Fig. 1.6. Potential diagram for electrons on the surface of liquid helium. The potential barrier V_0 and the image potential V_{image} confine the electrons as indicated by the charge distribution function $\psi(z)$, where z measures the distance above the surface

GaAs and GaAlAs can form a *quantum well* instead of a superlattice. The term, *heterostructure* is often used also for similar layered systems. Superlattices or heterostructures are not limited to the above semiconductor combination and, for the sake of distinction, GaAs/GaAlAs superlattices are classified as type I. The energy structure is different in InAs/GaSb, called type II, in that the conduction band of InAs is close to the valence band of GaSb. In GaAs/GaSbAs heterostructures belong to the same type II. Unlike type I, where both electrons and hole states are in the GaAs regions, type II has electrons in InAs and holes in GaSb or GaSbAs. The energy diagrams of these superlattices are illustrated in Fig. 1.5. ε_c and ε_v represent the conduction and valence band edges respectively. The shaded area in the subbands represents regions in which carriers are confined.

Other types of superlattices include the *polytype*, in which a third component is added [1.5]. In the NIPI type n(Si)- and p(Be)-doped GaAs layers are separated from each other by thin intrinsic (i) layers of the semiconductor. In the *spin type* HgMnSe/HgCdSe superlattice magnetic atoms such as Mn are used to produce an alternating energy structure in the presence of a magnetic field.

(c) Electrons on Liquid He

A completely different 2D electron system has been realized on the surface of liquid helium [1.6]. These electrons are subject to their own image forces near the surface of liquid helium, which acts as a dielectric. A schematic potential diagram is given in Fig. 1.6. The surface of liquid helium presents a barrier V_0 of more than 1 eV to electron transmission into the liquid. On the other hand, the image potential V_{image} attracts the electrons towards the liquid. Therefore, the electrons are localized as shown by $\psi(z)$ representing the charge distribution. The coordinate z is directed towards the exterior of the liquid.

In comparison with interface electrons in semiconductors there is no impurity on liquid helium so that a very pure 2D system can be realized. Their typical density is $10^9 \, \text{cm}^{-2}$.

There are some other types of 2D electron systems. Since the available density range of all these 2D electron systems is very wide, they provide as

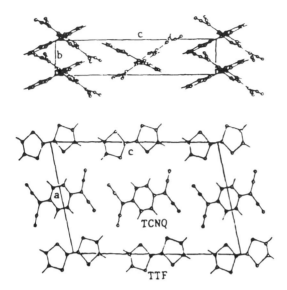

Fig. 1.7. Stacking structure of TTF–TCNQ

a whole a very good testing ground for many body theory. In addition, new and very unusual properties have been discovered, as will be discussed.

1.2.2 One-Dimensional Electron Systems

A 1D system was considered as a convenient but highly theoretical object which is remote from reality. Thus it has attracted no significant attention until recently, despite some important theoretical contributions [1.7].

The situation changed with the synthesis of TCNQ (tetracyanoquino-dimethane) salts by Du Pont in 1962, polyacetylene and other 1D conductors. In particular the appearance [1.8] of large conductivity peaks and the metal insulator transition in tetrathiafulvalene-tetracyanoquinodimethane (TTF–TCNQ) have caused considerable interest. The area of one-dimensional conductors is now one of the new exciting fields in solid state physics.

Among several charge transfer salts of TCNQ the best known case is TTF–TCNQ in which the cation TTF acts as a donor and the anion TCNQ as an acceptor. In TTF–TCNQ the large and almost flat TTF cations form a stack separated from the column of the TCNQ anions. At low temperatures TTF–TCNQ is an insulator. It shows a sharp metal-insulator transition at around 55 K, above which it is a metal. TTF–TCNQ is 1D in the sense that electrical conduction is high in the direction in which the flat TCNQ molecules are stacked while it is low in other crystal directions. The metallic state is caused by the overlap along the stacks of the highly directional π orbitals for the unpaired electrons on the acceptor (TCNQ) and on the donor (TTF). The stacking in TTF–TCNQ is schematically illustrated in Fig. 1.7.

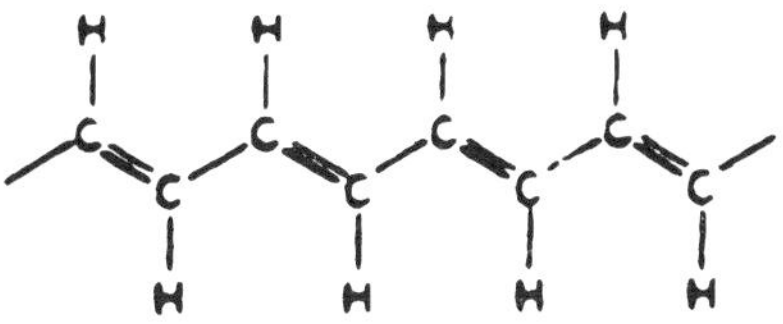

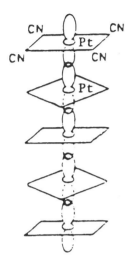

Fig. 1.8. Chemical structure of poly-acetylene

Fig. 1.9. Platinum salt KCP

There are many other conductors which are 1D in the same sense as in TTF–TCNQ but let us describe only three other typical types. First, mention must be made of polyacetylene (Fig. 1.8) as it is a representative organic conductor in the polyene family made up of conjugate double bonds which provide a conduction channel. The electronic structure of polyenes has long been known. Three out of four valence electrons of a carbon atom are in the sp^2 orbital, while the remaining valence electron forms a π bond. If there is no distortion of the chain, the overlapping π bonds results in conductivity. Polyacetylene $(CH)_x$ is the simplest conductive polymer among polyenes. The electrical properties of films of polyacetylene vary in a very wide range depending on doping from insulating through semiconducting to metallic. Polyacetylene has been used to fabricate lightweight rechargeable organic batteries [1.9].

Second, platinum atoms lie in stacks in salts called KCP, $K_2Pt(CN)_4X_{0.3} \cdot xH_2O$ with $X = Cl$ or Br. When surrounded by four CN molecules, each platinum atom provides two electrons to the d_z^2 band. Figure 1.9 illustrates this configuration. The overlap of d_z^2 orbitals gives rise to conductivity. Synthesized in 1842, these salts are actually the oldest known 1D conductors.

Third, trichalcogenides of group V transition metals such as $NbSe_3$ are generally built up from trigonal prismatic chalcogenides which are lined up

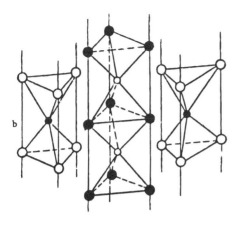

Fig. 1.10. Structure of trichalcogenides such as NbSe$_3$. Small (large) white and black circles represent Nb(Se) in different layers

along the b axis as in Fig. 1.10. The small white and black circles represent Nb in different layers in correspondence to the larger white and black circles for Se.

A number of other 1D materials have been discovered. Among them are (SN)$_x$, TSeF–TCNQ, NMP–TCNQ, trichalcogenides such as TaS$_3$, NbSe$_3$, and simple 2:1 salts (TMTSF)$_2$X and X can be PF$_6$, AsF$_6$, TaF$_6$, SbF$_6$, ClO$_4$, etc. In the last salt group a wide variety of properties have been observed, ranging from superconduction to insulation depending on the anion X. Another family of salts (TMTTF)$_2$X can also be superconducting, particularly under pressure.

In 1D conductors many interesting phenomena and properties have been observed. The discovery in 1980 of superconductivity in the selnium-based 1D organic compounds known as Bechgaard salts caused special excitement [1.10]. There are other aspects which make 1D uniquely important. For example any disorder in 1D which breaks the only available conduction channel is expected to cause localization, as we shall discuss in Chap. 9. Also, it is known in statistical mechanics that in one dimension no phase transition exists if the interaction is short ranged.

These 1D materials are generally metallic at high temperatures, in many cases down to 100 K. At low temperatures they generally become semiconductors or insulators due to an inherent instability characteristic of 1D. When a 1D system is made up of atoms with separation a each giving one conduction electron to the system, the electron wave function $\psi(x + a)$ must be the same as $\psi(x)$ and in fact is of the form $u_k(x)\exp(ikx)$. In the tight binding approximation, which is generally valid for these compounds, the energy is given by

$$\varepsilon_k = -2t\cos ka\,,$$

where t is the overlap integral between the nearest neighbouring sites. *Peierls* [1.7b] showed that at low temperatures the system is more stable if the lattice is slightly distorted and an energy gap is opened up at $k = 2k_F$. According to *Fröhlich* [1.7a] this gap increases with the strength of electron–phonon coupling. At low temperatures the electron wave number is either k_F or $-k_F$ so that the electrons couple strongly with phonons whose wave vector is $2k_F$. The gap increases also with the response function $F(q, \omega)$. In the RPA, this function is logarithmically divergent at $2k_F$ in 1D (Fig. 2.1). Therefore, electron–phonon coupling can be expected to play an important role for 1D conductors, and indeed it has been related to many interesting properties.

By using narrow gates, 2D electron systems such as Si MOSFETs can be made quasi one-dimensional [1.11]. At low temperatures, they show unique conductance fluctuations as we shall discuss in Chap. 10.

1.3 Characteristic Lengths

In this work we shall be concerned with electron systems but with appropriate modifications our basic description can be applied to systems of any other charged particles. Electron systems are very widely spread and require different theoretical approaches. Nevertheless, they can be classified generally in terms of several characteristic lengths.

Of the many length parameters the Bohr radius is the most fundamental one. It is given by $a_0 = \hbar^2/m_0 e^2 = 0.529 \times 10^{-8}$ cm. However, for electrons in condensed matter, an effective mass m and a dielectric constant κ must be used. That is, an effective Bohr radius

$$a_0^* = \hbar^2 \kappa / m e^2 \qquad (1.3.1)$$

represents a characteristic length. For inversion layer electrons in [100] Si, $a_0^* = 2.1$ nm with an average dielectric constant $\kappa = 7.7$ and $m = 0.19 m_0$.

If $m = 0.068 m_0$ as in the case of electrons in GaAs/GaAlAs and for $\kappa = 10$, $a_0^* = 7.78$ nm. These large values of the actual Bohr radius can be considered to effectively reduce the Coulomb interaction between the electrons because the Bohr radius is inversely proportional to e^2. In this respect it is worth noting that in units of $\hbar = 1$ and $2m = 1$,

$$a_0 = 2/e^2 .$$

That is, e^2 has the dimension of a reciprocal length.

For actual systems the Bohr radius must be compared with other length parameters. These include the average length per electron called the *Wigner–Seitz radius*, given by $(3/4\pi n)^{1/3}$ (for 3D), and the de Broglie thermal wavelength

$$\lambda = h/(2\pi m k_B T)^{1/2} .$$

The Wigner–Seitz radius is reduced to r_s in the atomic units in which $a_0 = 1$. The Debye screening length

$$1/\kappa_D = (k_B T / 4\pi n e^2)^{1/2}$$

includes both n and $k_B T$, which, however, represents the thermal energy rather than the de Broglie thermal wavelength. In condensed matter an effective screening constant must be used.

Although λ is an important length, r_s as a length parameter characterizes ground state properties and is related to $1/k_F$. Screening at low temperatures is characterized by the Thomas–Fermi screening length which is proportional to $1/k_F$.

The Wigner-Seitz radius for 2D is $(\pi n)^{-1/2}$. For $n = 10^{12}\,\mathrm{cm}^{-2}$ it is 5.642 nm. Its ratio to $a_0^* = 2.145$ nm is 2.63. This is the effective r_s for Si inversion layer electrons. In the case of GaAs/GaAlAs, the effective r_s becomes 0.725 for the same density and $\kappa = 10$.

The diversity of charged particle systems can be seen from the wide range of particle densities. In unit volume of gaseous plasmas the particle density ranges from 1 to 100 in interstellar gases; from 10^{10} to 10^{12} in the dense ionosphere, upper stellar atmosphere and tenuous laboratory plasmas; from 10^{14} to 10^{16} in the lower stellar atmosphere, and intense laboratory plasmas; from 10^{16} to 10^{18} in dense laboratory plasmas. Much higher densities are expected in stellar interiors, particularly in the white dwarfs and neutron stars. The white dwarfs are faint and small due to the gravitational collapse of stars of the size of the sun which takes place when the hydrogen fuel has been used up. They consist of ionized He gas and electrons and have mass density of order $10^6\,\mathrm{g/cm}^3$ at around 10^7 K in the center. Their electron density is around 10^{29} per unit volume. With this density the Fermi temperature is around 10^9 K so that the electrons are quantum mechanically degenerate even though their actual temperature is high. Since the electrons are nearly free, the ions can be considered to constitute an interesting OCP object.

The pulsars have been identified as rotating neutron stars. They are caused by the strong gravitational collapse of stars much larger than the sun into a radius of around only 10 km. Their outer crust consists of ionized iron nuclei and electrons and has mass density $1o^4$–10^7 g/cm^3 at temperatures around 10^7 K. With their high density, they form a nearly free electron gas. Hence, the outer crust may be treated as an ion OCP with negative charges in the background. Due to the large charge ze on the iron nuclei and to the high density, $\Gamma = (ze)^2 / r_s a_0 k_B T$ can be as large as 10^2. The outer crust has a thickness of only about 1 km followed by a neutron-rich inner crust.

In a magnetic field, the *magnetic length* defined by

$$l = (c\hbar/eH)^{1/2} \tag{1.3.2}$$

plays an important role. This represents the size of the lowest cyclotron orbit. In terms of H in tesla, it is

$$l = 25.66H^{-1/2} \tag{1.3.3}$$

so that for $H = 10$, $l = 81.14\,\text{nm}$. The quantity

$$1/(2\pi l^2) = eH/ch$$

represents the density of states per Landau level. The electron density relative to the density of states is expressed by the *filling factor* defined by

$$\nu = n/(eh/ch). \tag{1.3.4}$$

If $\nu < 1$, the lowest Landau level is only partially filled by electrons. Recent experiments on such low density electron systems have revealed very unusual properties which are caused by Coulomb interaction.

1.4 Fermi Liquid Theory

The linear specific heat and some other electronic properties of metals can be well described by an ideal electron gas model provided suitable effective masses are used. As a generalization of such a description, we can consider characterizing an electron system by introducing quasiparticles which are more or less free. Landau's Fermi liquid theory is based on this generalization. The basic idea behind this theory is that, if the interactions are switched on adiabatically, the energy states of the quasiparticles may still be classified in terms of the corresponding free-particle states. However, the energy of a quasiparticle may depend on the state of other particles and is considered to be a functional of the quasiparticle distribution function.

Let E be the total energy of a Fermion system. In the absence of excitation above the Fermi energy, it is equal to the ground state energy E_g. The quasiparticles in the excited states may still be labeled by momentum $\mathbf{p}$. If n_p quasiparticles are in the excited states $\mathbf{p}$ the energy E is given by

$$E = E_g + \sum_{\mathbf{p}} \varepsilon(p) n_p. \tag{1.4.1}$$

The entropy may be expressed as

$$S = -k \sum_{\mathbf{p}} [n_p \ln n_p + (1 - n_p) \ln(1 - n_p)]. \tag{1.4.2}$$

The equilibrium distribution $n_p(T)$ is obtained by imposing the condition that the free energy be minimum. The result is simply the ordinary Fermi distribution

$$n_p(T) = \frac{1}{e^{[\varepsilon(p)-\mu]/k_BT} + 1} \tag{1.4.3}$$

It is important to recognize that the energy $\varepsilon(p)$ may depend on electron spin. Although for the linear specific heat of metals this dependence does not come into play, it does for the magnetic susceptibility. To see this, let us evaluate the paramagnetic susceptibility. The quasiparticle energy is changed due to spin magnetic field coupling in accordance with

$$\delta\varepsilon(p,\sigma) = -\mu_B^*\boldsymbol{\sigma}\cdot\mathbf{H}. \tag{1.4.4}$$

where μ_B^* is the effective Bohr magneton of the quasiparticles which must be determined, and σ is their spin. The effecive Bohr magneton can depend on momentum, but in what follows let us assume that it is constant.

The energy change $\delta\varepsilon$ can be constructed from two terms. One is the direct term due to spin-magnetic field coupling and is given by

$$\delta_1\varepsilon(p,\sigma) = -\mu_B\boldsymbol{\sigma}\cdot\mathbf{H}. \tag{1.4.5a}$$

The other is the indirect term which is caused by the change in the distribution function. We express this energy for $\delta n = n - n_p$ in terms of an f-function such that

$$\delta_2\varepsilon(p,\sigma) = \sum_{\mathbf{p}',\sigma'} f(\mathbf{p}\sigma,\mathbf{p}'\sigma')\delta n(\mathbf{p}'\sigma'). \tag{1.4.5b}$$

For a weak magnetic field, δn can be replaced by

$$\delta n = \frac{\partial n_p(\varepsilon')}{\delta\varepsilon'}\delta\varepsilon(p,\sigma), \tag{1.4.6}$$

where $\delta\varepsilon(p,\sigma)$ can be expressed as in (1.4.4).

In general, the f-function may depend on spin. However, this dependence can be small for a weak magnetic field. In view of the symmetry between σ and σ', we expand f in the following form:

$$f(\mathbf{p}\sigma,\mathbf{p}'\sigma') = f_0(\mathbf{p},\mathbf{p}') + (\boldsymbol{\sigma}\cdot\boldsymbol{\sigma}')f'(\mathbf{p},\mathbf{p}') \tag{1.4.7}$$

we introduce this expression into (1.4.5), equate $\delta\varepsilon$ of (1.4.4) with the sum of (1.4.5a) and (1.4.5b) and perform the summation over σ'. We arrive at

$$\mu_B^* = \mu_B + 2\mu_B^*\sum_{\mathbf{p}'} f(\mathbf{p},\mathbf{p}')\frac{\partial n_p(\varepsilon')}{\partial\varepsilon'}. \tag{1.4.8}$$

This is a self-consistent equation to determine μ_B^*.

For low temperatures, we can use

$$\frac{\partial n_p(\varepsilon')}{\partial\varepsilon'} = -\delta(\varepsilon_F - \varepsilon'). \tag{1.4.9}$$

Moreover, the magnitude of $\mathbf{p}'$ or $\mathbf{p}$ can be replaced by p_F. Hence, one can rewrite $f'(\mathbf{p},\mathbf{p}')$ of (1.4.7) as

$$f'(\mathbf{p},\mathbf{p}') = \phi(p_F,\theta). \tag{1.4.10}$$

where θ is the angle between $\mathbf{p}$ and $\mathbf{p}'$. Using (1.4.9) and (1.4.10) in (1.4.8), we obtain

$$\mu_B^* = \frac{\mu_B}{1+s}. \tag{1.4.11}$$

Here

$$s = g(\varepsilon_F)\langle\phi(p_F, \theta)\rangle \qquad (1.4.12)$$

where $g(\varepsilon_F)$ is the density of states at the Fermi surface, and $\langle\cdots\rangle$ is the average of ϕ on the Fermi surface. Per unit volume, the susceptibility is then given by

$$\chi_p = \frac{\chi_p^0}{1+s} \qquad (1.4.13)$$

where χ_p^0 is the Pauli susceptibility:

$$\chi_p^0 = \mu_B^2 g(\varepsilon_F). \qquad (1.4.14)$$

According to (1.4.13) one can relate the susceptibility with the electronic specific heat through the density of states $g(\varepsilon_F)$. The parameter s modifies the Pauli susceptibility. Although s is generally expected to be small, it can be negative. If so, the paramagnetic susceptibility χ_p is enhanced over the Pauli susceptibility. In fact, in both Na and K, χ_p is enhanced; the experimental value of χ_p/χ_p^0 is 1.74 and 1.58, respectively. If s is close to -1, χ_p becomes very large. The system is then nearly ferromagnetic. The case of Pd is such an example.

The Landau theory of a Fermi liquid is very convenient in interpreting many-body effects. Moreover, it has been extended to nonequilibrium cases and has been successfully applied not only to electrons but also to liquid ^{3}He [1.12].

1.5 Tight Binding, Anderson and Hubbard Models

Usually, electrons are confined to a finite volume in the presence of oppositely charged particles and their properties depend directly or indirectly on the status of these charges. In accordance with the status of these counter charges, suitable models must be introduced in order to treat the electronic properties. For gaseous plasmas or ionic solutions the one-component plasma model (OCP) is often used in which the ions are smeared to form a cloud of positive charges so that the total system maintains charge neutrality. This is the basic and simplest model. It is also called the *jellium model*, particularly when electron correlations are emphasized.

In second quantization, the Hamiltonian of an OCP is given by

$$\mathcal{H} = \sum_{q\sigma}\varepsilon(q)a_{q\sigma}^\dagger a_{q\sigma} + \frac{1}{2V}\sum_{kpq}\sum_{\sigma\sigma'}u(q)a_{k+q,\sigma}^\dagger a_{p-q,\sigma'}^\dagger a_{p\sigma'}a_{k\sigma}, \qquad (1.5.1)$$

where $\varepsilon(q) = \hbar^2 q^2/2m$ and $u(q)$ is the Fourier transform of the Coulomb potential. In order to maintain charge neutrality, we require

$$u(0) = 0$$

This model can be used for electrons in metals but the electron–ion interaction may be important. For this interaction one can use

$$\mathcal{H}_{ei} = \sum_q V(q)\rho_q \tag{1.5.2}$$

where

$$\rho_q = \sum_{k\sigma} a_{k+q,\sigma} a_{k\sigma} \tag{1.5.3}$$

is the number density of electrons.

In general more explicit representations of ions are necessary to describe electronic properties in condensed matter. For metals the *tight binding model* is a typical example. In this model, the electrons are labeled by the ion index i, and can hop from one site to another. With their mutual interaction, their Hamiltonian is given by

$$\mathcal{H} = \mathcal{H}_0 + \mathcal{H}_1. \tag{1.5.4}$$

The first term represents hopping,

$$\mathcal{H}_0 = \sum_{(ij)} (w_{ij} a_{i\sigma}^\dagger a_{j\sigma} + c.c.). \tag{1.5.5}$$

The matrix element w_{ij} corresponds to the electron jump from the site j to i. Since w_{ij} depends usually on the difference $|i - j|$, one can express $\mathcal{H}_0$ in a Fourier transformed form

$$\mathcal{H}_0 = \sum_{k\sigma} w(k) a_{k\sigma}^\dagger a_{k\sigma}, \tag{1.5.6}$$

where the Fourier transforms are defined by

$$w(k) = \sum_j w(j) e^{ikj}, \quad a_{k\sigma} = \sum_j e^{-ikj} a_{j\sigma}. \tag{1.5.7}$$

The second term $\mathcal{H}_1$ is the electron electron interaction, which is of the form

$$\mathcal{H}_1 = \frac{1}{2} \sum_{(ijkl)} U_{ijkl} a_j^\dagger a_k^\dagger a_l a_i, \tag{1.5.8}$$

where U_{ijkl} is given by

$$U_{ijkl} = \int dr_1 dr_2 \psi_j^*(\mathbf{r}_1) \psi_i(\mathbf{r}_1) \phi(\mathbf{r}_1 - \mathbf{r}_2) \psi_k^*(\mathbf{r}_2) \psi_l(\mathbf{r}_2). \tag{1.5.9}$$

$\phi(r) = e^2/r$, and $\psi_i(r)$ is the wave function of the electron at site i. Obviously, the magnitude of U_{ijkl} depends on overlapping of the wave functions. Since overlapping of electrons at four different sites is generally small, U may be approximated by two-site Coulomb interaction for which $k = l$ and $i = j$. Moreover, the nearest neighbor interaction should be the strongest.

On the other hand, when spins are taken into consideration, two electrons with opposite spins may be on one site. Denoting their densities by $n_{j\uparrow}$ and $n_{j\downarrow}$ and assuming constant interaction parameters w, t and U, we arrive at the *Hubbard Hamiltonian* [1.13]

$$\mathcal{H} = w \sum_{i\sigma} n_{i\sigma} + t \sum_{(ij)\sigma} (a_{i\sigma}^{\dagger} a_{j\sigma} + c.c.) + U \sum_j n_{j\uparrow} n_{j\downarrow}, \qquad (1.5.10)$$

where (ij) represents nearest neighboring pairs. The same Hamiltonian was used also by *Gutzwiller* [1.13] for the study of d-electrons in a ferromagnet.

The Hubbard model is used effectively for narrow bands for which the parameter w, representing the band width, is often assumed to be small. The U term represents on-site electron–electron interaction such that if two electrons are on the same site, the energy goes up by U. Hence, U can be assumed to be large. If $t/U \ll 1$, this Coulomb interaction is very strong and, equivalently, hopping is negligible. In the limit $U \to \infty$ no two electrons can occupy the same site unless their spins are antiparallel. In this limit, each electron takes on its own site and the system is in a sense similar to the case of noninteracting electrons although the Fermi wave number k_F is different. If U is large but finite, a virtual process is possible in which electrons with antiparallel spins in adjacent sites hop to each other's site and back and can lower the energy. That is, the system can have an effective antiferromagnetic interaction between spins. If on the other hand $t \to 0$, the electrons are localized. This localization has nothing to do with the exclusion principle. The electrons are more or less uniformly spaced and behave classically. Note that this case represents the atomic limit. In the opposite case, where $U \to 0$, the model becomes the nearest-neighbor tight binding model.

Conduction electrons interact with phonons, i.e., vibrations of the lattice. Electron–phonon scattering can cause polaron effects. The electron is then expected to have a larger effective mass. Electron–phonon interactions lead to a novel type of ground state known as the charge-density wave state. *Peierls* was the first to discuss that the ground state of a 1D metal with electron–phonon coupling is sharacterized by a collective mode formed by electron–hole pairs at $2k_F$. The charge density associated with this collective mode is

$$\rho = \rho_0 + \rho_1 \cos(2\mathbf{k}_F \cdot \mathbf{r} + \phi)$$

where ρ_0 is the average electron density. A *charge density wave* (CDW) can move under the action of an electric field, thus causing a current. On the other hand, it can easily be pinned by defects. CDW states have been observed in 1D organic metals.

Electron–electron interaction via phonons gives rise to superconductivity according to the BCS theory. The conduction electrons can also interact with localized magnetic moments as in the case of the *Kondo effect* [1.14] and the *Anderson model* [1.15]. The latter is expressed by the Hamiltonian

$$\mathcal{H} = \mathcal{H}_0 + \mathcal{H}_{0d} + \mathcal{H}_c + \mathcal{H}_{sd};$$

$$\mathcal{H}_0 = \sum_{\mathbf{k}\sigma} \varepsilon(k) n_{\mathbf{k}\sigma}, \quad \mathcal{H}_{0d} = \varepsilon \sum_{\sigma} n_{\mathbf{d}\sigma}, \quad \mathcal{H}_c = U n_{\mathbf{d}\uparrow} n_{\mathbf{d}\downarrow}, \tag{1.5.11}$$

$$\mathcal{H}_{sd} = \sum_{\mathbf{k}\sigma} V_{\mathbf{k}} (a_{\mathbf{k}\sigma}^\dagger a_{\mathbf{d}\sigma} + c.c).$$

Here $\mathcal{H}_0$ and $\mathcal{H}_{0d}$ are the unperturbed energies of the conduction and localized electrons, respectively. The latter can be in the d states of the magnetic impurity atoms which are dissolved in a metal. $\mathcal{H}_c$ is the repulsive interaction between the localized electrons, to which a constant strength given by the parameter U is assigned. The last term $\mathcal{H}_{sd}$ represents the s-d interaction. This type of s-d interaction corresponds to a one-electron energy and is entirely different from the s-d exchange interaction.

The Anderson model is solvable in the absence of the correlation term $\mathcal{H}_c$. Since it describes the interaction of conduction and localized electrons, it is related to the Kondo problem.

Because there are so many interesting phenomena for which electrons are responsible, we shall not be able to discuss all the models and phenomena in detail. We intend in the following only to present a very basic description which is based more or less on the OCP model. Even within this model, some phenomena can be described by single-electron theory while others require many-electron theory. Depending on the particular property under consideration a suitable model and approach must be introduced.

2. Dielectric Function

An electron plasma can be described by its dielectric function. Hence the theoretical determination of this function is important.

2.1 RPA Dielectric Function

One of the most basic quantities for an electron system as an OCP is its dielectric function. This is deiermined as a linear response of this plasma to a test charge. Because of the charge neutrality, the Fourier transform $\varepsilon(\mathbf{q}\omega)$ of the dielectric function is defined in the q-ω space by [2.1]

$$\tilde{D}(\mathbf{q}, \omega) = \varepsilon(\mathbf{q}, \omega)\tilde{E}(\mathbf{q}, \omega), \tag{2.1.1}$$

where $\tilde{D}$ and $\tilde{E}$ are the Fourier transforms of the displacement and electric field vectors which depend on the densities of a test charge and induced charges as follows:

$$\text{div } \mathbf{D} = 4\pi V^{-1} \times (\text{ test charge})$$
$$\text{div } \mathbf{E} = 4\pi V^{-1} \times (\text{ test charge} + \text{induced charges}).$$

The dielectric function determines screening of the long range Coulomb potential, which in turn characterizes electronic properties of the plasma. Its explicit form has been evaluated in the random phase approximation (RPA). In this approximation, the density of the induced charges is evaluated as a linear response of the ideal case in which Coulomb interaction is absent.

We first introduce the free particle polarization function. In the RPA, it is defined by [2.1]

$$\chi(q, \omega) = -\frac{2}{(2\pi)^3} \int \frac{f(\varepsilon_{\mathbf{k+q}}) - f(\varepsilon_k)}{\varepsilon_{\mathbf{k+q}} - \varepsilon_k + \omega + i0} d\mathbf{k}, \tag{2.1.2}$$

where $i0$ represents a small imaginary number to be brought to zero after the integration, $f(\varepsilon_k)$ is the Fermi distribution function, and ε_k is the kinetic energy. As denoted, the function $\chi(q, \omega)$ depends on $q = |\mathbf{q}|$. The dielectric function is given by [2.1]

$$\varepsilon(q, \omega) = \frac{1}{1 - u(q)\chi(q, \omega)}. \tag{2.1.3}$$

This function is also dependent on q rather than $\mathbf{q}$, as indicated. In (2.1.3), the Fourier transform $u(q)$ of the Coulomb potential may be replaced by $u(q)/\varepsilon(q,\omega)$ because the electron–electron interaction takes place in the medium of the dielectric constant $\varepsilon(q,\omega)$. This replacement yields the self-consistent RPA expression

$$\varepsilon(q,\omega) = 1 + u(q)\chi(q,\omega). \tag{2.1.4}$$

The k integral in (2.1.2) corresponds to 3D. For 1D and 2D, $d\mathbf{k}/(2\pi)^3$ should be replaced by $dk/(2\pi)$ and $d\mathbf{k}/(2\pi)^2$ respectively. The Fourier transform $u(q)$ of the Coulomb potential is given by

$$u(q) = \begin{cases} 4\pi e^2/q^2 & \text{(3D)} \\ 2\pi e^2/q & \text{(2D)} \end{cases} \tag{2.1.5}$$

It cannot be defined in 1D in the same way. However, for TTF–TCNQ, its 3D form may be adopted because electrons in different stacks interact with each other.

The dielectric function is complex. In the RPA and for 3D, its real and imaginary parts, $\varepsilon_r(q,\omega)$ and $\varepsilon_i(q,\omega)$, are given by

$$\varepsilon_r(q,\omega) = 1 + \frac{2}{\pi a_0 k_F s^2}\left[1 + \frac{1}{2s}\left(\frac{(t+s^2)^2}{4s^2} - 1\right)\ln\left|\frac{s^2 - 2s + t}{s^2 + 2s + t}\right|\right.$$
$$\left. - \frac{1}{2s}\left(\frac{(t-s^2)^2}{4s^2} - 1\right)\ln\left|\frac{s^2 + 2s - t}{s^2 - 2s - t}\right|\right] \tag{2.1.6}$$

$$\varepsilon_i(q,\omega) = \begin{cases} \dfrac{t}{a_0 k_F s^3} & (0 \leq t \leq 2s - s^2), \\[2mm] \dfrac{2}{a_0 k_F s^3}\left[1 - \dfrac{(t-s^2)^2}{4s^2}\right] & (-s^2 + 2s \leq t \leq s^2 + 2s), \\[2mm] 0 & (t \geq 2s + s^2). \end{cases} \tag{2.1.7}$$

Here, $s = q/k_F$ and $t = \omega/k_F^2$, k_F being the Fermi energy ($\hbar = 1$, $2m = 1$). The dimensionless variables s and t represent respectively q and ω. Note in these equations that

$$4/(\pi a_0 k_F s^2) = q_T^2/q^2.$$

Here a_0 is the Bohr radius and

$$q_T = (6\pi ne^2)^{1/2}/k_F \tag{2.1.8}$$

is the *Thomas-Fermi screening constant* which is effective at low temperatures.

The function $\chi(q,\omega)$ depends on the dimensionality. In the RPA and in the static limit, it is real and is given by

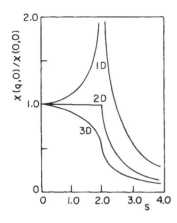

Fig. 2.1. Normalized polarization function vs $s = q/k_F$

$$\chi(q,0) = \begin{cases} \dfrac{k_F}{4\pi^2}\left[1 - \dfrac{1}{4}\left(s - \dfrac{4}{s}\right)\ln\left|\dfrac{s^2 + 2s}{s^2 - 2s}\right|\right], & \text{(3D)} \\[2ex] \dfrac{1}{2\pi}\begin{cases} \times 1, & s \leq 2, \\[1ex] \times\left[1 - \left(1 - \dfrac{4}{s^2}\right)^{1/2}\right], & s \geq 2, \end{cases} & \text{(2D)} \\[3ex] \dfrac{1}{2\pi q}\ln\left|\dfrac{s^2 + 2s}{s^2 - 2s}\right|, & \text{(1D)} \end{cases} \qquad (2.1.9)$$

where $s = q/k_F$ as before, and k_F is given by (1.1.5). At $q = 2k_F$, $\chi(q,0)$ diverges in 1D, shows a kink in 2D, and changes its curvature in 3D. These properties are illustrated in Fig. 2.1, in which $\chi(q,0)/\chi(0,0)$ is plotted against s.

The 3D expression for the dielectric function, first studied by *Lindhard* [2.2], is rather involved. Near $\omega = 0$, $\varepsilon_r(q,\omega)$ for 3D is positive as indicated by its limiting expression

$$\varepsilon_r(q,0) = 1 + 4/(\pi a_0 k_F s^2). \qquad (2.1.10)$$

This finite value corresponds to static screening, which manifests itself strongly for $q \to 0$. In fact, generally $\varepsilon_r(q,\omega)$ decreases as ω increases, becomes negative, and eventually approaches zero. In particular, in the long wavelength limit in which $q = 0$,

$$\varepsilon_r(0,\omega) = 1 - \omega_p^2/\omega^2. \qquad (2.1.11)$$

Here, ω_p, the *plasma frequency*, is defined by

$$\omega_p = (4\pi n e^2/m)^{1/2}. \qquad (2.1.12)$$

The imaginary part vanishes at $\omega = 0$, and is finite only for certain ω values. In fact, it reaches a maximum and then drops to zero again. In 2D, the RPA dielectric function is given by

$$\varepsilon(q,\omega) = 1 + \frac{u(q)}{2\pi}$$
$$\times \left(1 - \frac{1}{2s}\left\{([s+it/s)]^2 - 4]^{1/2} + [(s-it/s)^2 - 4]^{1/2}\right\}\right). \quad (2.1.13)$$

As in (2.1.6), the real and imaginary parts of this function can be written down. For simplicity, let us use the natural unit in which $\hbar = 1$ and $2m = 1$ and introduce

$$f_\pm = 1 - g_\pm^2, \quad g_\pm = (s \pm t/s)/2,$$
$$U = 2^{1/2} r_s/s = e^2/q.$$

The two parts assume different forms in accordance with $f_\pm$ as follows

(i) $f_+ \geq 0$ and $f_- \geq 0$

$$\varepsilon_r = 1 + U,$$
$$\varepsilon_i = (U/s)(f_-^{1/2} - f_+^{1/2}). \quad (2.1.14a)$$

(ii) $f_+ \leq 0, \ f_- \geq 0$

$$\varepsilon_r = \left[1 + U[1 - (-f_+)^{1/2}/s]\right],$$
$$\varepsilon_i = (U/s)(f_-)^{1/2}. \quad (2.1.14b)$$

(iii) $f_+ \geq 0, \ f_- \leq 0$

$$\varepsilon_r = 1 + U\left(1 - s^{-1}(g_- + 1)\left|\frac{1-g_-}{1+g_-}\right|^{1/2}\right),$$
$$\varepsilon_i = -(U/s)(1+g_+)\left|\frac{1-g_+}{1+g_+}\right|^{1/2}. \quad (2.1.14c)$$

(iv) $f_+ \leq 0, \ f_+ \leq 0$

$$\varepsilon_r = 1 + U\left[1 - \frac{1}{s}\left((1+g_-)\left|\frac{1-g_-}{1+g_-}\right|^{1/2} - (1+g_+)\left|\frac{1-g_+}{1+g_+}\right|\right)\right],$$
$$\varepsilon_i = 0. \quad (2.1.14d)$$

In spite of the functional differences between the RPA dielectric functions for 3D and 2D, the screening behavior is similar in that it is strong in the static, long-wavelength limit. There is also a certain similarity between the two cases when ε_r or ε_i is plotted against t. Figure 2.2 illustrates the 2D case for $r_s = 1$ and $q = 0.5k_F$. As a function of t, ε_r does not decrease immediately but stays constant for small t or ω, vanishes at $\omega_p = 1.5k_F^2$, and gradually approaches 1 for large ω. Here, $k_F^2 = \varepsilon_F$ is the ideal Fermi energy ($\hbar = 1, 2m = 1$). In the corresponding plot in Fig. 2.3 for 3D, ε_r is maximum at $t = 0$, and approaches 1 for large t after passing through a minimum. The imaginary part

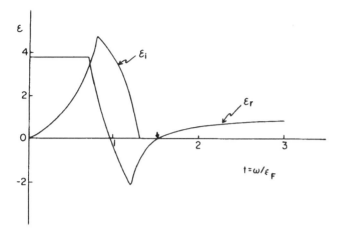

Fig. 2.2. Real and imaginary parts of the RPA dielectric function for 2D ($r_s = 1$ and $q = 0.5k_F$)

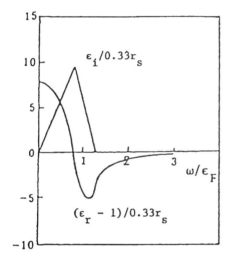

Fig. 2.3. Real and imaginary parts of the RPA dielectric function for 3D

ε_i is triangular in 3D. In this graph, the real and imaginary parts are scaled in terms of $0.33r_s$, which represents $2(4/9\pi)^{1/3}r_s/\pi = e^2/(\pi k_F)$. Hence, the curves are independent of r_s. In 2D ε_i is illustrated by a slightly modified triangle as in Fig. 2.2. This modification depends on q. Note that ε_i is zero at the origin and is finite only up to approximately $\omega = 1.3k_F^2$. The arrow indicates the location of ω_p.

As q increases, the real part becomes smaller and decreases as t increases from the origin. Such a variation is shown in Fig. 2.4 for the case $q = k_F$. As can be seen, the constant value of ε_r is close to one half of the case of $q = k_F/2$ in Fig. 2.2. The triangular shape of ε_i is still maintained but is flattened somewhat and stretched out. Significant changes in ω_r and ω_i take

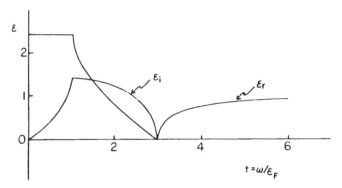

Fig. 2.4. 2D dielectric function for $q = k_F$ and $r_s = 1$

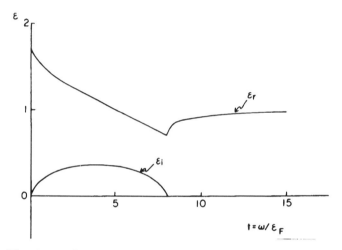

Fig. 2.5. 2D dielectric function for $q = 2k_F$ and $r_s = 1$

place at $q = 2k_F$ as shown in Fig. 2.5. Note in this graph that ε_r is still finite at the origin. The singularity in the real part is due to the plasmon. The imaginary part is now smooth and parabolic below the plasmon frequency, where it vanishes.

In general, the dielectric function $\varepsilon(\mathbf{q}, \omega)$ is related to the dynamic structure factor $S(\mathbf{q}, \omega)$ by

$$nu(q)S(\mathbf{q}, \omega) = -\Im\{1/\varepsilon(\mathbf{q}, \omega)\}, \tag{2.1.15}$$

where $u(q)$ is the Fourier transform of the Coulomb potential. The reason is that $1/\varepsilon(q, \omega)$ depends on the charge density n_q which in turn determines $S(\mathbf{q}, \omega)$ as the density–density correlation function. Note that the Coulomb interaction energy in the ground state can be expressed in terms of the Fourier transform n_q of the density such that

$$E_c = \sum_{\mathbf{q}\neq 0} \frac{1}{2} u(q)(n_{\mathbf{q}}^* n_{\mathbf{q}} - n)$$

$$= -\sum_{\mathbf{q}} \frac{1}{2\pi} \left(\int_0^\infty d\omega \Im\{1/\varepsilon(\mathbf{q},\omega)\} + \frac{2\pi n e^2}{q^2} \right) \qquad (2.1.16)$$

In these equations, n is the equilibrium density.

2.2 Beyond the RPA

The RPA dielectric function is only approximate and in general becomes a poor approximation as electron correlations increase. Essentially the same shortcoming of the RPA appears in some other quantities. Therefore, efforts have been made to improve the RPA results. We discuss in what follows some of these efforts on classical OCPs in 3D.

2.2.1 Monte Carlo Calculations

The recent remarkable developments in computer facilities have made numerical approaches to difficult problems very effective. Monte Carlo simulations have produced interesting results, especially for electron systems with strong correlations, which are helpful in improving analytical approaches. Important results based on Monte Carlo simulations of electron systems are numerous and will be mentioned from time to time throughout this book.

We shall be concerned in this chapter with improvements on the RPA dielectric function . For this purpose let us first discuss some relevant Monte Carlo results for the pair distribution function $\rho_2(r)$ because it is related to the dielectric function and determines the thermodynamic properties. The Fourier transform of $\rho_2(r)$ yields the static structure factor $S(q)$

$$S(q) = 1 + \frac{1}{n} \int [\rho_2(r) - n^2] e^{i\mathbf{q}\cdot\mathbf{r}} d\mathbf{r}. \qquad (2.2.1)$$

The static structure factor is obtained by integrating the dynamic structure factor over frequency,

$$S(q) = \int_0^\infty S(q,\omega) d\omega. \qquad (2.2.2)$$

A pioneering Monte Carlo work on the pair distribution function of an OCP was performed by *Brush* et al. [2.3a] in 1966 for systems containing from 32 to 500 particles in the range $0.05 \leq \Gamma \leq 100$. Later, Hansen [2.3b] and Slattery et al. [2.3c] reported improved Monte Carlo results for larger values of Γ. Figure 2.6 illustrates the radial distribution function of an OCP based on these Monte Carlo calculations. The curves for $\Gamma \leq 100$ illustrate the results of *Brush* et al., while the curve $\Gamma = 160$ represents those of *Slattery*

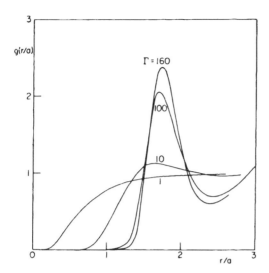

Fig. 2.6. Monte Carlo results for the radial distribution function

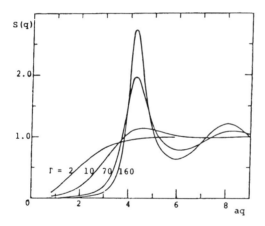

Fig. 2.7. Structure factor for several Γ [2.3e]

et al., who improved *Hansen*'s results for large Γ. The abscissa is scaled in terms of the *ion sphere radius* $a = (3/4\pi n)^{1/3}$. The radial distribution function for $\Gamma < 3$ increases monotonically from 0 to its asymptotic value 1. The Monte Carlo results in particular for $\Gamma = 0.1$ agree well with the Debye Hückel results. The radial distribution function for $\Gamma \geq 3$ exhibits oscillations indicating the development of short range correlations. These short range correlations are not well represented by the RPA. It is known that the RPA overestimates these. The static structure factor $S(q)$ reported by *Galam* and *Hansen* [2.3e] is shown in Fig. 2.7. In the abscissa, $a = (4\pi n/3)^{-1/3}$ is the ion-sphere radius. The general shape of these curves resembles that of $g(r)$.

Increasing oscillations of $S(q)$ are seen as Γ increases above 10. The function $S(q)$ will be used shortly. Numerical results on the Coulomb energy will be discussed toward the end of this section.

2.2.2 Hubbard's Attempt

In an attempt to improve the RPA by taking account of short-range exchanges, *Hubbard* [2.4] suggested in 1957 that (2.1.3) be replaced by

$$\varepsilon(q,\omega) = 1 + \frac{u(q)\chi(q,\omega)}{1 - u(q)G(q)\chi(1,\omega)} \tag{2.2.3}$$

where

$$G(q) = \frac{q^2}{2(q^2 + k_F^2)}. \tag{2.2.4}$$

The factor $1/2$ in $G(q)$ comes from the fact that only spin parallel electrons can have exchange interaction, and the q-dependent part $[q^2 + k_F^2]/q^2$ is called the *Thomas-Fermi dielectric function*. The appearance of the Fermi wave number k_F indicates that at low temperatures momentum exchange due to interaction takes place only near the Fermi surface.

Equation (2.2.4) is an approximation which becomes poor for large q. Hence, let us discuss more generally that a certain function $G(q,\omega)$, called the dynamic *local field correction*, must be introduced to improve the RPA dielectric function [2.5–7]. As indicated, this correction can depend not only on q but also on ω. However, in what follows we consider mostly the static case with the static local field correction $G(q)$, which may be different from that given by (2.2.4). In the static case, we shall attempt to interpret the origin of this function.

Before this attempt we remark on the simplest form of the static dielectric function $\varepsilon(q)$. In (2.2.3), the polarization function $\chi(q)$ for an ideal electron gas without Coulomb interaction is simply

$$\chi(q) = n\beta,$$

where $\beta = 1/k_B T$. Hence,

$$\varepsilon(q) = 1 + \frac{\kappa_D^2}{q^2 - \kappa_D^2 G(q)}, \tag{2.2.5}$$

where $\kappa_D^2 = 4\pi\beta n e^2$. Although we do not know the local field correction $G(q)$ for the static case, we can expect that it approaches 0 for $q \to 0$. Hence, $\varepsilon(q) \to \infty$. This indicates the crudeness of the approximation.

2.2.3 Phenomenological Consideration of $\varepsilon(\mathbf{q}, \omega)$

Let us now examine the dielectric function in a phenomenological way. We note that the dielectric function $\varepsilon(q, \omega)$ can be introduced generally such that the total potential $V = V(q, \omega)$ is proportional to an externally imposed potential $V_e = V_e(q, \omega)$, and that the total potential includes the Coulomb potential $u(q)$ due to the density of polarization charges $\langle \rho \rangle = \langle \rho(q, \omega) \rangle$. Accordingly we have

$$V = \frac{V_e}{\varepsilon} = V_e - u(q)\langle \rho \rangle. \tag{2.2.6}$$

The polarization charges are proportional to the external potential, the proportionality constant being the response function $\chi = \chi(q, \omega)$,

$$\langle \rho \rangle = \chi V_e. \tag{2.2.7}$$

However, a new response function $\chi_s = \chi_s(q, \omega)$ may be introduced such that $\langle \rho \rangle$ depends also on a local effective potential $v(q, \omega)$,

$$\langle \rho \rangle = \chi_s[V_e - v(q, \omega)\langle \rho \rangle]. \tag{2.2.8}$$

This yields

$$\langle \rho \rangle = \frac{\chi_s V_e}{1 + \chi_s v} \tag{2.2.9}$$

The function $v = v(q, \omega)$ is unknown and yet to be specified.

From (2.2.7) and (2.2.9), we find

$$\chi = \frac{\chi_s}{1 + \chi_s v}. \tag{2.2.10}$$

Also, (2.2.6) and (2.2.9) yield

$$\varepsilon = 1 + \frac{u\chi_s}{1 - \chi_s u(1 - \frac{v}{u})}. \tag{2.2.11}$$

Comparing this result with (2.2.3), we see that the function $G(q)$ may be interpreted as $(1 - v/u)$ in the static limit:

$$G(q) = 1 - v/u. \tag{2.2.12}$$

The denominator of (2.2.9) or (2.2.10) can be interpreted as a result of summing a series in ascending powers of an effective potential v.

If this potential is equated to u, then

$$\langle \rho \rangle = \chi_s(V_e/\varepsilon). \tag{2.2.13}$$

We learn that χ_s represents the response to the external potential screened by ε. Note in this case that

$$\varepsilon = 1 + u\chi_s. \tag{2.2.14}$$

This form may be compared with the self-consistent definition of ε which is obtained by replacing u in (2.2.6) by u/ε in the following form:

$$\varepsilon = 1 + u\chi. \tag{2.2.15}$$

On the other hand, if $v = 0$,

$$\chi = \chi_s, \quad \varepsilon = 1/(1 - u\chi_s). \tag{2.2.16}$$

If in this expression u is replaced by u/ε, we recover (2.2.14).

2.2.4 Self-Consistent Approach

The above phenomenological consideration does not provide a method of determining the dielectric function. In 1968 *Singwi* et al. [2.7a,b] proposed a self-consistent determination of the unknown function which is given by (2.2.12). As can be seen from (2.1.15) and (2.2.2), the static structure factor $S(q)$ is given in terms of the dielectric function such that

$$S(q) = -\frac{1}{nu(q)} \int \Im\left\{\frac{1}{\varepsilon(q,\omega)}\right\} d\omega. \tag{2.2.17}$$

Using a simplified equation for the pair distribution function, they proposed to determine $G(q)$ numerically from the integral:

$$G_{ST}(q) = -\frac{1}{n(2\pi)^3} \int \frac{\mathbf{q}\cdot\mathbf{k}}{k^2}[S(|\mathbf{q}-\mathbf{k}|) - 1]d\mathbf{k}. \tag{2.2.18}$$

and the dielectric function from

$$\varepsilon(q,\omega) = 1 + \frac{u(q)\chi}{1 - u(q)G(q)\chi}, \tag{2.2.19}$$

with the RPA polarization function χ. For an assumed $G(q)$, ε can be obtained from this expression. This ε is then introduced into (2.2.17) to determine $S(q)$, which is in turn used in (2.2.18). The resulting $G_{ST}(q)$ must agree with the one chosen initially. The process is repeated numerically until self-consistency has been reached.

It has been found that the above scheme falls short of reproducing the known classical results for small Γ and does not satisfy the compressibility sum rule [2.1]. Therefore, several modifications have been proposed. In fact, *Singwi* et al. [2.7c] proposed in 1970 to introduce the static dielectric function $\varepsilon(q)$ into the integrand of (2.2.18). Hence, the correction function is now determined from

$$G_{ss}(q) = -\frac{1}{n(2\pi)^3} \int \frac{\mathbf{q}\cdot\mathbf{k}}{k^2\varepsilon(q)}[S(|\mathbf{q}-\mathbf{k}|) - 1]\,d\mathbf{k}. \tag{2.2.20}$$

Vashishta and *Singwi* [2.7d] proposed the replacement

$$G \rightarrow \left(1 + \alpha n\frac{\partial}{\partial n}\right)G(q)$$

with a parameter α to be chosen so as to satisfy the compressibility sum rule, in order to achieve agreement with the classical OCP results for small Γ. However, these modifications do not produce significant improvements beyond the Debye–Hückel energy expression.

On the other hand, *Niklasson* [2.7e] derived a limiting formula

$$\lim_{q\to\infty} G(q,\omega) = \frac{2}{3}\left[1 - g(0)\right],\tag{2.2.21}$$

where $g(0)$ is the radial distribution function at $r = 0$. Note that for $g(0) = 1/2$ [See (3.3.11)],

$$\lim_{q\to\infty} G(q,\omega) = \frac{1}{3}.$$

Effects of short range correlations on $G(q)$ were considered by *Yasuhara* [2.7f].

Generally, the correction function $G(q)$ may depend on frequency ω and wave number q. However, let us confine ourselves to the static case and examine these attempts to determine the correction function. For this purpose we shall use the well-known *hypernetted chain*, to be abbreviated as HNC, *approximation* [2.8] to the radial distribution function. This approximation is correct to first order in density. In order to show how this approximation yields the radial distribution function, it is convenient to introduce first an unknown potential ϕ_H such that

$$g(r) = \exp\left[-\beta\phi_H(r)\right].\tag{2.2.22}$$

The HNC approximation amounts to using ϕ_H determined from the equation [2.8a]

$$-\beta\left[\phi_H(r_{12}) - \phi(r_{12})\right] = n\int C(r_{13})\left[g(r_{32}) - 1\right]d\mathbf{r}_3,\tag{2.2.23}$$

where the direct correlation function $C(r)$ is given by

$$C(r) = g(r) - 1 + \beta\left[\phi_H(r) - \phi(r)\right].\tag{2.2.24}$$

The gradient of the direct correlation function is given by

$$\nabla C = \nabla g + \beta\nabla(\phi_H - \phi),\tag{2.2.25}$$

where the arguments of the functions have been omitted for simplicity. The Fourier transform of (2.2.25) is

$$-\tilde{C}(q) = \beta u(q) + \frac{\beta}{n(2\pi)^3}\int \frac{\mathbf{q}\cdot\mathbf{k}}{q^2}u_H(k)\left[S(|\mathbf{q}-\mathbf{k}|) - 1\right]d\mathbf{k}$$
$$= \beta u(q)\left[1 - G(q)\right],\tag{2.2.26}$$

where $\tilde{C}(q)$, $u_H(q)$ and $u(q)$ are the Fourier transforms of $C(r)$, $\phi_H(r)$ and $\phi(r)$ respectively, and where in the last equality $G(q)$ is defined by

$$G(q) = -\frac{1}{n(2\pi)^3}\int \frac{\mathbf{q}\cdot\mathbf{k}}{q^2}\frac{u_H(k)}{u(q)}\left[S(|\mathbf{q}-\mathbf{k}|) - 1\right]d\mathbf{k}.\tag{2.2.27}$$

We note that $u_H(q)$ is given by the Fourier transform of the HNC equation (2.2.23) such that

$$u_H = u - \frac{1}{\beta}\tilde{C}(S-1). \qquad (2.2.28)$$

On the right side, the Fourier transform $\tilde{C}(q)$ can be obtained by the definition (2.2.24) of the HNC direct correlation function as

$$\tilde{C} = \frac{1}{n}(S-1) + \beta(u_H - u). \qquad (2.2.29)$$

Introducing this into (2.2.28) we find

$$u_H = u - \frac{1}{\beta n}\left(1 - \frac{1}{S}\right)(S-1). \qquad (2.2.30)$$

Elimination of u_H in (2.2.29) yields

$$\tilde{C} = \frac{1}{n}\left(1 - \frac{1}{S}\right). \qquad (2.2.31)$$

Using this in (2.2.26), we obtain

$$S = \frac{1}{1 + n\beta u(1-G)}. \qquad (2.2.32)$$

Hence,

$$G = 1 + \left(1 - \frac{1}{S}\right)\frac{q^2}{\kappa_D^2}, \qquad (2.2.33)$$

where $\kappa_D^2 = 4\pi n e^2/k_B T$ as before.

Equation (2.2.33) enables us to evaluate $G(q)$ if $S(q)$ is given. For small q, $G(q)$ approaches zero if $S(q)$ varies with q linearly. For large q, $S(q) \to 1$ so that $G(q) \to 1$. Since $S(q)$ is oscillatory for large Γ, $G(q)$ is expected to be also oscillatory. A numerical result [2.8b] based on the Monte Carlo $S(q)$ of *Brush* et al. and others support these expectations and also $G(q)$ is Γ dependent. However, since a local field correction is expected to be important for large q rather than for small q, such a result may not necessarily represent short-distance corrections. In any case, in neglect of its oscillatory behavior and dependence on Γ, $G(q)$ may be approximated by

$$g(q) = \begin{cases} aq/4, & aq \lesssim 4, \\ 1, & aq \gtrsim 4, \end{cases}$$

where a is the ion-sphere radius.

At large distances or small q, $S(q,0)$ might approach $S_D(q)$ given by

$$S_D(q) = \frac{q^2}{q^2 + \kappa_D^2}. \qquad (2.2.34)$$

This expression shows that $S(q)$ approaches zero proportionally to q^2. Note that when $S_D(q)$ is used for S in (2.2.33), $G(q) = 0$.

We remark that if a dielectric function ε is introduced such that

$$u_B = \frac{u}{\varepsilon} \qquad (2.2.35)$$

we obtain

$$\frac{1}{\varepsilon} = S - G(S - 1). \qquad (2.2.36)$$

This equation can be used to evaluate the dielectric function when S and G are given.

Note that (2.2.24) is not independent of (2.2.25). Hence, the function $G(q)$ must satisfy a self-consistent equation

$$-G(q) = \frac{1}{n(2\pi)^3} \int \frac{\mathbf{q} \cdot \mathbf{k}}{k^2} \{S(k) - G(k)[S(k) - 1]\} [S(|\mathbf{q} - \mathbf{k}|) - 1] \, d\mathbf{k}. \qquad (2.2.37)$$

Equation (2.2.37) was derived by *Chihara* [2.9] in a somewhat more complicated way. Since it is an integral equation and S is unknown, the determination of G requires further steps. Therefore, let us choose G in the integrand in a few convenient ways. Such choices are arbitrary but may be acceptable in view of the approximate nature of the original HNC equation. Sometimes, errors in an approximation are compensated by further approximations.

Let us comment on two special cases. First, G in the integrand is set equal to 1. We then recover (2.2.18) for G_{ST}. Second, if $G = 0$ in the integrand, we obtain

$$-G(q) = \frac{1}{(2\pi)^3 n} \int \frac{\mathbf{q} \cdot \mathbf{k}}{k^2} S(k) [S(|\mathbf{q} - \mathbf{k}|) - 1] \, d\mathbf{k}, \qquad (2.2.38)$$

which is the form used by *Totsuji* and *Ichimaru* [2.10]. In this equation, we could replace $S(k)$ in the integrand by $1/\varepsilon(q)$ in view of (2.2.36) for $G = 0$. The result of this replacement yields (2.2.20) for G_{ss}.

2.2.5 Excess Energy

Coulomb effects on the thermodynamic properties of a classical OCP may be judged from the internal energy. For this purpose, it is convenient to use the excess energy U_{ex}. For a classical OCP, it is defined by

$$U_{ex} = U - \frac{3}{2} N k_B T, \qquad (2.2.39)$$

where U is the internal energy. As can be expected, $U_{ex}/N k_B T$ is negative and varies approximately as

$$-\varepsilon \sim -\Gamma^{3/2}.$$

Here, ε is the plasma parameter. This expression may be compared with that given by (3.1.16).

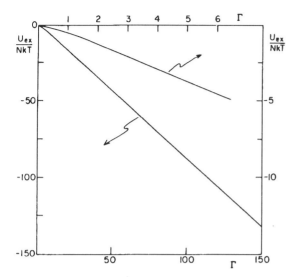

Fig. 2.8. Excess internal energy of a classical OCP [2.10]

Figure 2.8 illustrates the Monte Carlo results for the excess energy obtained by *Slattery* et al. [2.3]. The upper curve with the top and right coordinates represents U_{ex}/Nk_BT for a small region of Γ. The lower curve corresponds to the large scale given by the left ordinate and bottom abscissa. Except for the region below $\Gamma \lesssim 2$, U_{ex} appears to vary almost linearly with Γ. However, *Slattery* et al. reported an empirical equation given by

$$U_{ex}/Nk_BT = -0.89752\Gamma + 0.94544\Gamma^{1/4} + 0.17954\Gamma^{-1/4} - 0.80049.$$

There are some other numerical calculations of the excess energy. For instance, the energy has been determined [2.10] based on the Monte Carlo results of *Brush* et al. and the work of *Berggren* [2.11].

The radial distribution function $g(r)$ of a many-body system has been investigated diagrammatically. For classical systems, it is often expressed as

$$g(r) = \exp(-\beta\phi + \psi). \qquad (2.2.40)$$

The diagrams which contribute to $\psi = \psi(|\mathbf{r}_2 - \mathbf{r}_2|)$ can be classified by the presence and non-presence of nodes, which are the junctions that must be passed in going from $\mathbf{r}_1$, to $\mathbf{r}_2$, where $r = |\mathbf{r}_2 - \mathbf{r}_1|$. The nodal diagram contribution is given by a sum over the irreducible diagrams of arbitrary numbers of nodes. The HNC approximation amounts to neglecting those diagrams without nodes. Hence, there have been several attempts at improving the HNC approximation, for instance, by taking into consideration short range interaction and three-particle correlations [2.12].

2.3 Plasmon Dispersion

The condition that the dielectric function of an electron gas vanishes

$$\varepsilon(q,\omega) = 0, \qquad (2.3.1)$$

determines the relation. For 3D, the plasma frequency ω_p determined from this equation is well defined in the limit $q \to 0$ and is given by

$$\omega_p^2 = \frac{4\pi n e^2}{m}, \qquad n = \frac{N}{V}. \qquad (2.3.2)$$

The corresponding plasmon energy ε_p is approximately $12\,\mathrm{eV}$ for metallic electrons with density $n \sim 10^{23}\,\mathrm{cm}^{-3}$. It represents a quantum of the collective plasma oscillation with frequency of order $2 \times 10^{16}\,\mathrm{Hz}$.

Although the plasmon frequency is well defined, the plasmon dispersion relation determined by (2.3.1) depends on a particular form of the dielectric function which is subject to a certain approximation. In the most familiar random phase approximation (RPA), the dispersion relation for small q is given by

$$\hbar\omega = \hbar\omega_p + \alpha\frac{\hbar^2}{m}q^2. \qquad (2.3.3)$$

Hence, when plotted against q, it varies quadratically, starting with the plasmon energy $\hbar\omega_p$. The value of the constant α is

$$\alpha = \frac{3}{5}\frac{\varepsilon_F}{\hbar\omega_p}. \qquad (2.3.4)$$

It is proportional to the ratio of the average energy per electron to the plasmon energy. The numerical factor $3/5$ is characteristic of the RPA.

The value of α decreases when electron correlations are taken into consideration. For example, it has been shown that an exchange-correlation correction decreases α such that [2.13]

$$\alpha_{xc} = \alpha\left[1 - \left(\frac{\hbar\omega_p}{\varepsilon_F}\right)^2\right]. \qquad (2.3.5)$$

The correction amounts to a decrease in α by approximately 10%.

Experimentally, the plasmon dispersion relation can be obtained by electron and X-ray scattering experiments. In principle, these experiments measure the energy loss, which is related to the imaginary part of the inverse dielectric function. In a single relaxation time approximation, one can adopt a Lorentzian form for $\Im\{-1/\varepsilon\}$,

$$\Im\{-1/\varepsilon\} = \frac{\omega_p^2\omega/\tau}{(\omega^2 - \omega_p^2)^2 + (\omega/\tau)^2}. \qquad (2.3.6)$$

If the plasmon excitation is the only energy loss, the loss function can be expressed as

$$\Im\{-1/\varepsilon\} = \frac{\pi}{2}\omega_p\delta(\omega - \omega_p). \tag{2.3.7}$$

We obtain

$$\int_0^\infty \Im\{-1/\varepsilon\}\omega d\omega = \frac{\pi}{2}\omega_p^2. \tag{2.3.8}$$

This is what we expect from the sum rule of the dynamical structure factor

$$\int_0^\infty S(q,\omega)\omega d\omega = \frac{N\hbar^2}{2m}q^2, \tag{2.3.9}$$

because the loss function is connected with $S(q,\omega)$ by

$$-\Im[1/\varepsilon(q,\omega)] = \frac{4\pi e^2}{\hbar q^2 V}[S(q,\omega) - S(q,-\omega)]. \tag{2.3.10}$$

The quadratic variation in the long-wavelength limit of the plasmon dispersion relation has been confirmed by scattering experiments. However, experimental values of the proportionality constant α are generally smaller than the theoretical value given by (2.3.4) [2.14]. Moreover, experimental dispersion curves deviate from the q^2 proportionality as q increases. Indeed, even a maximum, followed by a dip, has been observed. Figure 2.9, from [2.14b], illustrates such deviations which start at around $q = q_c$. The ordinate is the plasmon energy normalized to its limiting value. The dashed curve represents the RPA result for Be, and the chain curve corresponds to the RPA with the exchange and correlation corrections. Such deviations are of course due to the inadequacy of the RPA for large q. This failure of the RPA is connected with that for $S(q,\omega)$. Although there are attempts to improve the RPA $S(q,\omega)$, for example by introducing local field corrections, further theoretical studies are still needed.

So far our discussions have been concerned with plasmons in bulk. Since the plasmon dispersion relation is important for the study of collective coupling of electrons, mention must be made of the case of two dimensions. It is crucial to recognize that a plasmon does not exist for $q \to 0$ in two dimensions. Hence, one of the most significant differences between bulk and surface electrons is exhibited in the plasmon dispersion relation in the $q \to 0$ limit.

For a classical 2D system of electrons in Si inversion or accumulation layers at a Si/SiO$_2$ interface, it has been shown that [2.15]

$$\omega_p^2 = \frac{ne^2}{2\kappa m}q, \tag{2.3.11}$$

where κ is the dielectric constant. If the dielectric constants of Si and SiO$_2$ are κ_s and κ_{ox} respectively, one can use the effective dielectric constant κ given by

$$\kappa = \frac{1}{2}[\kappa_s + \kappa_{ox}\coth(qd)]. \tag{2.3.12}$$

where d is the layer thickness. For GaAs/GaAlAs quantum wells, the dielectric constants of Ga and Al can be used instead of κ_s and κ_{ox} The essential

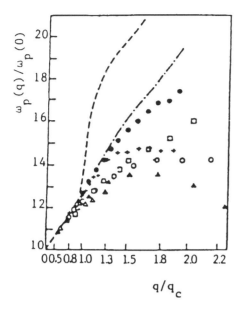

Fig. 2.9. Plasmon dispersion [2.14 b]

feature of the two-dimensional dispersion relation (2.3.11) is that ω_p vanishes in the limit $q \to 0$. This feature was observed by *Theis* et al. [2.16] in 1980, confirming the 2D character of inversion layer electrons. They found that in general the theoretical dispersion relation (2.3.11) is satisfactory.

However, several effects cause deviations from the theoretical relation [2.17]. First, the finite thickness of a sample causes a more complicated form for the effective dielectric constant. If the wave function behaves in the perpendicular direction such that

$$\psi(z) \sim z \exp(-zb/2),$$

where b is a parameter, the effective dielectric function is given by

$$\varepsilon(q, \omega) = \frac{\varepsilon_1}{G(q/b)}. \tag{2.3.13}$$

Here

$$G(x) = (1+x)^{-6} \left[\frac{1}{8}(33 + 54x + 44x^2 + 18x^3 + 3x^4)\right]$$
$$+ \frac{2\varepsilon_1}{\varepsilon_1 + \varepsilon_2 \coth(qd)}, \tag{2.3.14}$$

where ε_1 is the dielectric constant of Si or GaAs, and ε_2 is that for SiO_2 or GaAlAs layers. This finite thickness effect reduces the plasmon frequency below the above classical result.

Second, the electron correlations tend to reduce the plasmon frequency. For instance, for small q and with p_F for Fermi momentum one has a correction:

$$\delta\omega_p^c = -\omega_p \frac{q}{4p_F}. \tag{2.3.15}$$

Third, nonlocal quantum corrections to first order in q give rise to a correction

$$\delta\omega_p^q = \omega_p \frac{3q}{4q_s}. \tag{2.3.16}$$

That is, due to the quantum character of the electrons, the plasma frequency is increased. Here,

$$q_s = \frac{g_v}{2\pi\kappa a_0}, \tag{2.3.17}$$

where g_v is the valley degeneracy for Si inversion layers and a_0 is the Bohr radius.

The quantum effect is stronger in GaAs than in Si because of the small effective mass in the former case. Using the 2D plasmon dispersion in high mobility GaAs/GaAlAs heterostructures, *Batke* et al. [2.17] observed that the increase of 8% in the plasmon frequency takes place due to the quantum effect at q of order $2 \times 10^5 \, \mathrm{cm}^{-1}$. However, this increase is largely canceled by the thickness effect. Due to this cancellation, the classical result yields rather good agreement with experimental data. Nevertheless, for a small wave vector, the data points appear generally above the classical theoretical curve. Figure 2.10 depicts the data of *Batke* et al. together with theoretical plasmon dispersion relations for GaAs/GaAlAs heterostructures in the region of small q. The electron density is $6.7 \times 10^{11} \, \mathrm{cm}^{-2}$, the depletion layer density is $1.5 \times 10^{11} \, \mathrm{cm}^{-2}$. The dielectric constant of GaAs is 12.6, while that of GaAlAs is 11.0. The solid curve represents the classical dispersion relation. Its corrections are illustrated as follows: curve 1, the nonlocal quantum correction; curve 2, finite thickness correction; curve 3, nonlocal and finite thickness corrections; curve 4, all corrections.

In GaAs/GaAlAs superlattices of a large but finite number (~ 15) of periods, discrete plasmons have been observed by inelastic light scattering experiments [2.18]. These discrete modes demonstrate lifting of the degeneracy of pairs of modes with wave numbers q_i and $2\pi/d - q_i$ in the first Brillouin zone of the superlattice with period d. In the infinite superlattice, these modes are degenerate. Grating-assisted Raman scattering has been introduced as a new way of studying the dispersion of near surface modes. The plasma frequency of a finite superlattice has been expressed as [2.19]

$$\omega_s^2(q, q_i) = \omega_p^2(q) \frac{\sinh qd}{\cosh qd - \cos q_i d}, \tag{2.3.18}$$

where

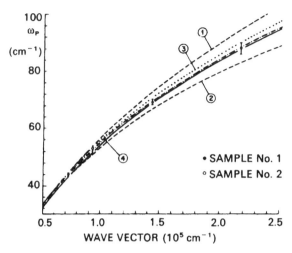

Fig. 2.10. Plasmon dispersion in GaAs/GaAlAs [2.17]

$$q_i = i\frac{2\pi}{Nd}. \tag{2.3.19}$$

N is the number of periods and $i = 1, 2, \ldots, N$.

In the presence of a magnetic field, the magnetoplasma frequency in the long-wavelength limit, i.e., $X = \mathbf{q} \cdot \mathbf{k_F} l^2 \ll 1$, is given by

$$\omega_{mp}^2 = \omega_p^2 + \omega_c^2, \tag{2.3.20}$$

where ω_c is the cyclotron frequency and l is the magnetic length. Magnetoplasmons have been observed [2.20] for a fairly wide range of electron densities which corresponds to liquid-like and solid phases.

Due to nonlocal quantum corrections to the above expression, the plasmon resonance interferes with the harmonics $N\omega_c(N = 2, 3, \ldots)$ of the cyclotron frequency, causing splitting of the magnetoplasmon dispersion when the magnetoplasmon dispersion crosses the harmonics $N\omega_c$. This splitting has been observed by *Batke* et al. The amount of this splitting is determined by $(qv_F/\omega_c)^2$, where v_F is the Fermi velocity, in a way similar to the case of the nonlocal quantum effect. The splitting increases with wave vector q, as expected from a quasiclassical nonlocal theory.

In the short-wavelength limit, i.e., $X \gg 1$, the magnetoplasmons are given by

$$\omega_{mp}(q) = N\left[\omega_c + F_N(X)\right], \tag{2.3.21}$$

where $n = 1, 2, \ldots$, and $F_N(X)$ is a positive, monotonically decreasing, and oscillatory function. Electron–electron interaction is strong in this region and affects the short-wavelength limit significantly. Moreover, there are spin and coupled magnetoplasmon-spin modes. New resonance lines have been observed in GaAs/GaAlAs heterostructures near the cyclotron frequency [2.21].

3. One-Component Plasmas at High Temperatures

Even though most of the properties of a classical electron plasma depend on the plasma parameter, quantum effects may become important at extremely high temperatures or high densities.

3.1 ε and Γ Series

The Coulomb potential between electrons is long-ranged. The presence of an electron at any point in a coordinate system of an OCP influences all other electrons, as well as positive charges, and causes a redistribution of charges. The positive charges which are attracted cause screening of the charge of the electron. As can be guessed, the larger the distance the more effective this screening. As a consequence, the electron is expected to exert an effective potential at large distances which decreases faster than the original Coulomb potential. This screening effect depends on the temperature and density of the electrons in the system. Since the redistribution of charges is involved, it is a many-body effect.

The screening was first theoretically treated by *Debye* and *Huckel* [3.1] in 1923 for an ionic solution. They used a self-consistent method in which an effective potential is determined by the Poisson equation. The Boltzmann distribution of charges with the effective potential gives the right side of this equation.

In 1950, *Mayer* [3.2] discussed how Debye screening could be interpreted graphically in a low density and high temperature limit of an OCP. Since then it has been a theoretical problem to find corrections to Debye screening. In what follows, we shall discuss how such corrections can be obtained using the methods which are outlined in the Appendix.

Let us first note that in a very dilute classical electron gas the pair distribution function is given approximately by the direct Coulomb interaction in the form

$$\rho_2(r) = n^2 \exp(-e^2/rk_{\mathrm{B}}T). \tag{3.1.1}$$

Collective couplings of the electrons cause deviations from this limiting expression. These deviations depend on the distance r of the two electrons in the pair distribution function and also on the density n of the entire electrons. As in the Appendix, Sect. A.2, the pair distribution function $\rho_2(r)$ can be evaluated in the chain diagram approximation. As in (A2.7), $\rho_2(r)$ is expressed in terms of the eigenvalues $\lambda_j(q)$ of the free electron propagator such that

$$\rho_2(r) = n^2 - \frac{1}{(2\pi)^3\beta} \sum_j \int \frac{u(q)\,[\lambda_j(q)]^2}{1 + \lambda_j(q)u(q)} \exp(i\mathbf{q}\cdot\mathbf{r})d\mathbf{q}. \tag{3.1.2}$$

Here, $u(q)$ is the Fourier transform of the Coulomb potential. The eigenvalues $\lambda_j(q)$ are given by (A2.4) in the Appendix. In the high temperature limit, the eigenvalues are given by

$$\lambda_j(q) = z\lambda^{-3}\beta\delta_{j0} , \tag{3.1.3}$$

where z is the absolute activity and λ is the de Broglie thermal wavelength. Note that at high temperatures and for low densities, $z\lambda^{-3}$ can be replaced by the electron density n. Therefore, (3.1.2) yields

$$\rho_2(r) = n^2 - \frac{n\kappa_D^2 e^{-\kappa_D r}}{4\pi r}. \tag{3.1.4}$$

Note that the Debye screening constant has been associated with the simple chain diagrams on which (3.1.2) is based.

The above result can be generalized to the case in which a representative electron pair is connected by an arbitrary number of chains which are independent of each other. The corresponding diagrams are called *watermelon diagrams*. The summation over such watermelon diagrams yields

$$\rho_2(r) = n^2 \exp\left[-\phi_D(r)/k_B T\right], \tag{3.1.5}$$

where $\phi_D(r)$ is the potential of average force given by

$$\phi_D(r) = \frac{e^2}{r} \exp(-\kappa_D r), \tag{3.1.6}$$

which is a screened Coulomb potential.

It is instructive to rewrite (3.1.5) in terms of the plasma parameter ε,

$$\rho_2(r) = n^2 \exp\left[-\varepsilon \exp(-x)/x\right], \tag{3.1.7}$$

where $x = \kappa_D r$ is a dimensionless distance variable. We learn that the Debye screening potential is proportional to ε and that the pair distribution function in the watermelon approximation contains terms of all orders in ε. In this respect, we remark that there are other types of graphs which also contribute terms of order ε^2 or higher. Hence, the above expression must be amended.

In order to derive the pair distribution function which is correct to order ε^2, it is convenient to replace each simple chain by an effective line and examine topologically different prototype graphs. A systematic estimate of the contributions of classical prototype graphs to the pair distribution function can be made in accordance with the following rules [3.3]:

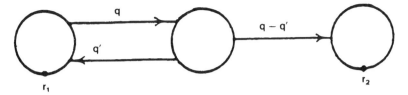

Fig. 3.1. $W_{31}(r)$ in (3.1.8)

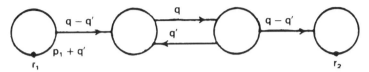

Fig. 3.2. $W_{42}(r)$ in (3.1.8)

1) Each effective line contributes a factor ε/κ_D^3.
2) Each component graph (unlabeled electron at a junction of the effective lines) yields a factor n.
3) Each independent momentum variable representing an effective line results in a factor κ_D^3.

Note that an effective line is represented by a single momentum variable because the integration over the coordinate of an electron at a junction gives rise to a conservation of the total momentum entering and leaving the junction. This conservation law is similar to Kirchhoff's law for the conservation of current in electric circuits. The above rules result from the necessary integrations over the coordinates of all the unlabeled electrons and over the intermediate temperatures representing the locations of the interaction lines.

The integrand of (3.1.2) indicates that a chain of electrons produces screening. In fact, since the eigenvalues $\lambda_j(q)$ are constant in the classical limit as in (3.1.3), the process of summing over chains amounts to replacing $u(q)$ by an effective potential $u(q)/[1 + n\beta u(q)]$.

Let us now investigate how the above rules can be used to extend the results based on simple chain diagrams. Because a series connection of chains yields only another chain and thus does not produce a new prototype graph, any correction to the chain diagram approximation should include parallel connections of effective lines.

In order to obtain the corrections to (3.1.7), let us write

$$\rho_2(r) = n^2 \exp\left[-\phi_D^*(r) + nW_{31}(r) + n^2 W_{42}(r) + \cdots\right]. \tag{3.1.8}$$

An application of the above rules shows that W_{ij} represents the contribution from the graphs with i effective lines and j junctions (unlabeled electrons at junctions of effective interaction lines). Thus, W_{31} corresponds to a prototype graph with 3 effective lines and 1 junction, as illustrated in Fig. 3.1, and W_{42} has 4 effective lines and 2 junctions as in Fig. 3.2. Because of the momentum conservation law, both have 2 independent momentum variables. The former contribution is estimated to be $(\varepsilon(\kappa_D^3)^3 n(\kappa_D^3)^2 = n\varepsilon^2/\kappa_D^3$. This is of order $n\varepsilon^2$ because

$$\kappa_D^3/(4\pi n) = \varepsilon. \tag{3.1.9}$$

The latter contribution is estimated as $(\varepsilon/\kappa_D^3)^4 n^2(\kappa_D^3)^2 = n^2\varepsilon^4/(\kappa_D^3)^2$, which is of order ε^2 due to (3.1.9). We can conclude that there are no other graphs of order $n^2\varepsilon^2$, because the more components in a connected graph, the more lines for their connections. As a consequence, its order becomes higher.

We can write down W_{31} and W_{42} as

$$nW_{31}(r) = n\int [-\phi_D^*(r_{13})]\,[-\phi_D^*(r_{32})]^2\,dr_3, \tag{3.1.10}$$

$$nW_{42}(r) = n\int [-\phi_D^*(r_{13})]\,[-\phi_D^*(r_{34})]^2\,[--\phi_D^*(r_{42})]\,dr_3dr_4. \tag{3.1.11}$$

Using the known Fourier transforms of the screened Coulomb potential and its square, these integrals can be evaluated in a straightforward way. Expressing (3.1.8) in the form

$$\rho_2(r) = n^2 \exp\left[-\Psi(r)/k_B T\right]. \tag{3.1.12}$$

we obtain

$$\begin{aligned}
\Psi/k_B T = \frac{\varepsilon}{x}e^{-x} + \frac{\varepsilon^2}{2x}&\left[e^{-x}\left(-\frac{3}{4}\ln 3 + \frac{x}{4}\ln 3 - \frac{1}{3}\right)\right.\\
&+ \frac{e^x}{4}\left[xE_i(-3x)\right] + \frac{e^{-x}}{4}\left[xE_i(-x)\right] + \frac{3e^x}{4}E_i(-3x)\\
&\left.- \frac{3e^{-x}}{4}E_i(-x) + \frac{1}{3}e^{-2x}\right] + \cdots,
\end{aligned} \tag{3.1.13}$$

where

$$E_i(-x) = -\int_x^\infty \frac{e^{-x}}{x}\,dx. \tag{3.1.14}$$

Equation (3.1.13) is correct to order ε^2.

We have seen how screening appears in the pair distribution function. Let us now evaluate several thermodynamic functions. First, we note that the energy of a classical system can be expressed in terms of the pair distribution function,

$$U = \frac{3}{2}Nk_B T + \frac{V}{2}\int \phi(r)[\rho_2(r) - n^2]dr. \tag{3.1.15}$$

We find that to order ε^2 the energy is given by

$$U = \frac{3}{2}Nk_B T - \frac{Nk_B T}{2}\left[\varepsilon + \varepsilon^2\left(\gamma - \frac{2}{3} + \frac{1}{2}\ln 3\varepsilon\right)\right], \tag{3.1.16}$$

where $\gamma = 0.57721$ is Euler's constant. The energy can of course be evaluated directly, without using the pair distribution function [3.4]. The second term in the curly brackets yields approximately 2% corrections when ε is of order 10^{-2}.

The free energy may be obtained from

$$F = \beta^{-1} \left(\int_0^\beta U d\beta + \text{constant} \right), \tag{3.1.17}$$

where the constant of integration can easily be determined from the known result at $\varepsilon = 0$. We arrive at

$$F = N \, k_B T \ln \left[\frac{n}{e} \left(\frac{h^2}{2\pi m k_B T} \right)^{3/2} \right]$$
$$- \frac{N k_B T}{6} \left[2\varepsilon + \varepsilon^2 \left(\gamma - \frac{11}{12} + \frac{1}{2} \ln 3\varepsilon \right) \right]. \tag{3.1.18}$$

Hence, the equation of state is given by

$$\frac{p}{n k_B T} = 1 - \frac{\varepsilon}{6} \left[1 + \varepsilon \left(\gamma - \frac{2}{3} + \frac{1}{2} \ln 3\varepsilon \right) \right]. \tag{3.1.19}$$

These results show that thermodynamic functions can be expressed in ascending powers of ε. Note that the terms of order ε^2 in these equations are negative for small ε. The ε series may alternate in sign. It is expected to converge for small values of ε. However, for consistency it becomes increasingly necessary to take quantum effects into consideration. Even within the classical framework, it is rather involved to evaluate analytically higher order terms in ε. Therefore, several Monte Carlo calculations have been utilized to obtain thermodynamic functions in wider ranges of ε or Γ [3.5]. For instance, fitting with Monte Carlo results for 128 particles, *Slattery* et al. [3.5a] obtained an empirical energy expression which can be used in the range $1 \leqq \Gamma \leqq 160$. Since this expression gives a very good fit, *DeWitt* and *Rosenfeld* [3.5b] tried to interpret its form by making use of a free energy which is good for systems with short-range potentials. They minimized the free energy with respect to the packing fraction $x = \pi n \sigma^3 / 6$, σ being an effective hard-sphere diameter. The fraction at the minimum point was then found to be proportional to $\Gamma^{-1/4}$ near $x = 1$. Although the point $x = 1$ is actually unphysical because the maximum value of $x = \pi/(3.2^{1/2}) = 0.740$, this proportionality leads us to a term with $\Gamma^{1/4}$. A related expression for the equation of state is

$$\frac{p}{n k_B T} = 1 - 0.29917\Gamma + 0.31514\Gamma^{1/4} + 0.05951\Gamma^{-1/4} - 0.26683. \tag{3.1.20}$$

The excess energy U_{ex} behaves like $\Gamma^{3/2}$ for $\Gamma \to 0$. It has been found numerically that this energy varies predominantly like Γ in the limit of large Γ. Therefore, *Hansen* proposed an expansion [3.5c]

$$\frac{U_{ex}}{N k_B T} = \Gamma^{3/2} \left(\frac{a_1}{(b_1 + \Gamma)^{1/2}} + \frac{a_2}{b_2 + \Gamma} + \frac{a_3}{(b_3 + \Gamma)^{3/2}} + \frac{a_4}{(b_4 + \Gamma)^2} \right). \tag{3.1.21}$$

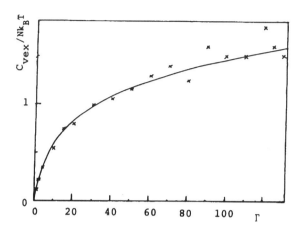

Fig. 3.3. Numerical results for the excess specific heat at constant volume [3.5c]

The optimum coefficients he obtained are

$$a_1 = -0.895929, \quad a_2 = 0.11340656, \quad a_3 = -0.90872827, \quad a_4 = -0.11614773;$$

$$b_1 = 4.6664860, \quad b_2 = 13.675411, \quad b_3 = 1.8905603, \quad b_4 = 1.0277554.$$

He obtained similar expressions for the specific heat and free energy. The excess specific heat which is due to Coulomb interaction varies as $\Gamma^{3/2}$ for small Γ and roughly as Γ for large Γ. His results on the excess specific heat at constant volume are illustrated in Fig. 3.3. The curve represents an expansion similar to (3.1.21), and the crosses are the Monte Carlo results. The excess specific heat increases considerably due to electron correlations. The graph shows also increasing fluctuations in the Monte Carlo results for large Γ.

The parameters ε and Γ characterize Coulomb effects in classical plasmas, and their values vary significantly in wide ranges of electron density and temperature. Their typical values are given in the following two examples.

1) Laboratory plasma

The number density of electrons is usually in the range $10^{10} - 10^{12}$ cm^{-3} in tenuous laboratory plasmas, and $10^{14} - 10^{16}$ cm^{-3} in intense ones. A dense laboratory plasma may have the following values:

$$n = 10^{14}\,\mathrm{cm}^{-3}, \quad T = 10^4\,\mathrm{K};$$

$$\kappa_D = 1.45 \times 10^4\,\mathrm{cm}^{-1}, \quad \varepsilon = 2.42 \times 10^{-3}, \quad \Gamma = 3.48 \times 10^{-2}.$$

2) Fusion plasma

For a thermonuclear fusion plasma, a density of 10^{16} cm^{-3} and a temperature 10^6 K may be required.

$$n = 10^{16}\,\mathrm{cm}^{-3}, \quad T = 10^6\,\mathrm{K},$$

$$\kappa_D = 1.45 \times 10^5\,\mathrm{cm}^{-1}, \quad \varepsilon = 2.42 \times 10^{-2}, \quad \Gamma = 1.1 \times 10^{-1}.$$

In these cases, ε or Γ is small, and the electrons can be considered weakly coupled. However, as the density increases, the electrons will be more and more strongly coupled. Also, in the case of ions, Γ can be relatively large. In fact, for ions with mass number A, charge ze and mass density ρ, their effective Γ parameter is given by

$$\Gamma = 2.29 \times 10^5 z^2 (\rho/A)^{1/3}/T. \tag{3.1.22}$$

The hydrogen plasma in a Jovian planet is considered to have a mass density of order a few $g\,cm^{-3}$. At 10^4 K, Γ is of order 23 so that the plasma is strongly coupled. In the outer crust of a neutron star, the ionized iron nuclei may have a mass density of order 10^6 $g\,cm^{-3}$. With $Z = 26$ they are strongly coupled. On the other hand, in the extreme case of white dwarf stars, which occur when stars of the size of the sun have used up their hydrogen fuel, the helium atoms are considered to be completely ionized, and the number density of electrons is around 10^{30} cm^{-3}. The temperature can be of order 10^7 K, but the high density causes the Fermi temperature of the electrons to be of order 10^{11} K. Hence, the electrons can be treated as a nearly free degenerate Fermi gas. Because of the light electron mass, $\Gamma = 0.022$ even though the number density is large. However, the helium ions in the interior of white dwarfs can be different. With a density of order 10^{23} $g\,cm^{-3}$, $z = 2$, and $A = 4$ and at 10^7 K, the effective Γ of ions is very large so that they are strongly coupled, although at such a high temperature quantum effects become important. In the opposite limit of low temperatures, the Γ parameter can no longer be useful. In fact, at low temperatures strong electron–electron coupling occurs at low densities.

3.2 Quantum Effects at High Temperatures

In an electron plasma, quantum effects reappear at extremely high temperatures for a reason which differs from the case of low temperatures. This somewhat peculiar aspect of an electron plasma has been shown in the phase diagram in Fig. 2.1. High temperature quantum effects occur because the de Broglie thermal wavelength of an electron decreases more slowly with temperature than the Landau length, which represents the average distance of closest approach. We have discussed this aspect already in Sect. 1.1, but let us examine a semiclassical case. Although quantum effects are relatively small in this case, we shall see how they begin to play a role.

Let us note that the Debye screening constant for a classical electron gas is given by

$$\kappa_D = 0.145(n/T)^{1/2}\,cm^{-1}.$$

Hence, for $n = 10^{16}$ cm^{-3} and $T = 10^2$ K, $\kappa_D = 1.45 \times 10^6$ cm^{-1}. For electrons at the same temperature, the de Broglie thermal wavelength is $\lambda = 7.45 \times 10^{-7}$ cm. The product $\lambda \kappa_D$ is a constant equal to 1.08. This value is smaller than the plasma parameter $\varepsilon = e^2\kappa_D/k_BT = 24.2$. However, at 10^5 K, $\kappa_D = 0.459 \times 10^5$, $\varepsilon = 7.66 \times 10^{-4}$ and the ratio $\lambda/l_L = \eta = \lambda\kappa_D/\varepsilon = 1.41$, as in Table 1.1. Hence, quantum effects become important.

Of course, in the opposite limit of very low temperatures, quantum effects are significant, but let us first investigate quantum effects at high temperatures. We attempt to find quantum corrections to the classical Debye screening constant. Such corrections in turn affect the thermodynamic properties.

For this purpose, let us evaluate the grand partition function Ξ based on the formula [see (A3.2)]:

$$\ln \Xi/\Xi_0 = \frac{V}{2(2\pi)^3} \sum_j \int d\mathbf{q}\{u(q)\lambda_j(q) - \ln[1 + u(q)\lambda_j(q)]\}. \qquad (3.2.1)$$

Here V is the total volume, Ξ_0 is the grand partition function of an ideal electron gas. As in (3.1.3), in the high temperature limit all the eigenvalues vanish except for λ_0. Hence, its contribution can be considered to be important even in the semiclassical case. For this reason, we evaluate the grand partition function based on (3.2.1) by splitting the j-sum into two parts, $j = 0$ and $j \neq 0$.

We note that the first approximation to the surviving eigenvalue is $z\lambda^{-3}\beta$ as in (3.1.3). Since this is an approximation to the $j = 0$ eigenvalue, let us introduce a new notation

$$\lambda_0^0 = z\lambda^{-3}\beta.$$

We have shown in (3.1.4) that λ_0^0 leads us to Debye screening. Hence, we can surmise that quantum corrections would emerge if we consider corrections to the above approximation. It is then appropriate to write λ_0 such that

$$\lambda_0 = \lambda_0^0 + \lambda_0^1,$$

and investigate the contribution from the correction term λ_0^1 which is given by [see (A2.4)]

$$\lambda_0^1 = \int_0^1 \{\exp[-\beta q^2 x(1 - x)]-1\}dx. \qquad (3.2.2)$$

For the present semiclassical case, we treat λ_0^1 as a small correction.

The part of the grand partition function which is associated with λ_0 is then given as

$$\begin{aligned}
\ln \Xi(j = 0)/\Xi_0 &= \frac{V\kappa^3}{2(2\pi)^3}4\pi \int_0^\infty \left\{\left[1 - x^2\ln\left(1 + \frac{1}{x^2}\right)\right]\right. \\
&\quad \left. + \left[\lambda_0^1 - x^2\ln\left(1 + \frac{\lambda_0^1}{1+x^2}\right)\right]\right\}dx \\
&= \frac{V\kappa^3}{12\pi}\left(1 + \frac{\pi}{3}(J_1 + J_2 + \ldots)\right)
\end{aligned}$$

$$= \frac{V\kappa^3}{12\pi}\left(1 - \frac{3}{8}\pi^{1/2}\zeta^{1/2} + \frac{\zeta}{4}\right.$$
$$\left. + \frac{z^{1/2}\pi^{1/2}}{8}(2^{1/2} - 1) + \dots\right), \tag{3.2.3}$$

where J_1 and J_2 are obtained from the terms given by λ_0^1 and $(\lambda_0^1)^2$ respectively, and

$$\zeta = \kappa^2\lambda^2/4\pi, \tag{3.2.4}$$

$$\kappa^2 = 4\pi\beta e^2 z/\lambda^3. \tag{3.2.5}$$

Note that κ differs from κ_D. We shall shortly determine z/λ^3 as a function of n and T, and express this parameter and also ζ in terms of κ_D. The first term on the right side of (3.2.3) yields contributions of order ε to thermodynamic functions, ε being the plasma parameter. The second and third terms represent quantum corrections for small ζ.

The eigenvalues $\lambda_j(j \neq 0)$ contribute terms of order $\zeta^{1/2}$ as follows:

$$\ln\Xi(j \neq 0) = \frac{V}{2(2\pi)^3}\frac{1}{2}\sum_j\int u^2(q)\lambda_j^2(q)d\mathbf{q}$$
$$= \frac{V\kappa^3}{8\pi^2}J_3, \tag{3.2.6}$$

where

$$J_3 = \frac{\pi^{3/2}(5 \times 2^{1/2} - 7)\zeta^{1/2}}{2^{5/2} \times 3}. \tag{3.2.7}$$

The grand partition function is given by the sum of (3.2.3) and (3.2.6),

$$\ln\Xi = \frac{Vz}{\lambda^3} + \frac{V\kappa^3}{12\pi}(1 + A\zeta^{1/2} + \frac{\zeta}{4}), \tag{3.2.8}$$

where

$$A = \pi^{1/2}(-\frac{3}{8} + \frac{2^{1/2} - 1}{8} + \frac{5 \times 2^{1/2} - 1}{2^{1/2}}) = 0.378. \tag{3.2.9}$$

The quantity z/λ^3 is determined by iteration in the form

$$\frac{z}{\lambda^3} = n - \frac{\kappa_D^3}{8\pi}(1 + \frac{4}{3}A\zeta_0^{1/2} + \frac{5}{12}\zeta_0), \tag{3.2.10}$$

where ζ_0 is now given by replacing κ in (3.2.4) by κ_D,

$$\zeta_0 = \kappa_D^2\lambda^2/4\pi.$$

We arrive at

$$\frac{p}{nk_BT} = 1 - \frac{\varepsilon}{6}\left(1 + 0.756\zeta_0^{1/2} + \frac{3}{4}\zeta_0\right). \tag{3.2.11}$$

This equation of state may be compared with the classical expression of (3.1.19). Note that $\kappa_D^2 = 4\pi n\varepsilon$ and $\kappa_D^3/24\pi = n\varepsilon/6$. Hence, the first term in the curly brackets represents the first classical correction to the ideal gas equation of state. For 10^5 K and $n = 10^{16}$, $\zeta_0^{1/2} = 3.03 \times 10^4$ so that the first quantum correction given by the second term in the brackets is only 1.15×10^{-4}. This is small compared to 1. On the other hand, $\varepsilon = 7.66 \times 10^{-4}$ is of the same order of magnitude as $\zeta_0^{1/2}$. Hence, when corrections of order ε^2 come into consideration, this quantum correction may not be neglected. Quantum corrections based on η have been evaluated by *de Witt* [3.6b] for small η and by *Hoffmann* and *Ebeling* for a wider range [3.6a]. The latter authors expressed the equation of state in the form

$$\frac{p}{nk_\mathrm{B}T} = 1 - \frac{1}{6}\varepsilon - \frac{\varepsilon^2}{12}\ln\varepsilon - \frac{\varepsilon^2}{12}\left(2\gamma + \ln 3 - \frac{4}{3} + 12g(x)\right), \qquad (3.2.12)$$

where $x = \eta/(2\pi)^{1/2}$ and

$$g(x) = \begin{cases} -\dfrac{3\pi^{1/2}}{4}x^3 + \dfrac{3}{2}x^2 - \dfrac{3}{4}\pi^{1/2}(1 + \ln 2)x + \ln x + \dfrac{\gamma}{2} \\ \qquad\qquad + \ln 3 + 0.411 + O(x^{-1}), \quad (x > 1), \\ 2\gamma + \ln 3 - \dfrac{4}{3} - \dfrac{1}{2}x^2 + \dfrac{1}{10}x^4 + \dfrac{2}{21}x^6 + O(x^8), \quad (x < 1). \end{cases} \qquad (3.2.13)$$

The function $g(x)$ decreases from 1 as x increases, as shown in Fig. 3.4. For 10^6 K, $g(x)$ is approximately -5 so that the last term in the round brackets of (3.2.12) is large. The point $x = 1$ corresponds roughly to 3×10^5 K, and the abscissa in Fig. 3.4 covers roughly the temperature range $10^4 - 10^7$ K. The region near $x = 1$ has been obtained by extrapolation from the two sides. Beyond this point, $g(x)$ deviates from 1 more significantly. However, at high temperatures, the corrections of the type given in (3.2.11) also become appreciable, depending on density. The η corrections in the above equation of state are independent of density, but may be used for low densities where ζ_0 is small. In (3.2.13), the terms of $O(x^{-1})$ can be given explicitly by

$$O(x^{-1}) = 6\pi^{1/2} \sum_{s=-4}^{\infty} (-)^s 2^{-s} \left[\Gamma\left(\frac{s}{2} + 1\right)\right]^{-1} x^{-s+3}$$

$$\times \left[\zeta(s-2) - \frac{1}{2}(1 - 2^{2-s})\zeta(s-1)\right],$$

where $\zeta(s)$ is Riemann's ζ-function.

The parameter η has been listed in Table 1.1 for several temperatures. For nuclear fusion plasmas, a typical temperature 10^6 K may be adopted. The corresponding η is 4.46 and $x = 1.78$. Hence, $g(x)$ is roughly of order -5. With the numerical factor 12 in the last term in (3.2.12), we find that this quantum correction is large. As the temperature rises further, such a quantum effect will become even more important.

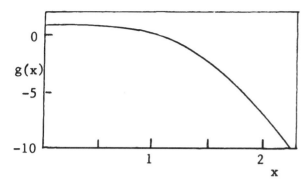

Fig. 3.4. $g(x)$ in (3.2.13) [3.6a]

3.3 Short-Distance Correlations

The chain-diagram approximation to the pair distribution function is useful to study long-range correlations. For short distances improvements on this approximation become necessary. The limiting form of the pair distribution function for short distances can be obtained by summing over ladder diagrams or, equivalently, by making use of the solutions of the Schrödinger equation for two electrons [3.7]. In fact, the short-distance divergence in the radial distribution function of a classical OCP can be eliminated by these methods.

For short distances, quantum effects are generally important. The direct and exchange graphs of a pair of electrons contribute to the pair distribution function as follows [3.7]:

$$\rho_{2D}(r) = n^2 \langle \mathbf{r}_1 \mathbf{r}_2 | e^{-\beta H} | \mathbf{r}_1 \mathbf{r}_2 \rangle$$
$$= n^2 2^3 (2\pi\beta)^{3/2} \int d\mathbf{k} e^{-2\beta k^2} \sum_{l,m} |R_{kl}|^2 Y_{lm}^*(\theta, \phi) Y_{lm}(\theta, \phi), \quad (3.3.1)$$

$$\rho_{2E}(r) = \frac{-1}{2} n^2 \langle \mathbf{r}_1 \mathbf{r}_2 | e^{-\beta H} | \mathbf{r}_2 \mathbf{r}_1 \rangle$$
$$\stackrel{!}{=} -\frac{1}{2} n^2 2^3 (2\pi\beta)^{3/2} \int d\mathbf{k} e^{-2\beta k^2} \sum_{l,m} R_{kl}(r)^2 Y_{lm}^*(\theta, \phi) Y_{lm}(\theta, \phi),$$

$$(3.3.2)$$

Here, units are such that $\hbar = 1$, $2m = 1$, H is the Hamiltonian of two electrons, and $R_{kl}(r)$ is the radial part of the eigenfunction given by

$$R_{kl}(r) = 2 \left(\frac{k}{a_0} \right)^{1/2} [\exp(\pi/ka_0) - 1]^{-1/2} \prod_{s=1}^{l} [s^2 + (ka_0)^{-2}]^{1/2}$$
$$\times \frac{(2kr)^l}{(2l+1)!} e^{-ikr} F\left(\frac{-i}{2ka_0} + l + 1, 2l + 2, 2ikr \right), \quad (3.3.3)$$

where

$$F(s,t,x) = \sum_{n=0}^{\infty} \frac{s(s+1)\dots(s+n-1)}{t(t+1)\dots(t+n-1)} \frac{x^n}{n!}$$

is the hypergeometric function. The radial wave function can be expanded in powers of r. To order r^3, the relevant contributions come from the s-waves ($l = 0$) and p-waves ($l = 1$) as follows:

$$\rho_{2\mathrm{D}}(r) = (2z)^2 \frac{(2\pi\beta)^{-3/2}}{(2\pi a_0)} \left\{ \left[1 + \frac{r}{a_0} + \frac{1}{2}\left(\frac{r}{a_0}\right)^2 + \frac{10}{9}\left(\frac{r}{2a_0}\right)^3 \right] I_1 \right.$$
$$\left. - \frac{1}{9}\left(\frac{r^3}{a_0}\right) I_2 \right\}, \tag{3.3.4}$$

$$\rho_{2\mathrm{E}}(r) = -\frac{(2z)^2}{2} \frac{(2\pi\beta)^{-3/2}}{(2\pi a_0)} \left\{ \left[1 + \frac{r}{a_0} + \frac{1}{3}\left(\frac{r}{a_0}\right)^2 + \frac{1}{18}\left(\frac{r}{a_0}\right)^3 \right] I_1 \right.$$
$$\left. - \left[\frac{2}{3}r^2 + \frac{4}{9}\frac{r^3}{a_0} \right] I_2 \right\}, \tag{3.3.5}$$

where $z = n\lambda^3$ and

$$I_1 = \int_0^{\infty} dk \frac{k\exp(-2\beta k^2)}{\exp(\pi/ka_0) - 1},$$
$$I_2 = \int_0^{\infty} dk \frac{k^3\exp(-2\beta k^2)}{\exp(\pi/ka_0) - 1}. \tag{3.3.6}$$

These integrals are difficult to evaluate analytically. However, in the cases for which

$$\delta = 2\beta^{1/2}/a_0 \tag{3.3.7}$$

is either very small or very large, $\rho_2(r)$ can be expanded in powers of r/a_0:

(1) $\delta \ll 1$

$$\rho_{2\mathrm{D}}(r) = n^2 \left\{ \left[1 + \frac{r}{a_0} + \frac{1}{2}\left(\frac{r}{a_0}\right)^2 + \frac{5}{36}\left(\frac{r}{2a_0}\right)^3 \right] \left[1 - \left(\frac{\pi}{2}\right)^{1/2}\delta + \frac{\pi^2}{12}\delta^2 \right] \right.$$
$$\left. - \frac{1}{12}\frac{r^3}{a_0\beta}\left[1 - \frac{2}{3}\left(\frac{\pi}{2}\right)^{1/2}\delta \right] \right\}, \tag{3.3.8}$$

$$\rho_{2\mathrm{E}}(r) = -\frac{n^2}{2} \left\{ \left[1 + \frac{r}{a_0} + \frac{1}{3}\left(\frac{r}{a_0}\right)^3 + \frac{1}{18}\left(\frac{r}{a_0}\right)^3 \right] \left[1 - \left(\frac{\pi}{2}\right)^{1/2}\delta + \frac{\pi^2}{12}\delta^2 \right] \right.$$
$$\left. - \left(\frac{1}{2}\frac{r^2}{\beta} + \frac{1}{3}\frac{r^3}{\beta a_0} \right) \left[1 - \frac{2}{3}\left(\frac{\pi}{2}\right)^{1/2}\delta \right] \right\}. \tag{3.3.9}$$

Combining these two results, we obtain

$$\rho_2(r) = \frac{1}{2}n^2 \left[1 - \left(\frac{\pi}{2}\right)^{1/2}\delta + \frac{1}{2}\frac{r}{a_0} + \frac{1}{2}\frac{r^2}{\beta} \right. $$
$$\left. - \frac{2}{3}\left(\frac{\pi}{2}\right)^{1/2}\frac{r^2}{a_0\beta^{1/2}} - \frac{1}{2}\frac{r^3}{a_0\beta} + \dots \right], \qquad (3.3.10)$$

Hence, $\rho_2(r)$ varies linearly with r starting with the following finite value at the origin:

$$\rho_2(0) = n^2(2\pi\beta)^{3/2}\frac{1}{\pi a_0}I_1$$
$$= \frac{n^2}{2}\left[1 - \left(\frac{\pi}{2}\right)^{1/2}\delta + \frac{\pi^2}{12}\delta^2 + \dots \right]. \qquad (3.3.11)$$

(2) $\delta \gg 1$

In this case, the denominator of (3.3.6) may be expanded as

$$\left[\exp\left(\frac{\pi}{2^{1/2}}\frac{\delta}{x}\right) - 1 \right]^{-1} = \sum_{s=1}^{\infty} \exp\left(-\frac{\pi}{2^{1/2}}\frac{\delta s}{x}\right).$$

We can show that both I_1 and I_2 are asymptotically proportional to $\exp[-3(\pi\delta)^{2/3}/2]$. Hence, an asymptotic value $\rho_2(0)$ is given by

$$\rho_2(0) = n^2[(\pi\delta)^{4/3}/3^{1/2}]\exp\left(-\frac{3}{2}(\pi\delta)^{2/3}\right). \qquad (3.3.12)$$

This formula is accurate for $r_s \gg 1$.

The above results have been obtained for Boltzmann statistics. The case of quantum mechanically degenerate electrons will be treated later where the limiting behavior of $g(0)$ will be illustrated for large and small r_s.

Instead of making use of the solutions of the two-body Schrödinger equation, one can improve the chain diagram formula of (3.1.2) for the pair distribution function of a dilute classical electron gas based on a short-distance consideration of the eigenvalues. We note that for short distances in which $r < q_0^{-1}$, where q_0 is a parameter which is of order κ_D the main contribution to the chain diagram result comes from the momentum region $q > q_0$. Hence, it becomes appropriate to split the q-integration of (3.1.2) into two parts. For $q \geqq q_0$ we use the correct λ_j in the numerator and neglect the denominator. For $q \leqq q_0$ we may use λ_0 only because of its dominance.

That is, the pair distribution function may be evaluated by

$$\rho_2(r) - n^2 = -\frac{4}{(2\pi)^3\beta}\left(\int\limits_{q\leqq q_0} \frac{\lambda_0^2 u(q)}{1 + 2\lambda_0 u(q)} - \sum_j \int\limits_{q\geqq q_0} \lambda_j^2 u(q) \right)e^{i\mathbf{q}\cdot\mathbf{r}}d\mathbf{q}. \quad (3.3.13)$$

In the second integral, the sum over j can be performed first. We arrive at

$$\rho_2(r)/n^2 - 1 = -\varepsilon\left[\frac{e^{-x}}{x} - \frac{\exp(-x/2\alpha)}{x} + \frac{2^{1/2}}{\alpha}\mathrm{erfc}\left(\frac{x}{2^{1/2}\alpha}\right) \right]. \quad (3.3.14)$$

Here, $x = \kappa_D r$, $\alpha = \beta^{1/2} \kappa_D$, and q_0 is a parameter which is expected to be between κ_D and $\beta^{1/2}$ but has been canceled out in the final result. The appearance of the de Broglie thermal wavelength λ indicates the necessity of quantum treatments at short distances. Equation (3.3.14) is a convenient interpolation formula which can be used for all distances for small ε.

4. Low Temperature Plasmas

Electron correlations are clearly exhibited in the pair distribution function and the correlation energy.

4.1 Asymptotic Correlations

The limiting behavior of the correlation function at low temperatures for both short and long distances involves interesting physics. These two limits are discussed separately since they require different theoretical approaches. In general, collective couplings between the electrons are important for long distances, while quantum effects are strong at short distances.

4.1.1 Long-Distance Correlations and the Effect of Impurities

It has been revealed by Knight shifts that impurity charges in metals cause a long distance effect at low temperatures. This effect is due primarily to the sharpness of the Fermi distribution coupled with the long distance Coulomb interaction between the impurity charges and electrons. Let us place one electron at the coordinate origin as an impurity and study how it correlates with other electrons at long distances by using the pair distribution function based on the chain diagram approximation [4.1]. For long distances the radial distribution function $g(r)$ may be determined by the chain diagram formula [see (A2.9)]

$$g(r) = -\frac{1}{n^2(2\pi)^3\beta} \sum_j \int \frac{u(q)[\lambda_j(q)]^2}{1 + \lambda_j(q)u(q)} \exp(i\mathbf{q} \cdot \mathbf{r}) d\mathbf{q}$$

$$= n^{-2}[\rho_2(r) - n^2 + I_2(r)],$$

(4.1.1)

where in the second equation $I_2(r)$ represents the ideal gas contribution due to quantum exchanges. This contribution is given simply by the Fourier transform of the Fermi distribution function, which is 1 for $k < k_F$. Hence, $I_2(r)$ is

$$I_2(r) = \left(\frac{1}{2\pi^2} \frac{\sin k_F r - k_F r \cos k_F r}{r^3}\right)^2.$$

(4.1.2)

Note that this function decreases rather slowly as r increases.

We are interested in $g(r)$, which represents the correlated part of the pair distribution function. It can be written as

$$g(r) = -\frac{3}{2\pi^2} \left(\frac{3}{2\pi}\right)^{1/3} \frac{r_s}{x} \int_0^\infty dy \int_0^\infty ds \frac{sF^2(s,y)\sin sx}{s^2 + s_0^2 F(s,y)}, \qquad (4.1.3)$$

where $F(s,y)$ is the essential part of the eigenvalue function which coincides with the polarization function and is given in (A2.15) and

$$s = q/k_F, \quad x = rk_F, \quad s_0^2 = 2/(a_0 k_F \pi) = 0.3317 r_s.$$

The right side of (4.1.3) is a Fourier sine transform of the complicated function resulting from the y-integration. The double integral is very difficult to evaluate, but it is known that, mathematically, the asymptotic behavior of such a Fourier transform at long distances is determined by the singularities of its integrand. Examining the original $\lambda_j(q)$, we discover that a singularity appears only in $\lambda_0(q)$ at $s = 2$. Hence, separating out this particular term, we study the behavior of a simpler Fourier transform defined by

$$g_0(x) = \frac{1}{x} \int_0^\infty \frac{sF^2(s,0)}{s^2 + s_0^2 F(s,0)} \sin(sx) ds, \qquad (4.1.4)$$

where

$$F(s,0) = 1 + \frac{1}{s}\left(1 - \frac{s^2}{4}\right) \ln\left|\frac{s+2}{s-2}\right| \qquad (4.1.5)$$

is the Lindhard function.

For small r_s, the integrand may be expanded in powers of s_0^2. In the first approximation, we find that there are three contributing terms. Assembling these, we obtain

$$g_0(x) = \frac{\pi}{2} \frac{e^{-s_0 x}}{x} + \frac{8 + s_0^2}{(4 + s_0^2)^2} \frac{\pi}{2x^4}(2x\cos 2x - \sin 2x)$$
$$- \frac{32\pi}{(4 + s_0^2)^3} \frac{\sin 2x}{x^4}\left(\gamma - \frac{3}{2} + \ln|x| - \ln 4\right). \qquad (4.1.6)$$

This represents the asymptotic form of $g_0(x)$ in the correlation function. The first term, which represents a screened Coulomb interaction, is short ranged. The second and third terms decrease algebraically and are long ranged and oscillatory.

A long distance effect was first studied theoretically by *Langer* and *Vosko* [4.2] by a different method in connection with a long-range impurity effect. However, their result has a wrong sign, misses $sin2x$ in the first term and $ln4$ in the second one. The origin of these terms and their roles have been clarified [4.1a].

A direct numerical evaluation of the double integral in (4.1.3) has also been carried out [4.1b]. Such a calculation shows that the correlation function plotted against x reaches the first peak at around $x = 3 \sim 4$, such that the larger r_s, the higher the peak and smaller the peak position. The correlation

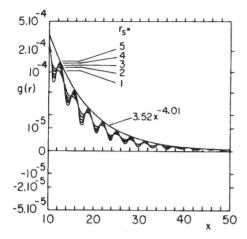

Fig. 4.1. Long distance behavior of the radial distribution function of an electron gas [4.1b]

function then oscillates. That is, the system is gas-like at high densities and liquid-like at low densities.

The oscillating pattern changes at long distances where the period becomes constant. Figure 4.1 illustrates oscillations of the radial distribution function for five values of r_s. The abscissa is the reduced distance variable $x = k_{\mathrm{F}}r$. These oscillations have the periodicity which is determined by k_{F}. The envelope function $g = 3.52x^{-4.01}$, obtained numerically, is shown for comparison.

4.1.2 Ladder Diagram Contribution

For low temperatures, the parameter r_s plays a primary role in the characterization of an electron system. In particular, for high densities, this parameter can be used as a small perturbational parameter. In fact, the limit $r_s \to 0$ corresponds to an ideal gas. As r_s increases, that is, as the density decreases, electron correlations become stronger and cause deviations from the quantum mechanically degenerate ideal case.

As in a classical electron gas, quantum effects are strong at short distances. This can be seen by examining the behavior of the pair distribution function in the short distance limit. In parallel with Sect. 3.3, the short distance correlations can be investigated in this limit based on the two electron density matrix or, equivalently, by summing ladder diagrams [4.3].

The contribution from the direct interaction graphs of two electrons to the pair distribution function is given by [4.3]

$$\rho_{2\mathrm{D}}(r) = n^2 \left\langle \mathbf{r}_1\mathbf{r}_2 | [1 - f(p_1)][1 - f(p_2)]\mathrm{e}^{-\beta H} | \mathbf{r}_1\mathbf{r}_2 \right\rangle . \qquad (4.1.7)$$

The trace can be taken in the momentum space representation by separating the center-of-mass coordinates from the relative coordinates. Thus, let us introduce

$$\mathbf{P} = \mathbf{p}_1 + \mathbf{p}_2, \quad \mathbf{p} = (\mathbf{p}_1 - \mathbf{p}_2)/2,$$

and note that for a degenerate electron gas, both $\mathbf{p}_1$, and $\mathbf{p}_2$ stay within the Fermi sphere.

In the absence of a quantum exchange between a given pair of electrons, the ideal gas contribution to the pair distribution function shall be denoted as $\rho_{2D}^{(0)}$. This is given by

$$
\begin{aligned}
\rho_{2D}^{(0)} &= 4 \left(\frac{4\pi}{(2\pi)^6} \right) \frac{32\pi}{3} k_F^3 \int_0^{k_F} \mathrm{d}p p^2 \left(1 - \frac{3}{2}\frac{p}{k_F} + \frac{1}{2}\frac{p^3}{k_F^3} \right) \\
&= \left(\frac{k_F^3}{3\pi^2} \right)^2 = n^2.
\end{aligned} \tag{4.1.8}
$$

This contribution is accompanied by the exchange contribution from the non-interacting electrons given by

$$
\begin{aligned}
\rho_{2E}^{(0)}(r) &= -\frac{2}{(2\pi)^6} \frac{32\pi}{3} k_F^3 \int_0^{k_F} \mathrm{d}\mathbf{p} \left(1 - \frac{3p}{2k_F} + \frac{p^3}{k_F^3} \right) e^{2i\mathbf{p}\cdot\mathbf{r}} \\
&= -\frac{1}{2\pi^4} k_F^6 \left(\frac{j_1(k_F r)}{k_F r} \right)^2 = -\frac{9}{2} n^2 \left(\frac{j_1(x)}{x} \right)^2.
\end{aligned} \tag{4.1.9}
$$

where $x = k_F r$ and

$$j_1(x) = \frac{\sin x - x \cos x}{x^2}.$$

On the other hand, the contribution from the interacting electrons can be obtained by using the solutions of the two-body Schrödinger equation as follows:

$$
\begin{aligned}
\rho_{2D}^L(r) = &\ 4 \left(\frac{32\pi}{3} \right) \frac{k_F^3}{(2\pi)^3} \int_0^{k_F} \mathrm{d}\mathbf{p} \sum_{lm} |R_{kl}(r)|^2 Y_{lm}^*(\theta,\phi) Y_{lm}(\theta,\phi) \\
&\times \left(1 - \frac{3p}{2k_F} + \frac{1}{2}\frac{p^3}{k_F^3} \right).
\end{aligned} \tag{4.1.10}
$$

Similarly, the contribution from the exchange graphs is

$$
\begin{aligned}
\rho_{2E}^L(r) = &\ -2 \left(\frac{32\pi}{3} \right) \frac{k_F^3}{(2\pi)^3} \int_0^{k_F} \mathrm{d}\mathbf{p} \sum_{lm} (-1)^l |R_{kl}(r)|^2 Y_{lm}^*(\theta,\phi) Y_{lm}(\theta,\Phi) \\
&\times \left(1 - \frac{3}{2}\frac{p}{k_F} + \frac{1}{2}\frac{p^3}{k_F^3} \right).
\end{aligned} \tag{4.1.11}
$$

For short distances, these contributions can be expressed in powers of r. To order r^3 and for $\alpha = (4/9\pi)^{1/3}$, they are given by

$$\rho_{2D}^L(r) = \frac{8k_F^6}{3\pi^3}\alpha r_s \int_0^1 dy \frac{y}{\exp(\pi\alpha r_s/y)-1}\left(1 - \frac{3}{2}y + \frac{1}{2}y^3\right)$$

$$\times\left[1 + \frac{r}{a_0} + \frac{1}{2}\left(\frac{r}{a_0}\right)^2 + \frac{5}{36}\left(\frac{r}{a_0}\right)^3 - \frac{1}{9}(k_F r)^2\frac{r}{a_0}y^2\right], \qquad (4.1.12)$$

$$\rho_{2E}^L(r) = -\frac{4k_F^6}{3\pi^3}\alpha r_s \int_0^1 dy \frac{y}{\exp(\pi\alpha r_s/y)-1}\left(1 - \frac{3}{2}y + \frac{1}{2}y^3\right)$$

$$\times\left[1 + \frac{r}{a_0} + \frac{1}{3}\left(\frac{r}{a_0}\right)^2 + \frac{1}{18}\left(\frac{r}{a_0}\right)^3 - \frac{2}{3}(k_F r)^2 y^2 - \frac{4}{9}(k_F r)^2\frac{r}{a_0}y^2\right].$$

$$(4.1.13)$$

Using these results we find that $\rho_2(0)$ is finite. In fact, the limiting value of $g(0) = \rho_2(0)/n^2$ is given by

$$g(0) = \begin{cases} \frac{1}{2} - \frac{3}{5}(\pi\alpha r_s), & (r_s \ll 1) \\ 12(\pi\alpha r_s)\int_0^1 dy y\left(1 - \frac{3}{2}y + \frac{1}{2}y^3\right)\exp(-\pi\alpha r_s/y), & (r_s \gg 1). \end{cases} \qquad (4.1.14)$$

Hence, for small r_s the radial distribution function decreases linearly with r_s starting from a finite value. However, the domain of such a linear variation is limited to r_s which is less than 0.6.

The asymptotic expression of $g(0)$ for large r_s is obtained as follows:

$$g(0) = 12\left\{\frac{1}{2}e^{-x}(1-x) - \frac{x^2}{2}E_i(-x) - \frac{3}{2}\left[\frac{1}{3}e^{-x}\left(1-x+\frac{x^2}{2}\right) - \frac{x^3}{3!}E_i(-x)\right]\right.$$

$$\left. + \frac{1}{2}\left[\frac{1}{5}e^{-x}\left(1-\frac{x}{4}+\frac{x^2}{4.3}-\frac{x^3}{4.3.2}+\frac{x^4}{4.3.2.1}\right) - \frac{x^5}{5!}E_i(-x)\right]\right\}, \qquad (4.1.15)$$

where

$$x = \pi\alpha r_s, \quad \alpha = (4/9\pi)^{1/3}, \quad r_s = 1/(\alpha k_F a_0),$$

$$E_i(-x) = e^{-x}\sum_{s=1}^{\infty}(-)^s\frac{(s-1)!}{x^s}.$$

Figure 4.2 illustrates $g(0)$ of the classical case discussed in the previous section in comparison with the present quantum mechanically degenerate case. Note that for a given r_s, $g(0)$ deviates from the limiting value of $1/2$ more strongly in the latter case, indicating that its short distance correlations are stronger. On the other hand, in both cases $\ln g(0)$ varies with r_s almost linearly as a function of r_s. In fact, the quantum case can be approximated by

$$g(0) = \frac{1}{2}\exp(-6\pi\alpha r_s/5).$$

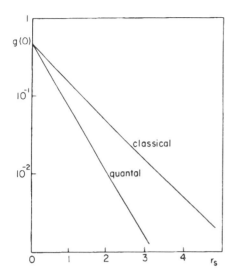

Fig. 4.2. Short distance limit of the radial distribution function [4.3a]

This approximation is particularly good for $r_s < 1$. It is remarkable that in both the classical and quantum cases $g(0)$ changes by approximately two orders of magnitude when r_s increases from 0 to 2 or 3. Such a strong variation indicates that electron correlations change significantly with r_s.

4.2 Correlation Energy in Three Dimensions

The strong binding of metals is represented by the cohesive energy which depends not only on the ions but also on the electrons. Among several relevant energies associated with the electrons is the *correlation energy*. This energy is the difference between the true ground state energy and that evaluated in the Hartree-Fock approximation and is defined by

$$\varepsilon_c = \varepsilon_g - \varepsilon_0 - \varepsilon_x, \tag{4.2.1}$$

where ε_g is the ground state energy, ε_0 is the kinetic energy and ε_x is the exchange energy. It is customary to express these energies per electron in the Rydberg unit which is $e^4 m/2\hbar^2$ and 13.61 eV. This energy value also represents the Coulomb energy at a distance of twice the Bohr radius a_0.

In what follows, we shall outline the derivation of the correlation energy for three dimensions by using a grand partition function approach. Although the correlation energy can be obtained directly without constructing the grand partition function, the present approach enables us to extend the calculation to finite temperatures.

4.2.1 Ideal Gas Contribution

The grand partition function Ξ_0 of free electrons is given by

$$\ln \Xi_0 = \frac{V}{4\pi^{3/2}\beta^{3/2}} \frac{1}{\Gamma(5/2)} \int dx \frac{x^{3/2}}{e^{x-\eta}+1}. \tag{4.2.2}$$

V is the total volume and $\eta = \beta k_F^2$ in the natural unit.

We note that k_F is the ideal Fermi wave number only for the noninteracting case. In the presence of Coulomb interaction we have to evaluate other contributions to the total grand partition function and then to renormalize k_F as a function of electron density.

For absolute zero, (4.2.2) yields

$$\ln \Xi_0 = \frac{2\eta^{5/2}V}{15\pi^2\beta^{3/2}}. \tag{4.2.3}$$

From this expression the kinetic energy per electron can be obtained easily,

$$\begin{aligned}
\varepsilon_0 &= \frac{3}{5}k_F^2 \\
&= \frac{3}{5}\left(\frac{9\pi}{4}\right)^{2/3}\frac{1}{r_s^2} = \frac{2.210}{r_s^2}.
\end{aligned} \tag{4.2.4}$$

Note that $r_s = r_0/a_0$, where r_0 is the radius of the sphere per electron defined by $4\pi r_0^3/3 = 1/n$. As is clear, the kinetic energy is important for small r_s. The kinetic energy is proportional to r_s^{-2} because it is determined by $k_F^2 \propto n^{2/3}$.

4.2.2 First-Order Exchange Contribution

At absolute zero, the contribution from the first-order exchange graphs to the grand partition function can be given by

$$\begin{aligned}
\ln \Xi_x &= \frac{V\beta}{16\pi^6}\int d\mathbf{p}d\mathbf{q}\, u(q)f(p)f(\mathbf{p}+\mathbf{q}) \\
&= \frac{Ve^2\eta^2}{4\pi^3\beta}.
\end{aligned} \tag{4.2.5}$$

The sign of the integral is positive because of the cancellation of the minus signs which originate from the presence of one interaction and one exchange in the exchange diagram for the propagator. The grand partition function is given by taking the trace of the propagator in the range between $\beta = 0$ and $\beta = 1/k_BT$.

The exchange energy can be derived by combining (4.2.3) and (4.2.5). It can actually be derived more directly without using the grand partition function. For absolute zero, it is

$$\varepsilon_x = -\frac{3}{2\pi}\left(\frac{9\pi}{4}\right)^{1/3}\frac{1}{r_s} = -\frac{0.916}{r_s} \tag{4.2.6}$$

For $r_s = 1$ the exchange energy is approximately one half the kinetic energy in absolute magnitude but negative. Its proportionality to r_s^{-1} indicates that it is of order $e^2 n^{1/3}$, which is the Coulomb energy at the average distance. For high densities, these two energies are rather large among the terms contributing to the ground state energy. In contrast, the correlation energy is not as large. However, it becomes increasingly important as r_s increases. Since r_s of actual metals is of order 2–3, it becomes necessary to evaluate the correlation energy correctly.

4.2.3 Ring Diagram Contribution

The kinetic and exchange energies are of order $(e^2)^0 = 1$ and (e^2) in terms of the Coulomb potential. To order e^4 the correlation energy is given by the ring and second order exchange diagram contributions. The former can be evaluated through (3.2.1). This expression indicates that in powers of $(u\lambda_j)$ the contribution starts with a term of order (e^4), but one can show that the entire contribution is actually of order (e^4). Here, $\lambda_j(q)$ is given by

$$
\begin{aligned}
\lambda_j(q) &= \frac{2}{(2\pi)^3} \int \frac{f(\mathbf{p}+\mathbf{q}) - f(p)}{p^2 - (\mathbf{p}+\mathbf{q})^2 + 2\pi i j/\beta} d\mathbf{p} \\
&= \frac{k_F}{(2\pi)^2} F(s, y).
\end{aligned}
\tag{4.2.7}
$$

The function $F(s, y)$, $s = q/k_F$ and $y = 2\pi j/\beta k_F^2$ is given in (A2.15). As we mentioned earlier, k_F has to be renormalized as a function of density in accordance with grand ensemble theory. It is related to the absolute activity z by

$$
z = \ln(\beta k_F^2).
\tag{4.2.8}
$$

The right side integral for $\lambda_j(q)$ in the first equality of (4.2.7) is essentially the same as that for the RPA polarization function. In fact, the change of variable from $2\pi j/\beta$ to $\omega - i0$ in the eigenvalue function results in the RPA dielectric function. In particular, when $y = 0$, (4.2.7) yields the RPA dielectric function for the static case.

The eigenvalue expression is rather involved. Nevertheless, two methods have been developed for the evaluation of the ring diagram contribution to the correlation energy. The first follows the pioneering work of *Gell–Mann* and *Brueckner* [4.4] based on a series expansion and a partial numerical integration [4.5]. The second method is a direct approach based on a numerical evaluation [4.6].

Let us discuss the first method [4.5]. We expand the ring diagram formula for the grand partition function in powers of $u(q)\lambda_j$ such that

$$
\ln \Xi_r = \frac{V}{16\pi^3} \sum_j \sum_{n=2} \frac{(-)^n}{n} \int d\mathbf{q} [u(q)\lambda_j(q)]^n.
\tag{4.2.9}
$$

The right side series may be summed based on the Mellin tranformation

$$\sum_{n=2}^{\infty}(-)^{n+1}\frac{x^n}{n} = \frac{1}{2\pi i}\int_c dv \frac{\pi x^v}{v\sin\pi v}, \quad (1 < c < 2).$$

In the limit $\beta \to \infty$, y is a continuous variable so that the sum over j in (4.2.9) can be replaced by integration. We can use a variable $t = y/2s$ and expand $\lambda_j(q)$ for absolute zero in powers of s^2,

$$\lambda_j(q) = \frac{k_F}{2\pi}[R(t) + s^2 R_1 + s^4 R_2(t) + \cdots],$$
$$R(t) = 1 - \tan^{-1}(1/t), \quad R_1(t) = -1/[12(1+t^2)^2], \quad (4.2.10)$$
$$R_2(t) = \frac{5t^2 - 1}{240(1+t^2)^4}, \quad \text{etc.}$$

This q expansion is expected to be valid at high density, where k_F is large, or equivalently, for small momentum transfer between the electrons. Since the Coulomb interaction $u(q) = 4\pi e^2/q^2$ is large at small momenta, the expansion makes sense. However, higher powers of $u(q)$ will be increasingly divergent in the limit $q \to 0$. Nevertheless, the total sum of the infinite series in $u\lambda_j$ is expected to yield a finite result since it behaves properly in the original integrand of $\ln \Xi_r$. The q-expansion enables us to evaluate the ring diagram contribution to the correlation energy in a series. Note that the terms with powers $n = 2, 3, \cdots$ correspond respectively to e^4, e^6, etc. The small correction due to the difference between the original and expanded eigenvalues can be evaluated numerically.

For absolute zero, the ring diagram contribution has been evaluated to order e^6 as follows:

$$\ln \Xi_r = \frac{V\beta k_F^3 e^4}{2\pi^5}[0.602 - \frac{1}{3}(1 - \ln 2)\ln \Lambda_0]$$
$$+ \frac{V\beta k_F^2 e^6}{\pi^6}[0.115 - 0.0422\ln \Lambda_0], \quad (4.2.11)$$

where

$$\Lambda_0 = 4(4/9\pi)^{1/3} r_s/\pi.$$

Equation (4.2.11) enables us to evaluate the ring energy up to order r_s.

4.2.4 Direct Calculation of the Ring Energy

On the other hand, there is a direct method to evaluate the ring diagram contribution for absolute zero without using such a q-expansion [4.6]. We note that the ring energy is given by

$$\frac{U_r}{V} = -\frac{e^4 k_F^3}{8\pi^4}\int_{-\infty}^{\infty} dy \int_0^{\infty} ds \frac{F^2 + 3yFF'}{s^2 + (e^2/\pi k_F)F}$$
$$= -\frac{e^4 k_F^3}{4\pi^4} A_3\left(\frac{e^2}{\pi k_F}\right), \quad (4.2.12)$$

where $F = F(s,y)$ and the function A_3 is a special case of a more general function defined by

$$A_s(x) = \frac{1}{2\pi}[A(x) + sB(x)],$$

$$A(x) = \int_0^\infty dy \int_0^\infty ds \frac{F^2}{s^2 + xF}, \quad B(x) = \int_0^\infty dy \int_0^\infty ds \frac{yFF'}{s^2 + xF}. \quad (4.2.13)$$

The notation $A_s(x)$ has been introduced because the ring diagram contribution to the number density can be expressed simply as

$$n_r = \frac{1}{V}\frac{\partial \ln \Xi_r}{\partial \ln z} = -\left(\frac{e^4 k_F}{4\pi^4}\right) A_5(e^2/\pi k_F). \quad (4.2.14)$$

The functions $A_3(x)$ and $A_5(x)$ can be evaluated numerically. n_r is necessary for the renormalization of k_F as a function of n. This renormalization and the final result for the correlation energy require the second-order exchange contribution.

4.2.5 Second-Order Exchange Contribution

The second-order exchange graphs are classified into two types called "regular" and "anomalous" depending on whether the two interaction lines cross each other or not. Figure 4.3 illustrates these two types; the dotted lines represent interaction and the arrows indicate the direction of propagation. The contribution from the regular second-order exchange graphs to the grand partition function is given by

$$\ln \Xi_{r2x} = \frac{\beta V e^4}{64\pi^7} \int d\mathbf{p}\,d\mathbf{q}_1 d\mathbf{q}_2 \frac{f(\mathbf{p} + \mathbf{q}_1 + \mathbf{q}_2)f(p)}{q_1^2 q_2^2(\mathbf{q}_1 \cdot \mathbf{q}_2)}$$
$$\times [1 - f(\mathbf{p} + \mathbf{q}_1)][1 - f(\mathbf{p} + \mathbf{q}_2)]$$
$$= -\frac{V e^4 \eta^{3/2}}{12\pi^2 \beta^{1/2}} I, \quad (4.2.15)$$

where

$$I = \frac{1}{3}\ln 2 - \frac{3}{2\pi^2}\zeta(3) = 0.04836. \quad (4.2.16)$$

The exact evaluation of the regular second-order diagrams was achieved by *Onsager* et al. [4.7].

The anomalous diagrams contribute as follows

$$\ln \Xi_{a2x} = \frac{\beta^2 V}{(2\pi)^9} \int d\mathbf{q}_1 d\mathbf{q}_2 d\mathbf{p}\, u(q_1)u(q_2)f(p)[1 - f(p)]f(\mathbf{p} + \mathbf{q}_1)f(\mathbf{p} + \mathbf{q}_2)$$
$$= \frac{V e^4 \beta p_F^3}{4\pi^4}. \quad (4.2.17)$$

The corresponding term in the correlation energy is canceled out by a term from $\ln \Xi_r$.

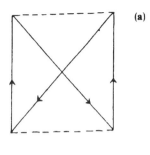
(a)

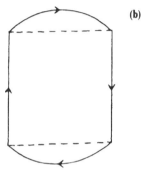
(b)

Fig. 4.3. (a) Regular and (b) anomalous second-order exchanges

4.2.6 Correlation Energy

Using the above results and including the unknown third-order exchange contribution, we can express the correlation energy in a series form as [4.5]

$$\varepsilon_c = -0.094 + 0.062 \ln r_s + 0.086 r_s - 0.013 r_s \ln r_s - \frac{24}{\pi^3} X_3 r_s. \qquad (4.2.18)$$

The last term is of order r_s and has a factor X_3 which represents the third-order exchange contribution. The second-order exchange contribution is constant and is combined with the corresponding term of the ring energy.

The reason for including the unknown term X_3 in (4.2.18) is that it is expected to cancel at least partly the preceding term $0.086 r_s$. Without the X_3 term, the above series becomes positive above around $r_s = 1$. Since the correlation energy is expected to be negative with magnitude decreasing monotonically to zero for $r_s \to 0$, the role played by the X_3 term is important. It can even completely cancel its preceding term $0.086 r_s$. The solid curve in Fig. 4.4 for $-\varepsilon_c$ has been obtained for this case so as to extrapolate smoothly to the numerical results of *Stevens* and *Pokrant* [4.8] for low densities given by dots.

The ground state energy ε_g can be obtained in accordance with (4.2.1) once ε_c is calculated. However, for a wide range of r_s, the high density series for ε_c cannot be used. On the other hand, the system is expected to form a crystalline lattice for a sufficiently large r_s, as discussed in [4.9]. A Padé approximation [4.9] to interpolate the gas phase energy to the lattice energy

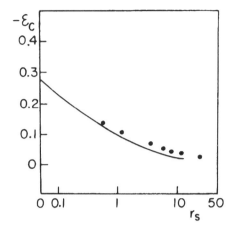

Fig. 4.4. Correlation energy of an electron gas [4.5c]

has shown that the ground state energy is minimum at $r_s = 3.76$ with a minimum value of -0.163 Ryd.

4.3 Correlation Energy of 2D Electrons

The density of 2D electrons in Si inversion layers can be varied in a very wide range. In the case of electrons in GaAs/GaAlAs quantum wells, the filling factor $\nu = n/(eH/ch)$ of Landau levels can also be changed in a wide range through a magnetic field. Therefore, these 2D electron systems provide many-body theory with an important testing ground.

The ground state energy of 2D electrons can be expressed as

$$\varepsilon_g = \frac{1}{r_s^2} + \varepsilon_x + \varepsilon_c, \tag{4.3.1}$$

where the right side members correspond respectively to the kinetic, first-order exchange and correlation energies. These energies can be evaluated in several ways, but the use of the electrostatic analog of the dielectric function found by *Isihara* and *Ioriatti* [4.10] facilitates the calculations. Through this analog, not only the exact exchange energy at finite temperatures but also the exact high and low density series for the ring energy can be obtained.

Therefore, let us start with the dielectric function given by

$$[\varepsilon(q, \omega)]^{-1} = 1 - u(q)\lambda(q, \omega), \tag{4.3.2}$$

where $u(q) = 2\pi e^2/q$ is the Coulomb potential and $\lambda(q, \omega)$ is related to the eigenvalues of the free-electron propagator given by

$$\lambda_j(q) = -\frac{2}{V\beta} \sum_{j,l} [(i\omega_j - \varepsilon_{k-q}^0 + \mu)(i\omega_{j+l} - \varepsilon_k^0 + \mu)]^{-1}, \tag{4.3.3}$$

where V is the total area, ε_k^0 is the kinetic energy, and μ is the chemical potential. By setting $2\pi j/\beta = -i\omega + \eta$ with $\eta \to 0+$ and $\beta \to \infty$, we obtain the polarization function $\lambda(q, \omega)$. We can easily show also that an alternative form of $\lambda_j(q)$ is

$$\lambda_j(q) = \frac{2}{(2\pi)^2} \int d\mathbf{p} f(p)[1 - f(\mathbf{p}+\mathbf{q})]$$
$$\times \int_0^\beta \exp\{\alpha[p^2 - (\mathbf{p}+\mathbf{q})^2]\} \exp(2\pi i j \alpha(\beta) d\alpha. \tag{4.3.4}$$

Here $f(p)$ is the Fermi distribution function.

Equation (4.3.4) can be transformed into an interesting form in terms of the new variables defined by [4.10]

$$\rho = 1/q, \quad z = \pi j(\beta q^2 p_F), \quad a = 1/(2p_F)$$
$$Q = 1/(2\pi)^2, \quad \eta = \beta p_F^2, \quad p = p_F r/\rho. \tag{4.3.5}$$

We obtain

$$\lambda(\rho, z) = Q \int_0^\infty \frac{2\pi r dr}{\exp(\eta r^2/\rho^2) + 1} \int_0^\infty \frac{dk}{a} \sin ka J_0(kr) e^{-k|z|}. \tag{4.3.6}$$

This equation suggests that it is appropriate to introduce

$$\phi(r, z) = Q \int dk \frac{\sin ka}{ka} e^{-k|z|} J_0(kr)$$
$$= \frac{Q}{a} \sin^{-1}\left(\frac{2a}{r_1 + r_2}\right). \tag{4.3.7}$$
$$r_1^2 = z^2 + (a+r)^2, \quad r_2^2 = z^2 + (a-r)^2.$$

$\phi(r, z)$ can be interpreted as the electrostatic potential due to a conducting disk of radius a with uniform charge Q. At absolute zero, $\lambda(\rho, z)$ represents the total electric flux passing through a flat disk at position z from the conducting disk at $z = 0$. ρ is the radial variable which is perpendicular to z. We note that the potential at $z = 0$ is given by

$$\phi(r, 0) = \begin{cases} \dfrac{Q}{a} \sin^{-1}\left(\dfrac{a}{r}\right) = \dfrac{k_F}{\pi^2} \sin^{-1}\left(\dfrac{1}{2k_F r}\right), & r > a, \\ \dfrac{\pi Q}{2a} = \dfrac{k_F}{2\pi}, & 0 < r < a. \end{cases} \tag{4.3.8}$$

The exchange energy can be obtained as

$$\varepsilon_x = \frac{1}{2(2\pi)^2 n} \int \Lambda(q) u(q) dq, \tag{4.3.9}$$

where

$$\Lambda(q) = \frac{1}{\beta} \sum_j \lambda_j$$
$$= \frac{2k_F q^2}{\pi} \int_0^\rho \phi(r, 0) 2\pi r dr. \tag{4.3.10}$$

We arrive at

$$\varepsilon_x = -\frac{8\sqrt{2}}{3\pi r_s}. \tag{4.3.11}$$

The above method can be extended to finite temperatures. For instance, the exact exchange specific heat has been found to be [4.10]

$$c_v = c_v^0 \left(1 + \frac{r_s}{\sqrt{2\pi}}(0.5802 - \ln \eta_0)\right), \tag{4.3.12}$$

where $\eta_0 = \beta k_F^2$.

The ring diagram contribution to the ground state energy is determined by $\lambda_j(q)$. In view of the above electrostatic analog, we employ an oblate spheroidal coordinate system in which

$$\rho = \cosh \xi \sin \phi,$$

$$|z| = \sinh \xi \cos \phi,$$

where $0 \leqq \phi \leqq \pi/2$ and $0 \leqq \xi \leqq \infty$. The eigenvalue transformed into this coordinate system is very simple [4.10],

$$\lambda(\phi) = \frac{1}{2\pi}(1 - \cos \phi). \tag{4.3.13}$$

In fact, it is only a cosine function and is independent of ξ. The ring energy is then given by

$$\varepsilon_r = -\frac{16}{\pi r_s^2} \int_0^{\pi/2} d\phi \int_0^\infty d\xi \frac{\cosh^2 \xi - \sin^2 \phi}{(\cosh \xi \sin \phi)^5} \left[\frac{r_s}{\sqrt{2}} \cosh \xi \sin \phi (1 - \cos \phi)\right.$$

$$\left. - \ln \left(1 + \frac{r_s}{\sqrt{2}} \cosh \xi \sin \phi (1 - \cos \phi)\right)\right]. \tag{4.3.14}$$

It is possible to perform the ξ integration. As a result, the ring energy is given in terms of a ϕ-integral in a finite domain. A very interesting aspect of this calculation is that the analytical behavior of the integrand of the ϕ-integral changes at $r_s = \sqrt{2}$. That is, depending on whether r_s is smaller or larger than $\sqrt{2}$, we obtain different energy expressions. These correspond to the following high and low density expressions:

$$\varepsilon_r = \begin{cases} -0.6137 - 0.1726 \ln r_s + 0.8653 r_s + O(r_s^2 \ln r_s), & r_s < \sqrt{2} \\ \\ -\dfrac{1.2935}{r_s^{2/3}} + \dfrac{1.2004}{r_s} - \dfrac{0.14018}{r_s^{4/3}} + O\left[\left(\dfrac{\ln r_s}{r_s}\right)^2\right], & r_s > \sqrt{2} \end{cases} \tag{4.3.15}$$

The ground state energy is given by

$$\varepsilon_g = \frac{1}{r_s^2} - \frac{1.2004}{r_s} + 0.2287 + \varepsilon_r, \tag{4.3.16}$$

where the third term is the second-order exchange energy. This energy has been evaluated by *Isihara* and *Ioriatti* [4.11] exactly in a relatively simple manner in comparison with the 3D case, but since it is basically mathematical, the derivation is omitted here.

Note that the first-order exchange energy in the second term is exactly canceled by the second term in ε_r in the low density case. Thus, the Hartree-Fock approximation is invalid for $r_s > \sqrt{2}$.

Although (4.3.16) represents the exact high and low density series, we must remember that for large r_s short distance correlations become important. As in Sect. 4.1.2, limiting short correlations can be treated either by summing ladder diagrams or by treating the Schrödinger problem for two electrons [4.12]. The correlation energy can also be evaluated for finite temperatures [4.13].

So far we have treated the case of a single band. By extending the theory one can treat the so-called *valley occupancy phase transition* in Si inversion layers [4.14]. The ground state in the [100] direction is doubly degenerate. Although these two so-called "valleys" are completely equivalent to each other and equally populated at high densities, an increase in electron correlations at low densities can lift this valley degeneracy. This can be shown from the energy difference between the single-valley and two-valley states, which is found to be given by

$$\Delta \varepsilon_g = \varepsilon_0 (1 - 0.14 r_s^{2/3} - \cdots) \qquad (4.3.17)$$

where $\varepsilon_0 = 1/2r_s^2$ is the kinetic energy of the double-valley equal-occupancy state. It is interesting to observe that this difference depends only on high-order correlation terms, and is independent of the first-order exchange energy and of the first term in the low density series of the correlation energy. It is clear that this difference is positive for small r_s, but it changes sign at a large r_s. A precise calculation shows that this takes place at $r_s = 8.011$. For Si inversion layers with an average dielectric constant of 7.8 and an effective mass of $0.2m$, this translates into a critical density of $1.2 \times 10^{11} \, \mathrm{cm}^{-2}$, which is not very far from a typical experimental value of around $5 \times 10^{11} \, \mathrm{cm}^{-2}$.

5. Electron–Hole Liquids

An electron–hole system condenses into a liquid state and exhibits additional very interesting phase transitions.

5.1 Excitons

The valence band in elemental semiconductors such as Ge or Si is separated from the conduction band by an energy gap. Therefore, at absolute zero they do not conduct. However, upon illumination by laser light with energy larger than the band gap, valence electrons can be excited to the conduction band, leaving holes behind. These holes are mobile, as are conduction electrons. On the other hand, a conduction electron and a hole attract each other through a Coulomb potential and can form a neutral particle called an *exciton* [5.1]. Their typical lifetime is of the order of a few microseconds, but under certain conditions unusually long lifetimes of order 1 ms in Ge and 10 μs in Si can be achieved. Their finite lifetime is due to recombination or a transfer of energy to the lattice. When recombination takes place, a certain amount of energy is released in the form of light, so that it becomes possible to make observations through luminescence experiments.

Excitons are also formed in some other semiconductors, but in what follows we shall be concerned with Si and Ge. In both cases the energy gap is referred to as an *indirect bandgap* because a change in wave vector **k** is required for an electron to make a transition from the conduction band to the valence band. The excitons and free charge carriers move within the host crystal, which serves as a neutralizing medium. They possess very interesting properties, which have been revealed since 1970. Some of these properties are summarized below [5.1–9].

(a) Excitonic Levels

The exciton is similar in structure to the hydrogen atom or more appropriately to positronium because the masses of a hole and an electron are not much different. Therefore, the exciton is a very simple and basic particle. Its energy structure has been revealed by spectroscopic studies with photon energy below the band gap, and indeed a spectral series similar to the Rydberg series of hydrogen (but with a smaller effective Rydberg energy) has been observed in Cu_2O and other semiconductors [5.2].

(b) Mobility

The excitons are highly mobile under an applied force or due to their thermal energy. Since they are neutral, stress, instead of an electric field, is used to induce motion [5.3]. In Si or Ge, stress reduces the energy gap between the valence and conduction bands. Therefore, the excitons can be moved toward the region of higher stress by applying a stress gradient. Their drift velocity v is expected to be proportional to an applied force F, the proportionality constant being the mobility μ.

$$\mu = v/F \tag{5.1.1}$$

On the other hand the drift momentum is $mv = F\tau$, where τ is a scattering time and m is the exciton mass. For motion along the [100] axis this mass is approximately given by the sum of transverse masses m_{et} and m_{0t} of the electron and hole under [011] stress. With $m_{et} = 0.19m_0$ and $m_{ht} = 0.26m_0$, $m = 0.45m_0$ approximately. The scattering time is given by

$$\tau = \mu m. \tag{5.1.2}$$

Therefore, a measurement of the drift velocity will determine the scattering time. Using ultrapure Si and pulsed laser light, *Tamor* and *Wolfe* [5.3] determined the exciton velocity and found that the data can be fitted to

$$\tau = 8.8 \times 10^{-10} T^{-3/2}. \tag{5.1.3}$$

The $T^{-3/2}$ proportionality and the magnitude are expected theoretically from a simple acoustic-phonon scattering model. At 1.3 K, their data lead to an equivalent charged-particle mobility $e\tau/m = 4 \times 10^6\,cm^2/Vs$ which is indeed very high. The diffusion coefficient may be determined from the Einstein relation $D = \tau k_B T/m$. This yields $D = 300T^{-1/2}\,cm^2/s$ so that at 11 K, $D = 90\,cm^2/s$. This value has been confirmed by an independent experiment.

(c) Bose Condensation

The above description has been based on classical statistics. In fact luminescence experiments at low densities indicate that generally excitons at much low densities follow classical statistics. However, excitons are lighter than the hydrogen atom and obey Bose statistics so that quantum effects, especially a Bose–Einstein condensation, can be expected at sufficiently high density [5.4]. The condensation temperature in an ideal Bose gas is given by

$$n = 2.612 \frac{(2\pi m k_B T)^{3/2}}{h^3}, \tag{5.1.4}$$

where n is the number density. Since the transition temperature is proportional to $n^{2/3}/m$ and the reduced mass m of excitons is small, the condensation temperature should be much higher than in liquid helium. Although the λ transition and the peculiar properties of He II have been associated with Bose condensation, liquid helium is not an ideal Bose gas but a high density liquid with strong interactions. Therefore, it is desirable to investigate more dilute and ideal systems and in this respect exciton gases step into the limelight.

Excitons in Cu_2O have several desirable characteristics for Bose condensation [5.4]. First, the excitons in Cu_2O seem to interact with each other repulsively. Second, the effective masses of electron and hole in these excitons are nearly equal to each other, resulting in a low effective mass and a high critical temperature. Third, para-excitons, i.e., excitons with antiparallel spins, have long lifetimes of order 10 μs. A line-shape analysis of a free exciton gas in Cu_2O at 1.5 K by *Hulin* et al. [5.4] indicates a gradual evolution from a classical regime at low density to a highly quantum statistical one with zero chemical potential at high density. It would be exciting if superfluidity were found in the condensed phase, as in the case of liquid helium. On the other hand, it is conceivable that excitons mediate superconductivity as phonons do for ordinary superconductivity [5.9].

(d) Metal–Insulator Transition

As the density of excitons increases, the excitons may be ionized so that the system becomes metallic. While this transition has been treated in several ways, the Mott criterion based on a consideration of screening is simple. In the metallic state, the Coulomb potential between an electron and a hole is screened by nearby charges in the form

$$\phi(r) = -\frac{e^2}{\kappa r}\exp(-r/r_{TF}), \tag{5.1.5}$$

where κ is the dielectric constant of the system. Near absolute zero, the screening constant is given by the Thomas–Fermi screening length r_{TF} defined by

$$r_{TF} = \hbar\left(\frac{\kappa}{8me^2}\right)^{1/2}\left(\frac{\pi}{3n}\right)^{1/6}, \tag{5.1.6}$$

where m is the electron mass, with which the hole mass is assumed to be identical. If classical statistics can be used for a finite temperature, the screening constant is given by the Debye length

$$r_D = \left(\frac{\kappa k_B T}{8\pi n e^2}\right)^{1/2}. \tag{5.1.7}$$

Here, the denominator factor is 8 instead of 4 because the system has two components.

The potential $\phi(r)$ is the long range attractive Coulomb potential for very small carrier density n. Hence, a pair of free carriers may not exist. It is more stable if they form a neutral exciton. However, if the carrier density is increased, the screening due to nearby charges will become more and more effective. The critical density may be obtained from the Schrödinger problem for a pair of particles with the above Coulomb potential such that the binding energy falls to zero. It is found that at this density the screening length is close to the exciton Bohr radius

$$r_{TF}(n_c) = 0.84 a_{\mathrm{ex}}. \tag{5.1.8}$$

At this point, the metallic plasma will become a neutral exciton gas. That is, a metal–insulator (Mott) transition has taken place [5.5].

(e) Electron–Hole Liquid

The excitons can condense into a liquid state in the form of droplets when their density exceeds a critical value. Such a condensation, which was first predicted by *Keldysh* [5.6] in 1968, has indeed been observed [5.7–9]. Rayleigh scattering has shown liquid droplets of about $2\,\mu$m in size. These droplets consist of approximately 10^7 electron–hole pairs and are massive so that they are expected to move only slowly. However, luminescence experiments have shown that they can drift fairly large distances by absorbing phonons and transferring the phonon momentum to carriers in the droplets. We say that the droplets diffuse in a "phonon wind" [5.8].

Let use now describe some basic aspects of the exciton luminescence in connection with the electron–hole liquid (EHL) formation. When the density of excitons is small, a sharp luminescence line is observed. The top two curves in Fig. 5.1 represent this line observed for Si by *Gourley* and *Wolfe* [5.9]. At lower temperatures two peaks appeared; one of these, marked as Ex, corresponds to free excitons and the other, Ex_2, to exciton molecules. Finally, at 1.4 K a broad line representing EHL was observed. That is, both of these excitons condensed into an EHL with density $3.3 \times 10^{18}\,\mathrm{cm}^{-3}$. This is a plasma-like droplet in which the electrons and holes move freely. The small circles represent theoretical line-shape fits from which the molecular binding energy of 1.53 meV was obtained. On the other hand, the relative intensity of the two peaks as a function of temperature resulted in a thermodynamic binding energy of 1.46 meV.

The free exciton line is due to recombination of free electrons and holes. A schematic energy diagram is given in Fig. 5.2. Here, the two parabolic curves represent the conduction and valence bands with a separation of $\varepsilon'_{\mathrm{gap}}$ which is renormalized from the band gap $\varepsilon_{\mathrm{gap}}$ due to the presence of other interacting electrons and holes. These bands are filled to their respective Fermi levels. An electron with energy ε_e and a hole with energy ε_h recombine, transfer energy $\hbar\omega_{\mathrm{ph}}$ to the lattice and emit a photon of energy $h\nu$. The condensate chemical potential is given by

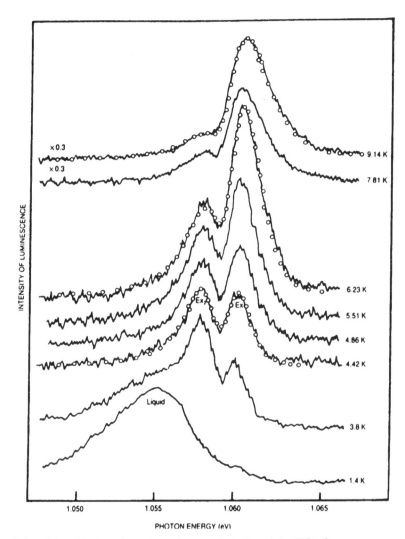

Fig. 5.1. Observation of an electron–hole liquid in Si [5.9]

$$\mu = \varepsilon'_{\text{gap}} + \varepsilon_{\text{F}}, \tag{5.1.9}$$

where ε_{F} is the sum of the electron and hole Fermi energies. The exciton threshold ε_{ex} is the sum of the chemical potential and the work function ϕ,

$$\varepsilon_{\text{ex}} = \mu + \phi. \tag{5.1.10}$$

The lower edge of the broad condensate line corresponds to the band gap $\varepsilon'_{\text{gap}}$. Its upper edge is given by $\varepsilon'_{\text{gap}} + \varepsilon_1$, where ε_1, the energy required to add an exciton, is given by the ground state energy per exciton such that

$$\varepsilon_1 - \varepsilon'_{\text{gap}} = E_0(N) - E_0(N-1) = E_0/N. \tag{5.1.11}$$

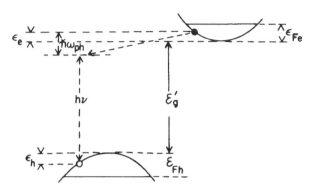

Fig. 5.2. Schematic energy diagram of an indirect gap semiconductor such as Ge or Si

The energy of a small but macroscopic number N of excitons, may be expressed as

$$E = NF(n) + N\varepsilon'_{\text{gap}}. \qquad (5.1.12)$$

Here, the function F depends only on the density n of excitons and is expected to be minimum at a certain density. This minimum corresponds to the ground state energy E_0 of the excitons. The excitons are expected to condense into a liquid state when their binding energy exceeds the ground state energy:

$$E_0/N < \varepsilon_{\text{ex}}. \qquad (5.1.13)$$

The binding energy of an exciton is several hundred times smaller than that of a hydrogen atom. The energy is characterized by an effective Bohr radius defined by

$$a_x = \kappa\hbar^2/m_x e^2 = \frac{a_0\kappa}{m_x}m_0, \qquad (5.1.14)$$

where the reduced mass m_x is given in terms of the optical masses of electrons and holes,

$$m_x^{-1} = m_{\text{oe}}^{-1} + m_{\text{oh}}^{-1}, \qquad (5.1.15)$$

and κ is the (high frequency limit of the) dielectric constant. κ is 15.4 in Ge and 11.4 in Si, and the reduced masses are 0.046 and 0.123 respectively. Thus, the excitonic effective Bohr radius is 177 A in Ge and 49 A in Si. These large values indicate that the exciton wave function is spread over many lattice constants so that a description in terms of effective masses and screened charges becomes possible. At the same time, the large values of the Bohr radius result in small binding energies, because these energies are expected to be of order $1/(\kappa a_x)$ Ryd. Actual binding energies are more complicated due to band structures and anisotropy in masses. Experimentally, the binding energy is 14.7 meV in Si and 4.15 meV in Ge. These values are rather small, indicating that the excitons can easily be ionized as the temperature rises. The binding energy of the EHL is determined by the ground state energy, which is the minimum average energy of the electron–hole system. The Hamiltonian of an electron–hole system consists of the kinetic energy and the Coulomb interaction energy. The determination of the ground state energy parallels that of a simple electron gas except for appropriate modifications in the present case with two components and multiple bands (valleys). In fact, at the bottom of the conduction band there are four equivalent valleys in Ge and six in Si. The top of the valence band is two-fold degenerate in both cases. However, an application of stress changes these degeneracies. The ground state energy is expressed in units of the excitonic Rydberg,

$$R = m_x e^4/(2\hbar^2 \kappa^2). \tag{5.1.16}$$

In m_x defined by (5.1.15), the optical mass of the electron is of the form

$$m_{oe}^{-1} = \frac{1}{3}(m_{le}^{-1} + 2m_{te}^{-1}). \tag{5.1.17}$$

There are two hole bands with "heavy" and "light" effective masses, m_{Hh} and m_{Lh}. The optical mass for the holes is given by

$$m_{oh}^{-1} = \frac{1}{2}(m_{Hh}^{-1} + m_{Lh}^{-1}). \tag{5.1.18}$$

In addition, the density of states mass m_d is defined by

$$m_d = (m_l m_t^2)^{1/3}, \tag{5.1.19}$$

when the energy surface is represented by an ellipsoid of revolution.

For systems with g equivalent valleys in the conduction band and for $\nu_h = 1 + (m_{Lh}/m_{Hh})^{3/2}$, the kinetic and exchange energies are given in units of the excitonic Rydberg R by

$$\varepsilon_0 = \frac{2.21}{r_s^2}\left(\frac{m_x}{g^{2/3}m_{de}} + \frac{m_x}{\nu_h^{2/3}m_{Hh}}\right), \tag{5.1.20}$$

$$\varepsilon_x = -\frac{0.916}{r_s}[g^{-1/3}\Phi(\gamma_e) + \Psi(\gamma_h)]. \tag{5.1.21}$$

Here, r_s is defined in terms of a_x in accordance with $(4\pi/3)(r_s a_x)^3 = 1/n$ and is from around 0.5 to 2.5. Φ and Ψ which are functions of $\gamma = m_t/m_l$ are of order 1. The parameter γ is a measure of the anisotropy of an ellipsoidal band for which the function Φ was evaluated by *Combescot* and *Nozières* [5.10a, b]. The above expressions show that increasing degeneracy lowers the Hartree–Fock energy. The anisotropy in masses is also favorable to stabilize the electron–hole plasma.

While the Hartree–Fock energy can be given in a straightforward way, the evaluation of the correlation energy for low densities faces the same difficulty as in the case of a single component electron gas. In fact, it is desirable to treat the case of r_s of order 1 or 2 which is not small. Hence, various modifications and improvements of a high density approach have been attempted, including a small momentum expansion, Hubbard approximation, and a self-consistent numerical integration. Since the essence of these approximations have already been given in Chaps. 2 and 4 for the simpler case of an electron gas, we shall not give any details, which are, in the present case, somewhat complicated by the complex nature of the band structure. The effective masses and other basic band parameters of Si and Ge are listed in Table 5.1. The effective masses are in units of the bare electron mass. We have given in (5.1.11) the relation between the energy gap and the ground state energy. The latter has been determined experimentally as [5.11]

$$E_0/N = -\ 6.0\,\text{meV (Ge)}$$
$$-\ 22.2\,\text{meV (Si)}.$$

Table 5.1. Band parameters for Ge and Si [5.10c]

	m_{le}	m_{te}	m_{oe}	m_{de}	m_{Lh}	m_{Hh}	ε_x [meV]	a_x [Å]
Ge	1.58	0.082	0.12	0.22	0.042	0.347	2.65	177
Si	0.9163	0.1905	0.259	0.32	0.154	0.523	12.85	49

The corresponding equilibrium electron densities are around $2.4 \times 10^{17}\,\text{cm}^{-3}$ for Ge and $33 \times 10^{17}\,\text{cm}^{-3}$ in Si. Despite basic difficulties, theoretical values close to these experimental values have been obtained [5.10, 12–15]. It has been found theoretically that the band structure and anisotropy affect the exchange and correlation energies separately, but in combination their net effects are rather small due to cancellation. Moreover, numerical calculations have shown that the ground state energy and equilibrium density decrease rapidly when a uniaxial strain is applied.

5.2 Phase Diagram

At low density and high temperature, excitons are in a gas phase and are neutral. As their density increases, a first-order phase transition into a metallic liquid state takes place below a critical temperature. An experimentally determined phase diagram of Si taken from [5.16a] is illustrated in Fig. 5.3. The black dots show that a free exciton gas coexists with liquid droplets at around 15 K in temperature and $10^{17}/\text{cm}^3$ in density. The crosses represent the metal–insulator transition. The triangles represent the data of *Hammond* et al. [5.16b] for the condensed phase. Note that on the high density side of the phase diagram, the EHL density varies as T^2, starting with $n_0 = 3.3 \times 10^{18}\,\text{cm}^{-3}$ at 0 K. The critical constants are

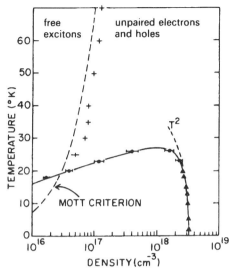

Fig. 5.3. Phase diagram of an electron–hole liquid. Above the curve is a gas of either free excitons or of unpaired particles. The dashed line represents the Mott criterion for the metal–insulator transition, and the circles are corresponding data [5.16a]

$$T_c = 27 \pm 1 \, \text{K}$$

$$n_c \cong 1.1 \times 10^{18} \, \text{cm}^{-3}.$$

We remark that theoretical values close to these experimental T_c and n_c have been reported. For instance, $28\,\text{K}$, $0.8 \times 10^{18}\,\text{cm}^{-3}$ [5.10b], $20.8\,\text{K}$, $1.2 \times 10^{18}\,\text{cm}^{-3}$ [5.14]. The theoretical coexistence curve in Fig. 5.3 was obtained by scaling the result of a phenomenological droplet theory [5.15] to the measured $T_c = 27\,\text{K}$ and $n_0 = 3.3 \times 10^{18}\,\text{cm}^{-3}$ at $0\,\text{K}$. The dashed curve represents the Mott criterion. For the classical regime, it is given by

$$n_M(T) = 1.5 \times 10^{15} T,$$

when the Debye screening length in (5.1.7) is used for (5.1.8). At absolute zero, it gives $n_M(0) = 1.3 \times 10^{14}\,\text{cm}^{-3}$. The experimental points on the gas side of the coexistence curve were determined with respect to the predicted gas density $1.25 \times 10^{17}\,cm^{-3}$ at $23\,\text{K}$ and under the assumption that the density varies linearly with T. The theoretical values were also based on some assumptions, approximations and adjustments. While it is beyond the scope of this section to discuss all these in detail, let us outline some of the theoretical features.

First, in order to derive a phase diagram, the free energy must be obtained for low but finite temperatures. The simplest approach can be made based on Fermi liquid theory, which states that the energy varies in proportion to T^2, as in an ideal gas. Theoretically, this adoption is reasonable because the

Fermi temperatures of the electrons and holes are higher than the critical temperature. In fact, T_F for Ge is around $45\,K$ for electrons and $29\,K$ for holes. Moreover, the experimental EHL density n shows such a quadratic variation. Hence, under the assumption of thermal equilibrium the free energy is expanded in powers of T^2 such that

$$f(n,T) = \varepsilon(n) - \frac{1}{2}\gamma(n)(k_B T)^2, \tag{5.2.1}$$

where $\varepsilon(n)$ is the ground state energy for a given condensate density n and

$$\gamma(n) = (\pi/3n)^{2/3}\hbar^{-2}(\nu^{2/3}m_{de} + m_{dh}). \tag{5.2.2}$$

For Si, an experimental value $\gamma = 1.6 \times 10^3\,\mathrm{eV}^{-1}$ and the following experimental temperature variations have been reported [5.16]:

$$\begin{aligned}
n(T) &= 3.33 \times 10^{18} - 1.5 \times 10^{15}T^2 \,\mathrm{cm}^{-3}, \\
\varepsilon_F(T) &= 22.2 - 8 \times 10^{-3}T^2 \,\mathrm{meV}, \\
\mu(T) &= 1.0887 - 5.9 \times 10^{-3}T^2 \,\mathrm{meV}, \\
\phi(T) &= 8.2 + 5.9 \times 10^{-3}T^2 \,\mathrm{meV}.
\end{aligned} \tag{5.2.3}$$

Here, n is the liquid density, ε_F the sum of the Fermi energies, μ the chemical potential and ϕ the work function.

The ground state energy of an electron–hole plasma is given by the Hartree–Fock and correlation energies. While the Hartree–Fock energy is well defined, it is the correlation energy which involves theoretical difficulties. *Combescot* and *Nozières* [5.10] used a high density approach with a finite momentum correction similar to that discussed in Sect. 4.2.3. *Vashishta* et al. [5.14] assumed a phenomenological form

$$\varepsilon_{xc}(r_s) = \frac{a + br_s}{c + dr_s + r_s^2}$$

for the exchange-correlation energy.

The free energy $f(n,T)$ may be expanded about the equilibrium density to second order. Approximating $\gamma(n)$ by $\gamma(n_0)$, we obtain

$$f(n,T) = \varepsilon(n_0) + \frac{1}{2}\varepsilon''(n_0)(n - n_0^2) - \frac{1}{2}\gamma(n_0)(k_B T)^2.$$

Minimization of f yields

$$n = n_0 + \frac{\gamma'(n_0)}{2\varepsilon''(n_0)}(k_B T)^2.$$

Hence, the quadratic variation of n in the EHL regime depends on $\varepsilon''(n_0)$. The sum of the Fermi energies, ε_{Fe} and ε_{Fh}, of electrons and holes in EHL also varies quadratically,

$$\varepsilon_F(T) = \varepsilon_F(0)\left(1 - \frac{\pi^2}{12}\frac{(k_B T)^2}{\varepsilon_{Fe}\varepsilon_{Fh}} + \frac{\gamma'(n_0)}{3n_0\varepsilon''(n_0)}(k_B T)^2\right) \tag{5.2.4}$$

Near absolute zero, the chemical potential of the EHL liquid is given by

$$\mu = \mu(0) - \frac{1}{2}\gamma(n_0)(k_B T)^2. \tag{5.2.5}$$

For the low density phase, a semiclassical expression for the chemical potential of the exciton may be used. With ν_{ex} as a degeneracy factor, this is

$$\mu_{ex} = \varepsilon_{ex} + k_B T \ln \left[\frac{n_{ex}}{\nu_{ex}} \left(\frac{2\pi\hbar^2}{m_x k_B T} \right)^{3/2} \right]. \tag{5.2.6}$$

This chemical potential is equated to the EHL chemical potential of (5.2.5) so that

$$n_{ex}(T) = \nu_{ex} \left(\frac{m_x k_B T}{2\pi\hbar^2} \right)^{3/2} \exp \left(\frac{\phi}{k_B T} - \frac{1}{2}\gamma k_B T \right), \tag{5.2.7}$$

where

$$\phi(T) = \mu_{EHL}(T) - \varepsilon_{ex}$$
$$\sim \varepsilon_0(n_0) - \varepsilon_{ex}. \tag{5.2.8}$$

The T dependence of ϕ has been neglected in (5.2.7).

It is clear from the experimental phase diagram of Fig. 5.3 that the critical region is rather flat and not parabolic. Hence, the above quadratic approximation becomes increasingly poor towards T_c. Nevertheless, the critical point has been determined from the familiar set of equations

$$\partial\mu/\partial n = 0, \quad \partial^2\mu/\partial n^2 = 0.$$

The critical constants thus determined have been found to agree with experiment.

Reinecke et al. [5.15] used a classical droplet theory of condensation. In this theory, droplets of size l are specified by a surface energy which is proportional to $l^{2/3}$. The unit surface energy $\sigma(T)$ is assumed to vary quadratically with T in such a way that

$$\sigma(T) = \sigma_0(1 - T^2/T_c^2). \tag{5.2.9}$$

Hence, $\sigma(T_c) = 0$. It is convenient to use T_c and σ_0 as theoretical parameters. For the former in Ge, the value $T_c = 6.5$ K has been obtained experimentally. The latter parameter σ_0 may be used so as to reproduce an experimental coexistence curve such as obtained by *Thomas* et al. [5.17].

The density of excitons in the gas phase is given by a sum of Boltzmann factors, $\exp(-f_l/k_B T)$, over l representing the size of droplet l. Along the coexistence curve, f_l consists of surface energy and entropy. For the latter, they adopted a form $l^{-\tau}$ with $\tau = 2.2$. The surface area was given by $a = 4\pi[3/4\pi n_0(T)]^{2/3}$, where $n_0(T)$ is the density due to single-particle excitations across the electron and hole Fermi energies. Hence, the density in the gas phase along the coexistence line was expressed as

$$n_G(T) = \frac{n_c}{\zeta(\tau - 1)} \sum_{l=1}^{\infty} \frac{\exp[-\sigma(T)al^{3/2}/k_B T]}{l^\tau}, \tag{5.2.10}$$

where Riemann's ζ function enters as a normalization factor and n_c is the critical density. Since $\sigma(T_c) = 0$, $n_G(T_c) = n_c$.

On the liquid side, the density was assumed to be given simply by the difference between the density n_s of the electron–hole system due to single particle excitations and n_G,

$$n_L(T) = n_s(T) - n_G(T). \tag{5.2.11}$$

$n_s(T)$ is expected to vary quadratically with temperature. For Ge, its form may be chosen from the experimental liquid density for $T \ll T_c$,

$$n_s(T) = 2.38 \times 10^{17}(1 - 0.0072T^2)\text{cm}^{-3}. \tag{5.2.12}$$

If complete symmetry exists between the gas and liquid phases at T_c, the critical density is given by

$$n_L(T_c) = n_G(T_c) = n_c = n_c(T_c)/2. \tag{5.2.13}$$

In Ge, $T_c = 6.5\,\text{K}$ experimentally. Hence, if an extrapolated use of (5.2.11) is made in (5.2.13), we obtain $n_c = 0.83 \times 10^{17}$ in comparison with the experimental value $(0.8 \pm 0.2) \times 10^{17}\,\text{cm}^{-3}$ of *Thomas* et al. [5.17].

5.3 2D Electron–Hole Plasma

The ground state energy of an electron–hole system is sensitive to band anisotropy [5.17, 18]. This has been demonstrated by diffusion experiments under stress [5.17]. For this reason, and also because of its actual realization in laboratories, the limiting case of two dimensions becomes interesting. Moreover, there are several advantages to studying 2D systems. First, the available density range of electrons is very wide. Second, electron correlations are relatively strong. The correlation energy has been obtained in a wide range as discussed in Sect. 4.3. Third, in the presence of a magnetic field, there is a possibility of Bose condensation.

Let us consider an ideal case in which the electron and hole have the same mass. In terms of the effective Bohr radius a_x with a reduced mass $m_x = 1/2m_e$, the density parameter r_s is given by

$$r_s = (\pi n)^{-1/2} a_x^{-1}, \tag{5.3.1}$$

where $n = n_e = n_h$ is the number density of electrons or holes. The energy is measured in units of the effective Rydberg defined by (5.1.16). The ground state energy of an electron–hole plasma in 2D may be expressed approximately in the form [5.19]

$$\varepsilon_g = r_s^{-2} - \frac{2.4008}{r_s(1 + 0.381r_s)} + \varepsilon_r, \tag{5.3.2}$$

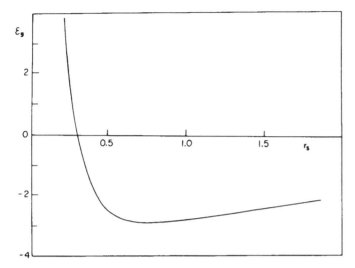

Fig. 5.4. Ground state energy of a 2D electron–hole plasma [5.19b]

where the first term is the kinetic energy, the second term represents an approximate exchange contribution obtained by combining the first- and second-order exchange energies such that the result is convergent for all r_s, and the last term is the ring energy. One can show that ε_r, is given exactly by [5.19]

$$\varepsilon_r = \frac{1}{2\pi i} \int_{c-i\infty}^{c+i\infty} \left(\frac{r_s}{2^{1/2}}\right)^{-s} \frac{4\pi^{1/2}\Gamma[(1+s)/2]}{(s-2)\sin \pi s \Gamma(1+s/2)} ds$$

$$\times \int_0^{\pi/2} \frac{1}{[\sin \phi(1-\cos \phi)]^{s-2}} \left(\frac{1}{\sin^5 \phi} - \frac{s+1}{s+2}\frac{1}{\sin^3 \phi}\right) d\phi, \quad (5.3.3a)$$

where $0 < c < 1$ and $\Gamma(x)$ is the Γ function. Accordingly, we arrive at

$$\varepsilon_r = \begin{cases} -4.891 - 5.523 r_s \ln r_s + 20.03 r_s + \dots, & (r_s < 2^{-3/2}) \\ -4.1064 r_s^{-2/3} + 2.4008 r_s^{-1} - 0.4105 r_s^{-4/3} + \dots, & (r_s > 2^{-3/2}). \end{cases} \quad (5.3.3b)$$

The ground state energy is illustrated in Fig. 5.4 as a function of r_s. Its minimum is -2.78 Ryd at $r_s = 0.7$. The minimum is approximately four times lower than the corresponding 3D value due to stronger correlations in 2D. In comparison, the binding energy of an electron–hole droplet is -1 Ryd in 3D and is -4 Ryd in 2D. If the 3D binding energy is used for comparison, the system is expected to be in a metallic state for r_s larger than approximately 0.35. Since actual 2D systems have always certain widths, the use may not be unreasonable. In fact, for these systems the ordinary Coulomb potential, instead of the logarithmic one, has been adopted, even though the electrons are confined nearly two-dimensionally. On the other hand, if the 2D binding energy is used for comparison, an electron–hole pair state is favored and no metallic state is expected. However, in the presence of band anisotropy, deviations from this conclusion can occur. In fact, such an anisotropy has been shown to lower the energy significantly [5.18].

The minimum in Fig. 5.4 is lower than some other theoretical results [5.20]. This is due to the improved treatment of the ring and exchange energies discussed above. A further improved result can be obtained by taking into consideration the ladder diagram contribution. For actual 2D systems, more sophisticated models than the above idealized one are necessary. For instance, electrons and holes may be in different layers. The ground state properties of electron–hole systems in two parallel planes and their condensation into pancake-like droplets have been treated theoretically [5.21].

On the experimental side, it has been found in Si/SiO$_2$ that the EHL luminescence is quenched and a new luminescence line appears if a bias electric field is increased from zero [5.22]. This line is observed only when a 2D hole (electron) gas is present at the [100] ([110] or [111]) plane and is sensitive to the bias electric field. It has been attributed to 2D carriers. Although in an ideal equilibrium situation, charge carriers are induced at the surface only by the applied electric field and the carriers of opposite sign are removed from the interface, excess charges may be created at the surface due to laser light. These charges attract opposite charges in the bulk to the surface, giving rise to recombination. The 2D systems become uniquely different from the 3D case in the presence of a perpendicular magnetic field. Therefore, let us discuss magnetic excitons for the most interesting case, in which the magnetic field is so strong that the magnetic length $l = (c\hbar/eH)^{1/2}$ is smaller than the effective Bohr radius a_0^* in the conduction or valence bands, and the Rydberg energy is less than the cyclotron energy,

$$l \ll a_0^*, \quad R \ll \omega_c.$$

The temperature is of course assumed to be low. Under these conditions, the electrons and holes are in the ground Landau level if their numbers do not exceed the capacity of this level. A natural density parameter is the Landau level filling factor ν defined by

$$\nu = 2\pi l^2 n.$$

For a given ν, what is the ground state of an electron–hole system? What happens if the density of excitons is increased? In what follows, we address these questions for the simple case of the electrons and holes having the same effective mass.

We first note that in the Landau gauge, which is asymmetric, the ideal electron ($\alpha = 1$) or hole ($\alpha = 2$) Green's function G_α can be expanded in terms of the Landau eigenfunctions $\phi_k^n(y)$ such that

$$G_\alpha^0(y, y'; p_x, \omega) = \sum_{n=0}^\infty \phi_k^n(y)\phi_k^n(y')g_{\alpha n}^0(\omega). \tag{5.3.4}$$

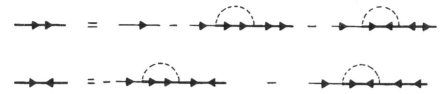

Fig. 5.5. Propagations in a self-consistent Hartree-Fock approximation

Here,

$$\phi_k^n(y) = \frac{\exp[-(y/l + kl)^2/2]}{(l\pi^{1/2}2^n n!)^{1/2}} H_n(y/l - kl), \tag{5.3.5}$$

$$g_{\alpha n}^0 = [i\omega_+^- (n + 1/2)\omega_c - \mu]^{-1}, \tag{5.3.6}$$

In (5.3.4), the sum over all Landau levels appears formally. However, under the present conditions in which only the lowest Landau level is partially occupied, the sole contribution from the ground level is relevant.

In the presence of Coulomb interaction, it is appropriate to introduce electron–electron, electron–hole and hole–hole propagators to be denoted by $G_{\alpha\alpha'}$ where α and α' are either 1 for an electron or 2 for a hole. These Green's functions form a matrix. The diagonal element $G_{11}(G_{22})$ represents electron (hole) propagation, and the off-diagonal elements G_{12} or G_{21} represents electron–hole pairing. We are interested in determining these Green's functions in a free-particle-like form. For this purpose, we adopt a self-consistent Hartree-Fock approximation [5.23], which is schematically illustrated in Fig. 5.5. In this approximation, the temperature Green's functions are determined by

$$G_{\alpha\alpha'}(y, y'; p_x, \omega) = \delta_{\alpha\alpha'} G_\alpha(y, y'; p_x, \omega)$$

$$- \sum_{\alpha''=1}^{2} k_B T \lim_{\tau \to 0} \sum_{\omega'} \int dy_1 dy_2 dp'_x u(y_1 - y_2, p_x - p'_x)$$

$$\times G_\alpha(y, y_1; p_x, \omega) G_{\alpha\alpha''}(y_1, y_2; p'_x, \omega') G_{\alpha''\alpha'}(y_2, y'; p_x, \omega') \exp(i\omega'\tau), \tag{5.3.7}$$

where $u(y, p)$ is the Coulomb interaction

$$u(y, p) = \frac{1}{2\pi} \int_\infty^\infty \frac{2\pi e^{*2} e^{iqy}}{(p^2 + q^2)^{1/2}} dq$$

$$= 2e^{*2} K_0(yp).$$

e^* is the effective charge, which may include the dielectric constant of the system.

We seek the solutions of the coupled equations (5.3.7) in a form similar to (5.3.4) because we are interested in free particles. When only the lowest Landau level is partially occupied and in the absence of level mixing, we use

$$G_{\alpha\alpha'}(y, y'; k, \omega) = \phi_k^0(y)\phi_k^0(y')g_{\alpha\alpha'}(\omega). \tag{5.3.8}$$

We find from (5.3.7) that the corresponding $g_{\alpha\alpha'}(\omega)$ is determined by

$$g_{\alpha\alpha'} = \delta_{\alpha\alpha'}g_\alpha^0 - \sum_{\alpha''=1,2} g_\alpha^0 \Delta_{\alpha\alpha''}g_{\alpha''\alpha'}. \tag{5.3.9}$$

This equation is nonlinear in g because $\Delta_{\alpha\alpha'}$ includes $g_{\alpha\alpha'}$. However, $\Delta_{\alpha\alpha'}$ orginating from (5.3.7) represents an average coupling energy of an $\alpha-\alpha'$ pair due to Coulomb interaction. Hence, we treat it as if it is a given parameter.

In view of electron–hole symmetry, we define

$$\Delta_{11} = -\Delta_{22} = \varepsilon, \quad \Delta_{12} = \Delta_{21} = \Delta. \tag{5.3.10}$$

We also introduce

$$\xi = -\varepsilon - \mu, \quad \eta = (\xi^2 + \Delta^2)^{1/2}. \tag{5.3.11}$$

μ is the chemical potential and Δ represents an energy shift. In this notation, (5.3.7) leads us to the following consistency equations:

$$\varepsilon = \frac{\varepsilon_0}{2}\left(1 - \frac{\xi}{\eta}\tanh\frac{\eta}{2k_BT}\right), \tag{5.3.12}$$

$$1 = \frac{\varepsilon_0}{2}\frac{1}{\eta}\tanh\frac{\eta}{2k_BT_c}, \tag{5.3.13}$$

where

$$\varepsilon_0 = (\pi/2)^{1/2}e^{*2}/l. \tag{5.3.14}$$

e^* is the effective charge, which may include the dielectric constant of the system.

As mentioned earlier, Δ has been considered as a parameter. In fact, for an electron–hole pair, it can be adopted as an order parameter. The condition that $\Delta = 0$ yields a critical temperature. That is, the plasma state corresponds to $\Delta = 0$, while the paired state to $\Delta \neq 0$.

In terms of the parameters ε and Δ, the Green's functions are given in a matrix form such that

$$G_{\sigma\sigma'} = \frac{\phi_{ok}(y)\phi_{ok}(y')}{\omega^2 + \xi^2 + \Delta^2}\begin{pmatrix} -i\omega - \xi & \Delta \\ \Delta & -i\omega + \xi \end{pmatrix}. \tag{5.3.15}$$

This is not an explicit solution of (5.3.7) but only a formal one in the sense that the parameters depend on the Green's functions. Nevertheless, the right side is in the form of a free particle propagator with ε and Δ representing respectively a deviation of the chemical potential and a gap.

The total number N of particles is determined by

$$N = \lim_{\tau\to 0}\frac{k_BTL}{2\pi}\sum_\omega \int dy dp_x G_{11}(y,y;p_x,\omega)e^{i\omega\tau}, \tag{5.3.16}$$

where L is the length of the system. From this equation, the filling factor ν is given by

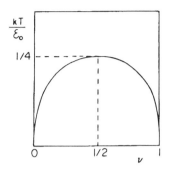

Fig. 5.6. Theoretical phase diagram for a 2D electron–hole liquid

$$\nu = \frac{1}{2}\left(1 - \frac{\xi}{\eta}\tanh\frac{\eta}{2k_{\mathrm{B}}T}\right). \tag{5.3.17}$$

Hence, (5.3.12) yields

$$\varepsilon = \nu\varepsilon_0. \tag{5.3.18}$$

Therefore, the ground state consists of noninteracting magnetic excitons. The lack of interaction is due to the assumed symmetry between the electron and hole. However, since there is no Bose condensation in two-dimensional ideal Bose gases, it is necessary to relax the strict two dimensionality somewhat. Since actual 2D systems have more-or-less a certain width, such a relaxation can be meaningful.

The chemical potential μ_p of the plasma state is determined by setting $\Delta = 0$ in η of (5.3.17). The result is

$$\mu_p = -\varepsilon_0 - k_{\mathrm{B}}T\ln(\nu^{-1} - 1). \tag{5.3.19}$$

The critical temperature is given also by setting $\Delta = 0$ in η in (5.3.13),

$$1 = \frac{\varepsilon_0}{2\eta_0}\tanh\frac{\eta_0}{2k_{\mathrm{B}}T_{\mathrm{c}}}.$$

When the excitons are in equilibrium with the plasma, the chemical potential in (5.3.13) must be equal to μ_p of (5.3.19). Hence, we find that the critical temperature varies with ν such that

$$T_{\mathrm{c}} = \frac{\varepsilon_0}{2k_{\mathrm{B}}}\frac{1 - 2\nu}{\ln[(1-\nu)/\nu]}. \tag{5.3.20}$$

Figure 5.6 illustrates the phase diagram determined by this equation. The ordinate represents $k_{\mathrm{B}}T$ in units of ε_0. When the temperature is reduced at a certain ν, a condensation takes place below the critical temperature given by $\varepsilon_0/4k_{\mathrm{B}}$, which occurs at $\nu = 1/2$. The left side of the curve represents a gas phase, while the right side is a liquid phase. Below T_{c} and under the curve is a coexistence region. In the condensed phase, the electron and hole share the same quantum number k, which determines the center $y_0 = kl^2$ of the cyclotron motion, and move in the same direction. Note that l is very small in the present case. Therefore, it is possible to associate the pair in the condensed phase with Bose condensation rather than a Cooper pair in superconductivity.

The ground state has been treated in a more rigorous manner. When transitions to higher Landau levels are neglected, the ground state is an ideal exciton gas whose long wave properties coincide with those of a 2D Bose gas. The energy dispersion relation is given by

$$\varepsilon_x(k) = -\varepsilon_0 \exp(-k^2 l^2/4) I_0(k^2 l^2/4), \qquad (5.3.21)$$

where I_0 is a modified Bessel function. Hence, there is no plasmon at $k = 0$. Furthermore, it has been shown that inclusion of higher Landau levels leads to a weak repulsive interaction. Hence, the spectrum of the low frequency excitations becomes acoustic and superfluidity may be expected as in the case of liquid helium.

6. Correlation in a Magnetic Field

Electron correlations are important in the magnetic response of an electron plasma, especially at low temperatures.

6.1 Spatial Correlation in a Magnetic Field

According to Van Leeuwen's theorem, a magnetic field does not affect the statistical properties of a classical electron gas. However, in 1930 Landau showed that free electrons do show a diamagnetic response to an applied field due to quantization of energy levels. This Landau diamagnetism comes from the orbital motion of free electrons and appears even when Boltzmann statistics is used instead of Fermi statistics. On the other hand, the coupling of the spin and the magnetic field results in the Pauli paramagnetism. We shall investigate the magnetic properties of an electron gas in this chapter, and first discuss how a magnetic fields affects the pair distribution function.

In the presence of a uniform magnetic field H in the z direction, the Hamiltonian of a system of electrons is given by

$$\mathcal{H} = \frac{1}{2m} \sum_i \left[\left(\mathbf{p}_i + \frac{e}{c}\mathbf{A} \right)^2 - \frac{g}{2}\mu_B \sigma_i \cdot \mathbf{H} \right] + \sum_{i<j} \frac{e^2}{r_{ij}}, \tag{6.1.1}$$

where the vector potential $\mathbf{A} = (-Hy, 0, 0)$, σ denotes electron spin, g is the Landè factor and μ_B is the Bohr magneton. The energy levels and the eigenfunctions of the Hamiltonian in the absence of Coulomh interaction are given by

$$\varepsilon_{n,\sigma} = (n + \frac{1}{2})\hbar\omega_c + \frac{p_z^2}{2m} \pm \frac{g}{4}\hbar\omega_c,$$

$$\tag{6.1.2}$$

$$\psi_{n,\sigma} = \frac{A_n}{2\pi} \exp[i(k_x x + k_z z)] \exp[-(y - y_0)^2/2l^2] H_n[(y - y_0)/l]|\sigma\rangle,$$

where H_n is the Hermite polynomial, $|\sigma\rangle$ is a spin state, $p_i = \hbar k_i$ for $i = x, y, z$, and

$$\omega_c = \frac{eH}{mc}, \quad y_0 = \frac{cp_x}{eH}, \quad A_n^2 = \frac{1}{\pi^{1/2}2^n n! l^2}, \quad l^2 = \frac{c\hbar}{eH}. \tag{6.1.3}$$

In these expressions, we have explicitly displayed $\hbar$ and the electron mass m. The magnetic length l defined here is a basic unit of length for electrons in a magnetic field.

The propagator for a free classical electron can be constructed as [6.1],

$$K_0(\mathbf{r}_2\beta_2, \mathbf{r}_1\beta_1) = Z \sum_{n,\sigma} \exp[-(\beta_2 - \beta_1)\varepsilon_{n,\sigma}]\psi_{n,\sigma}(\mathbf{r}_2)\psi^*_{n,\sigma}(\mathbf{r}_1)$$

$$= \frac{Z}{4\pi^{3/2}l^2} \frac{\cosh(gsa^2/2)}{s^{1/2}\sinh(sa^2)}$$

$$\times \exp\left(-\frac{z^2}{4s} - \frac{(x^2+y^2)}{4l^2}\coth(sa^2) + i\phi\right), \qquad (6.1.4)$$

where the capital letter Z has been used for the absolute activity in order to distinguish it from the coordinate z. That is, z in the exponential function is the z-component of the vector $\mathbf{r} = (x, y, z)$. This vector and other quantities are defined by

$$\mathbf{r} = \mathbf{r}_2 - \mathbf{r}_1, \quad s = (\beta_2 - \beta_1), \quad \phi = (x_1 - x_2)(y_1 + y_2)/2l^2,$$

$$a^2 = \hbar\omega_c/2 = \mu_B H.$$

Note in the natural unit that $1/l^2 = a^2$ and s has the dimension of length squared.

The above propagator, which we denote simply as $K_0(\mathbf{r}, s)$, is exact and consists of three factors. The first factor depends only on the magnatic field. The second factor represents the sum of the contributions from the two spin states. The last exponential factor shows that the propagation is free in the direction parallel to the magnetic field, but is regulated by the magnetic length in the perpendicular direction. The magnitude of the phase ϕ is determined by the curvature of the classical orbit.

In the quantum mechanically degenerate case, the propagator is given by

$$K(\mathbf{r}_2\beta_2, \mathbf{r}_1\beta_1) = \sum_{n,\sigma}[1 - f(\varepsilon_{n\sigma})]Ze^{-(\beta_2-\beta_1)\varepsilon_{n\sigma}}\psi_{n\sigma}(\mathbf{r}_2)\psi^*_{n\sigma}(\mathbf{r}_1)$$

$$= \sum_{n,\sigma} f(\varepsilon_{n\sigma})\exp[(\beta-\beta_2+\beta_1)\varepsilon_{n\sigma}]\psi_{n\sigma}(\mathbf{r}_2)\psi^*_{n\sigma}(\mathbf{r}_1), \quad (6.1.5)$$

where $f(\varepsilon)$ is the Fermi distribution function. The right hand sum can be obtained effectively by making use of the following relation which holds for absolute zero:

$$\frac{1}{2\pi i} \int_{c-i\infty}^{c+i\infty} \frac{e^{x(\varepsilon_F - \varepsilon)}}{x} dx = \begin{cases} 1, & \varepsilon \leq \varepsilon_F, \\ 0, & \varepsilon > \varepsilon_F, \end{cases}$$

where the path c must be chosen to the right of all the singularities of the integrand. The application of the above formula to (6.1.5) enables us to express the propagator of the degenerate case as

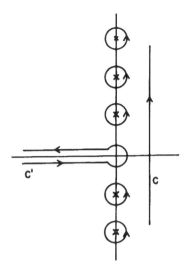

Fig. 6.1. Path of integration for (6.1.6)

$$K(\mathbf{r}_2\beta_2, \mathbf{r}_1\beta_1) = \frac{1}{2\pi i} \int_{c-i\infty}^{c+i\infty} \frac{e^{x\varepsilon_F}}{x} K_0(\mathbf{r}, x - \beta + s)\mathrm{d}x. \tag{6.1.6}$$

The path c can be divided into two parts, one surrounding the branch point on the real axis and the other encircling the poles at $(x - \beta + s)a^2 = il\pi$ as in Fig. 6.1. Accordingly, one can write

$$K(\mathbf{r}_2\beta, \mathbf{r}_1 0) = K_1(\mathbf{r}_2\beta, \mathbf{r}_1 0) + K_2(\mathbf{r}_2\beta, \mathbf{r}_1 0). \tag{6.1.7}$$

We assume that the system is quantum mechanically degenerate so that the Fermi energy is sufficiently larger than the field energy. In this case the integrand for K_1 can be expanded in powers of a dimensionless parameter γ defined by

$$\gamma = \hbar\omega_c/2\varepsilon_F = \mu_B H/\varepsilon_F.$$

We arrive at

$$K_1(\mathbf{r}_2\beta, \mathbf{r}_1 0) = \frac{Zk_F^3 e^{i\phi}}{\pi^2}$$
$$\times \left(\frac{j_1(k_F r)}{k_F r} - \frac{\gamma^2}{24} \left\{ \left[1 - 3 \left(\frac{g}{2} \right)^2 \right] \cos k_F r + k_F^2(x^2 + y^2)\frac{\sin k_F r}{k_F r} \right\} \right). \tag{6.1.8}$$

The contribution from the poles yields K_2,

$$K_2(\mathbf{r}_2\beta, \mathbf{r}_1 0) = \frac{Ze^{i\phi}}{4\pi^{3/2}l^3} \sum_{s=1}^{\infty} \frac{(-)^s}{(s\pi)^{3/2}} \left\{ \sin \left(\frac{\pi s}{\gamma} + \frac{z^2}{4\pi s l^2} - \frac{\pi}{4} \right) \cos \frac{g\pi s}{2} \right.$$
$$\times \left[J_0[k_F(x^2 + y^2)^{1/2}] + \frac{\gamma}{2} k_F(x^2 + y^2)^{1/2} \right.$$

$$\times \left(\frac{z^2}{4(\pi s)^2 l^2} + \frac{x^2 + y^2}{12 l^2} \right) J_1 [k_{\mathrm{F}} (x^2 + y^2)^{1/2}] \Big]$$

$$- \frac{\gamma}{s\pi} \left(\frac{3}{2} \cos \frac{g\pi l}{2} + \frac{g\pi l}{2} \sin \frac{g\pi l}{2} \right) k_{\mathrm{F}} (x^2 + y^2)^{1/2}$$

$$\times \cos \left(\frac{\pi s}{\gamma} + \frac{z^2}{4\pi s l^2} - \frac{\pi}{4} \right) J_1 [k_{\mathrm{F}} (x^2 + y^2)^{1/2}] \Big\} . \quad (6.1.9)$$

The contribution K_2 oscillates sinusoidally. The constant phase $\pi/4$ is characteristic of three dimensions. The amplitudes of the two oscillatory terms include Bessel functions J_0 and J_1 and are smaller for higher harmonics.

In terms of the propagator, the pair distribution function can be obtained. For example, at high temperatures, the pair distribution function of an ideal electron gas can be approximated by

$$\rho_2(r) - n^2 = -\frac{1}{2} K(\mathbf{r}_2 \beta, \mathbf{r}_1 0) K(\mathbf{r}_1 \beta, \mathbf{r}_2 0)$$

$$= -\frac{1}{2} \left(\frac{2Z}{\lambda^3} \right)^2 \left(\frac{\alpha \cosh(g\alpha/2)}{\sinh \alpha} \right)^2 e^{-2\pi z^2/\lambda^2} e^{-(x^2 + y^2)/2l^2 \tanh \alpha},$$

$$(6.1.10)$$

where $r = |\mathbf{r}_2 - \mathbf{r}_1|$, λ is the de Broglie thermal wavelength and α is a dimensionless parameter defined by

$$\alpha = \mu_{\mathrm{B}} H / k_{\mathrm{B}} T. \qquad (6.1.11)$$

The spatial variation of the pair distribution function is primarily related to the orbital motion rather than the spin states. Note that the distribution is anisotropic and is Gaussian. In the limit $\alpha \to 0$, the above formula is reduced to what *Uhlenbeck* and *Gropper* [6.2] obtained.

A very different expression is obtained for the pair distribution function at low temperatures. If the density of electrons is high and the parameter γ is small, it is given approximately by

$$\rho_2(r) - n^2 = -\frac{1}{2} \left(\frac{k_{\mathrm{F}}^3}{\pi^2} \right)^2 \left[\frac{j_1(k_{\mathrm{F}} r)}{k_{\mathrm{F}} r} - \frac{\gamma^2}{2} \left(\cos k_{\mathrm{F}} r + k_{\mathrm{F}}^2 (x^2 + y^2)^{1/2} \frac{\sin k_{\mathrm{F}} r}{k_{\mathrm{F}} r} \right) \right.$$

$$+ \frac{\gamma^{3/2}}{4\pi} J_0 [k_F (x^2 + y^2)^{1/2}]$$

$$\left. \times \sum_{s=1}^{\infty} \frac{(-)^s \cos(g\pi s/2) \sin(\pi s/\gamma + z^2/4\pi s l^2 - \pi/4)}{s^{3/2}} \right]^2 . \quad (6.1.12)$$

The oscillations in the pair distribution function have two sources. In the first two terms with the Fermi wave number k_{F} patial oscillations are due to the confinement in momentum space. In the last term, the magnetic length l couples with z because the location of the poles of the integrand of K_2 depends on α. This occurs because the de Broglie thermal wavelength

$\lambda = (4\pi\beta)^{1/2}$ is scaled by the magnetic length l. Therefore, the interference of two electron waves at distance z in the direction parallel to the magnetic field is also scaled by l. In addition to the appearance of the natural length unit l, the interference effect oscillates sinusoidally. It is not damped because the propagation is free in that direction. On the other hand, the wave vector of an electron in the perpendicular direction is confined to the orbit which is determined by the intersection of the Fermi surface and a plane normal to the field. That is, the electron moves along the circle which is given by slicing the Fermi sphere by the plane. Hence, the electron interference gives rise to damped oscillations which depend on k_F of the pair distribution function.

Let us now investigate long distance correlations in the presence of Coulomb interaction. Since the electron spin appears in (6.1.4) only through the constant factor $\cosh(gsa^2/2)$, it makes sense to study the case $g = 0$, because then when spin must be included we need only replace the absolute activity Z by $Z\exp(gsa^2/2)$ or by $Z\exp(-gsa^2/2)$ depending on the spin direction.

For studying long distance correlations mediated by Coulomb interaction, the chain diagram formula for the pair distribution function may be used. In order to find out how some of the classical OCP results are modified due to a magnetic field, let us use a semiclassical approach. The effective propagator representing the unit of a chain can be constructed following the process given by (A2.1) of the Appendix. For Boltzmann statistics and in the presence of a magnetic field the function which corresponds to $G_1(q; \beta_2 - \beta_1)$ in (A2.2) is given by

$$G_1(q, \beta_2 - \beta_1) = \frac{Z\alpha}{\lambda^3}(\sinh\alpha)^{-1}\exp\left[-\frac{s(\beta - s)}{\beta}q_z^2 - \frac{1}{t}(q_x^2 + q_y^2)\right], \quad (6.1.13)$$

where

$$t = \alpha^2\frac{\sinh\alpha}{\sinh sa^2\sinh(\beta - s)a^2}.$$

The eigenvalues are obtained as follows:

$$\begin{aligned}\lambda_j(q) &= \int_0^\beta dsG(q, s)e^{2\pi ijs/\beta} \\ &= \frac{Z\alpha\beta}{\lambda^3\sinh\alpha}\left(1 + \frac{\beta\alpha^2}{90}(q_x^2 + q_y^2) + \cdots\right)\delta_{j0}.\end{aligned} \quad (6.1.14)$$

Therefore, the use of (A2.7) results in [6.1]

$$\begin{aligned}\rho_2(r) - n^2 = -\frac{\alpha Z\kappa^2}{\lambda^3}&\left\{\frac{e^{-\kappa r}}{4\pi r}\left(1 - \frac{\alpha^2}{6}\right) + \frac{\lambda^2\alpha^2}{90(2\pi)^4}\right. \\ &\times\left[-\frac{e^{-\kappa r}}{r^3} - \frac{\kappa e^{-\kappa r}}{r^2} + 3z^2\left(\frac{e^{-\kappa r}}{r^5} + \frac{\kappa e^{-\kappa r}}{r^4} + \frac{\kappa^2 e^{-\kappa r}}{3r^3}\right)\right] \\ &\left.-\frac{\exp(-r^2/4\beta)}{4\pi r} + \frac{1}{4\pi}\left(\frac{2}{\beta}\right)^{1/2}\mathrm{erfc}\left(\frac{r}{(2\beta)^{1/2}}\right)\right\}, \quad (6.1.15)\end{aligned}$$

where the last two terms in the braces have been obtained in consideration of the necessary short distance quantum corrections, and where the new screening constant is defined by

$$\kappa^2 = 4\pi Z a^2 \beta^2 / \lambda^3 \sinh\alpha. \tag{6.1.16}$$

For dilute systems, $Z = n\lambda^3(1 - \varepsilon/2)$, where ε is the plasma parameter. In the presence of a magnetic field, this relation is modified such that

$$\frac{Z}{\lambda^3} = n\frac{\sinh\alpha}{\alpha}\left[1 - \frac{\varepsilon}{2}\left(1 + \frac{(\alpha\kappa_D\lambda)^2}{1080\pi^3}\right) + \frac{3}{8}\varepsilon^2 + \frac{\varepsilon\lambda\kappa_D}{8\sqrt{2}}\right].$$

The screening constant κ is related to the Debye screening constant κ_D by

$$\kappa^2 = \kappa_D^2\left[1 - \frac{1}{2}\varepsilon\left(1 + \frac{(\alpha\kappa_D\lambda)^2}{1080\pi^3}\right) + \frac{3}{8}\varepsilon^2 + \frac{1}{8\sqrt{2}}\varepsilon\lambda\kappa_D\right]. \tag{6.1.17}$$

As defined in (6.1.11), α is the ratio of the field energy to the thermal energy. Equation (6.1.17) shows that the magnetic field reduces the screening constant below Debye's value. Although it approaches κ_D in the high temperature limit, it becomes exponentially small in the opposite limit or in high magnetic fields.

It is interesting to derive the equation of state by making use of the above results. To order ε we arrive at

$$\frac{p}{nk_BT} = 1 - \frac{\varepsilon}{6}\left[1 + \frac{(\alpha\kappa_D\lambda)^2}{1080\pi}\left(\frac{1}{2\pi^2} - \frac{1}{5}\right) + \frac{3\kappa_D\lambda}{8\sqrt{2}}\right], \tag{6.1.18}$$

where ε is the plasma parameter, the first term in the square brackets is the classical OCP result correct to order ε, the second term represents the high temperature magnetic field effect, and the third term is a short range quantum correction to the classical OCP result. We can conclude that both the magnetic field and the Coulomb interaction decrease the electron pressure in the first approximation. In contrast to the complicated form of the correlation function, these two effects appear in the equation of state in a relatively simple way.

The magnetic susceptibility is given by

$$\chi = -\frac{1}{3}n\beta\mu_B^2\left(1 - \frac{3}{8}\varepsilon^2 - \frac{\varepsilon\lambda^2\kappa_D^2}{900\pi^3}\right). \tag{6.1.19}$$

The absolute magnitude of this diamagnetic susceptibility falls short of the ideal gas value by an amount which depends on ε and the ratio of the de Broglie thermal wavelength to the Debye screening length.

We have made a semiclassical approach to investigating a magnetic field effect on a classical OCP. Note that the first approximate term in (6.1.19) which is proportional to the plasma parameter ε includes the de Broglie thermal wavelength. For electron density $10^{14}\,\mathrm{cm}^{-3}$ and at 10^4 K, the Debye screening constant $\kappa_D = 1.448 \times 10^3\,\mathrm{cm}^{-1}$, $\lambda = 7.45 \times 10^{-8}$ cm, $\varepsilon = 2.42 \times 10^{-3}$, $\alpha = 0.67 \times 10^{-4}$ if $H = 10^4$ gauss, and $\alpha\lambda\kappa_D = 7.23 \times 10^{-9}$. Thus, the field correction is not very large, particularly in comparison with the product $\lambda \kappa_D = 1.08 \times 10^{-4}$ which represents a quantum correction. Yet the magnetic susceptibility is due to such a small perturbation of the energy. In the next section, we shall treat the case of low temperatures where even more interesting field effects can be found.

6.2 Magnetic Response in Three Dimensions

The magnetic response of an electron gas can be studied for various cases by using the propagator given in the previous section. The advantage of this approach is that both para- and diamagnetic susceptibilities can be derived simultaneously. Since the propagator takes various forms under different conditions one can expect that the response varies with regions of phase space and magnetic field. As is well known, the para- and diamagnetic susceptibilities of an ideal electron gas are inversely proportional to temperature at high temperatures but are constant at low temperatures. In spite of this difference, in both cases the former is three times the latter when the Landé factor $g = 2$. Such an ideal gas behavior is interesting, but in what follows our main concern will be the effects of Coulomb interaction on magnetic properties. Let us study the effects case by case in a systematic way.

6.2.1 Exchange Effect at High Temperatures

To study the exchange effect on the magnetic properties of a dilute electron gas at high temperatures, we construct the grand partition function of a noninteracting electron gas. To first order in absolute activity Z, its logarithm is given by the trace of the classical propagator K_0

$$\ln \Xi_0 = \int K_0(\mathbf{r}\beta, \mathbf{r}0)d\mathbf{r}$$
$$= V(2Z/\lambda^3)\alpha \cosh(g\alpha/2)/\sinh\alpha, \tag{6.2.1}$$

where $\alpha = \beta\omega_c/2$ and Z is the absolute activity.

In the classical limit, the exchange effect on the susceptibility can be found by using (6.1.10) in (A3.1). We obtain

$$\ln \Xi_x/\Xi_0 = V\beta e^2 \left(\frac{2Z}{\lambda^3}\right)^2 \frac{\alpha \cosh(g\alpha^2/2)}{\sinh\alpha}$$
$$\times \int d\mathbf{r}\frac{1}{r}\exp[-z^2/2\beta - (\omega_c/4)(x^2 + y^2)\coth\alpha]$$
$$= \frac{1}{4}V\beta^2 e^2 \left(\frac{2Z}{\lambda^3}\right)^2 \left(\frac{\alpha \cosh(g\alpha/2)}{\sinh\alpha}\right)^2 \{\cosh^{-1}B(\alpha)/[B^2(\alpha) - 1]^{1/2}\},$$
$$\tag{6.2.2}$$

where

$$B(\alpha) = 2\alpha \coth \alpha - 1.$$

Note that $\mathbf{r} = (x, y, z)$, and z here and in the integrand must be distinguished from the absolute activity Z.

The magnetic susceptibility is given by

$$\chi = \frac{1}{\beta HV} \left(\frac{\partial \ln \Xi}{\partial H} \right)_Z.$$

The absolute activity Z can be determined as a function of electron density by using the number density relation. We arrive at

$$\chi = \frac{ne}{cH} \left(\frac{g}{2} \tanh g\alpha/2 - L(\alpha) + \frac{\beta}{4} \kappa_D^2 \frac{\partial}{\partial \alpha} [\cosh^{-1} B/(B^2 - 1)^{1/2}] \right), \quad (6.2.3)$$

where

$$L(\alpha) = \coth \alpha - 1/\alpha$$

is Langevin's function.

The susceptibility χ consists of the paramagnetic and diamagnetic parts. Per unit volume, it is given by

$$\begin{aligned}
\chi &= \chi_p + \chi_d, \\
\chi_p &= n\beta\mu_B^2(g/2)^2, \\
\chi_d &= -\frac{1}{3} n\beta\mu_B^2(1 + \kappa_D^2\lambda^2/12\pi).
\end{aligned} \quad (6.2.4)$$

Here, μ_B is the Bohr magneton. These results are correct to order e^2. To this order, the paramagnetic susceptibility, which is identified by the g-factor, is independent of Coulomb interaction, while the diamagnetic susceptibility is increased by the exchange interaction. This increase is a many-body effect which is enhanced as the temperature is reduced. The combination of parameters in the form $n\beta\mu_B^2$ is characteristic of the classical case. In the natural unit, $\mu_B = e/c$ so that $n\beta\mu_B^2 = n\beta(e/c)^2$. Since $4\pi\beta = \lambda$, one finds that the susceptibilities are determined by the average Coulomb energy at $n^{-1/3}$ multiplied by a dimensionless factor $(n^{1/3}\lambda)^2$.

Both χ_p and $|\chi_d|$ increase as functions of $\alpha = \mu_B H/k_B T$, but the latter keeps increasing even after χ_p has slowed down its increase. As a result the diamagnetic part becomes increasingly important relative to the paramagnetic contribution. The exchange contribution increases first but decreases after reaching a maximum. This contribution is approximately 10^{-3} of χ_p.

6.2.2 Ideal Electron Gas at Low Temperatures

The magnetic response of an ideal and quantum mechanically degenerate electron gas can be studied through the grand partition function given by (6.2.1). It can be written as

$$\ln \Xi_0 = \sum_{s=1}^{\infty} (-)^{s+1} z^s A_s/s, \tag{6.2.5}$$

where we now have used the regular notation z for the absolute activity, and where

$$
\begin{aligned}
A_s &= \text{Tr}\{K_0(\mathbf{r}_1\beta, \mathbf{r}_s 0) K_0(\mathbf{r}_s\beta, \mathbf{r}_{s-1}0) \ldots K_0(\mathbf{r}_2\beta, \mathbf{r}_1 0)\} \\
&= \frac{V a^2}{4\pi^{3/2}} \left(\frac{1}{s\beta}\right)^{1/2} \frac{\cosh(gs\beta a^2)}{\sinh s\beta a^2}.
\end{aligned} \tag{6.2.6}
$$

For the evaluation of the magnetic susceptibilities, one needs to evaluate the grand partition function to order a^4. We find

$$\ln \Xi_0 = \frac{V}{4\pi^{3/2}\beta^{3/2}} F_{3/2}(\eta) + \left(\frac{g^2}{4} - \frac{1}{3}\right) \frac{V\beta^{1/2} a^4}{8\pi^{3/2}} F_{-1/2}(\eta) + O(a^8), \tag{6.2.7}$$

where

$$z = e^\eta = \exp(\beta k_F^2) = \exp(\beta \varepsilon_F),$$

$$F_s(\eta) = \frac{1}{\Gamma(s+1)} \int_0^\infty dx \frac{x^s}{e^{x-\eta}+1}.$$

At low temperatures, one can us he expressions

$$F_{3/2}(\eta) = 8\beta^{5/2} k_F^5/15\pi^{1/2} + \ldots,$$

$$F_{-1/2}(\eta) = 2\beta^{1/2} k_F/\pi^{1/2} + \ldots.$$

Hence, the internal energy is obtained as

$$
\begin{aligned}
U_0 &= -\left(\frac{\partial \ln \Xi_0}{\partial \beta}\right)_{z,V} \\
&= \frac{3V}{8\pi^{3/2}\beta^{5/2}} F_{3/2}(\eta) - \left(\frac{g^2}{4} - \frac{1}{3}\right) \frac{V a^4}{16\pi^{3/2}\beta^{1/2}} F_{-1/2}(\eta) + O(a^8) \\
&= \frac{V k_F^5}{5\pi^2} \left[1 - a^4 \left(\frac{g^2}{4} - \frac{1}{3}\right) \frac{5}{8k_F^4} + \ldots\right].
\end{aligned} \tag{6.2.8}
$$

The magnetic susceptibility is given by

$$
\begin{aligned}
\chi^0 &= \lim_{H \to 0} \frac{1}{V\beta H} \left(\frac{\partial \ln \Xi_0}{\partial H}\right)_{V,T,z} \\
&= \frac{k_F}{4\pi^2} \frac{e^2}{mc^2} \left(\frac{g^2}{4} - \frac{1}{3}\right),
\end{aligned} \tag{6.2.9}
$$

where the electron mass has been restored in order to show that the susceptibility is dimensionless. This susceptibility is determined by the Coulomb energy at $1/k_F$. It is dependent only on electron density through k_F, independently of temperature. It consists of the paramagnetic and diamagnetic parts, the former being three times the latter if $g = 2$. In the natural unit $e^2/c^2 = \mu_B^2$. Hence, χ^0 is proportional to $k_F \mu_B^2$ or to $(N\mu_B H/V H^2)(\mu_B H/k_B T)$. We learn that the field energy in units of the thermal energy is the key factor. In contrast, according to (6.2.4) for the classical case, the susceptibility χ is proportional to $(n^{1/3}\mu_B^2)(n^{1/3}\lambda)^2$ or to $(N\mu_B H/V H^2)(\mu_B H/\varepsilon_F)$. Thus, the ratio of the field energy of the electrons to the electromagnetic energy in volume V is the determining factor. However, the appearance of the de Broglie thermal wavelength λ or the Fermi energy ε_F indicates that the response is not purely classical.

6.2.3 Exchange Effects at Low Temperatures

The above results can be derived also by taking the trace of (6.1.6), which can be written as

$$K(\mathbf{r}_2\beta, \mathbf{r}_1 0) = \frac{\pi}{2\pi i} \int_{c-i\infty}^{c+i\infty} dx \frac{e^{x\beta\varepsilon_F}}{x \sin \pi x} K_0(\mathbf{r}_2 x\beta, \mathbf{r}_1 0). \qquad (6.2.10)$$

The grand partition function of an ideal electron gas at low temperatures is given by

$$\ln \Xi_0 = \mathrm{Tr}\{K(\mathbf{r}\beta, \mathbf{r}0)\}. \qquad (6.2.11)$$

As in Fig. 6.1, the path of integration consists of the cut which encircles the origin and the contour integrals around the poles. We obtain

$$\ln \Xi_0 = \frac{2V\beta k_F^5}{15\pi^2} \left\{ 1 + \frac{5}{8}\pi^2 (\beta\varepsilon_F)^{-2} + \frac{15}{8}\left[\left(\frac{g}{2}\right)^2 - \frac{1}{3}\right]\gamma^2 + \frac{15}{4}\gamma^{3/2}(\beta\varepsilon_F)^{-1} \right.$$
$$\left. \times \sum_{s=1}^{\infty}(-)^{s+1}\frac{\cos(g\pi s/2)\cos(\pi s/\gamma - \pi/4)}{s^{3/2}\sinh(\pi^2 s/\alpha)} \right\}. \qquad (6.2.12)$$

From this formula, the susceptibility of an ideal electron gas at low temperatures is derived,

$$\chi_0 = \frac{k_F}{4\pi^2}\frac{e^2}{mc^2}\left[\left(\frac{g}{2}\right)^2 - \frac{1}{3} \right.$$
$$\left. + \frac{\pi}{\beta\varepsilon_F\gamma^{3/2}} \sum_{s=1}^{\infty}(-)^{s+1}\frac{\cos(g\pi s/2)\sin(\pi s/\gamma - \pi/4)}{s^{1/2}\sinh(\pi^2 s/\alpha)} \right]. \qquad (6.2.13)$$

Here the electron mass m has been restored. Note that the right side consists of the paramagnetic, diamagnetic and oscillating parts. The oscillating part was first derived by *Lifshitz* and *Kosevich* in 1955 [6.3].

The above method can easily be extended to evaluate the exchange contribution. The corresponding grand partition function is obtained from

$$\ln \Xi_x = \frac{\beta}{2}\int\int d\mathbf{r}_1 d\mathbf{r}_2 \frac{\phi(r)}{2}K(\mathbf{r}_1\beta, \mathbf{r}_2 0)K(\mathbf{r}_2\beta, \mathbf{r}_1 0), \qquad (6.2.14)$$

where $\phi(r)$ is the Coulomb potential. The evaluation of the integral is similar to the case of (6.2.11) except that each of the propagators K is determined by the branch cut and the residues at the poles. In combination with the free particle part, the oscillating grand partition function can be expressed by

$$
\ln \Xi^{\text{osc}} = \frac{V k_{\text{F}}^3}{2\pi^2} \gamma^{3/2} \sum_{s=1}^{\infty} \frac{(-)^{s+1} \cos(g\pi s/2) \cos(\pi s/\gamma - \pi/4)}{s^{3/2} \sinh(\pi^2 s/\alpha)}
$$

$$
\times \left(1 - \frac{2^{1/2} s k_{\text{F}} I_s}{\pi^{1/2} a_0 \omega_c} \right), \tag{6.2.15}
$$

where in terms of a new dimensionless variable **r** replacing $\mathbf{r}k_{\text{F}}$ the quantity I_s is

$$
I_s = \int d\mathbf{r} \frac{1}{r} \frac{\sin(\gamma z^2/4\pi s)}{r^{3/2}} J_{3/2}(r) J_0[(x^2 + y^2)^{1/2}].
$$

The term with this integral represents the exchange contribution. Note that this integral vanishes in the limit $\varepsilon_{\text{F}} \to \infty$. Formally, $\ln \Xi_0$ includes a similar integral with cosine in place of sine in its integrand. However, when the absolute activity Z is renormalized, the corresponding term is canceled out [6.4]. Therefore, in (6.2.15) only the term with I_s has been retained. Since its oscillation is in phase with the ideal gas oscillation, the two oscillations could be combined as in (6.2.15).

The oscillating susceptibility is now given by

$$
\chi^{\text{osc}} = \frac{k_{\text{F}}^5 k_{\text{B}} T \gamma^{3/2}}{2\pi \mu_{\text{B}} H^3} \sum_{s=1}^{\infty} (-)^{s+1} e^{-s/\omega_c \tau} \frac{\cos(g\pi s/2) \sin(\pi s/\gamma - \pi/4)}{s^{1/2} \sinh(\pi^2 s/\alpha)}, \tag{6.2.16}
$$

where

$$
\tau = (2/\pi)^{1/2} k_{\text{F}} I_s / a_0. \tag{6.2.17}
$$

According to (6.2.16) the amplitudes of de Haas–van Alphen (dHvA) oscillations decrease because of the exchange interaction. The period depends on the Fermi energy so that the determination of the frequency is important for Fermi surface studies. In fact the experimental analyses of dHvA oscillations were focused on frequency until around 1969. However, it has been realized that the amplitude of dHvA oscillations provides additional information concerning scattering of electrons. Note that the most contributing term $s = 1$ suggests that a plot of $\ln[\chi^{\text{osc}} H^{3/2} \sinh(\pi^2/\alpha)(k_{\text{B}}T)^{-1}]$ against $1/H$ is a straight line. *Dingle* [6.5] arrived at a similar exponential variation of the amplitude based on a phenomenological consideration of collision broadening of Landau levels. Such a variation has been confirmed by experiment starting around 1969 [6.6]. The phase of the oscillations has also been found experimentally to agree with the free-electron formula [6.7].

6.2.4 Correlation Effects

The exchange contribution which we have considered is related to the ring-diagram contribution, and in fact their combination is necessary to remove a

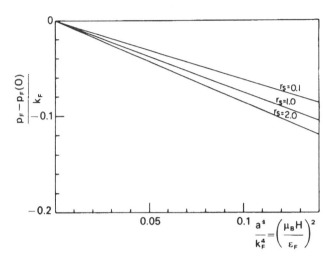

Fig. 6.2. Dependence of the actual Fermi momentum on magnetic field

certain divergence. Their combined contribution to the grand partition function is expressed by [6.8]

$$\ln \Xi_{x+r} = \frac{1}{(2\pi)^3} \frac{V}{2} \int d\mathbf{q} \left(u(q)D - \sum_{j=-\infty}^{\infty} \ln[1 + u(q)\lambda_j(q)] \right), \quad (6.2.18)$$

where

$$D = \frac{\alpha}{4\pi^{3/2}} \sum_{s=1}^{\infty} (-)^{s+1} z^s \left(\frac{1}{s\beta} \right)^{1/2} \frac{\cosh(gs\alpha/2)}{\sinh(s\alpha)}. \quad (6.2.19)$$

The eigenvalue $a_j(q)$ of the unit propagator of a ring diagram is associated with the process expressed by

$$\sum_{s=1}^{\infty} \sum_{j=0}^{s-1} (-)^{s+1} z^s K_0(\mathbf{r}', j\beta + \beta_2 - \beta_1; \mathbf{r}_1 0)$$

$$\times \phi(|\mathbf{r}_2 - \mathbf{r}'|) K_0(\mathbf{r}_1, s\beta - j\beta + \beta_1 - \beta_2; \mathbf{r}'0).$$

Its explicit expression is found to be

$$\lambda_j(q) = \frac{\alpha}{(2\pi)^3} \sum_{s=1}^{\infty} \sum_{j=0}^{s-1} (-)^{s+1} z^s \left(\frac{\pi^3}{s\beta} \right)^{1/2} \frac{2\cosh(gs\alpha/2)}{\sinh(s\alpha)} \int_0^1 dt e^{2\pi ijt}$$

$$\times \exp \left[-\beta q_z^2 (j+t) \left(1 - \frac{j+t}{s} \right) \right.$$

$$\left. -\beta(q_x^2 + q_y^2) \frac{\sinh[(j+t)\alpha]\sinh[(s-j-t)\alpha]}{\alpha\sinh(s\alpha)} \right]. \quad (6.2.20)$$

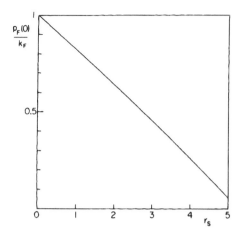

Fig. 6.3. Reduction of the actual Fermi momentum at zero field due to electron correlation

This is somewhat complicated, but can be expanded in powers of $a^2 = \mu_B H$. The first field-dependent term is found to vary like a^4. For the evaluation of the constant susceptibility, no higher terms are necessary. The quantity D can be expanded similarly. The grand partition function and the thermodynamic functions can then be evaluated analytically [6.8].

In the process of deriving the thermodynamic functions, the actual Fermi momentum p_F has to be renormalized in terms of the ideal Fermi momentum, which is a density parameter. We find that the actual Fermi momentum is reduced by the magnetic field in approximate proportion to a^4/k_F^4, which is the square of the ratio of the field energy $\mu_B H$ to the ideal Fermi energy, the proportionality constant being dependent on r_s. For small r_s, this reduction increases with r_s as in Fig. 6.2. Here, p_F is the actual Fermi momentum. Its zero field value $p_F(0)$ can deviate from the ideal Fermi momentum k_F in the presence of Coulomb interaction. Note here that the Fermi wave number plays the role of the Fermi momentum. The ratio $p_F(0)/k_F$ decreases as a function of r_s as in Fig. 6.3. This means that Coulomb interaction tends to destroy quantum degeneracy. Since the magnetic field causes a similar effect, the two effects combined vary with r_s as in Fig. 6.3.

The internal energy can be written as

$$\varepsilon = \varepsilon^{(0)} + \varepsilon^{(H)}, \tag{6.2.21}$$

where the first term is the energy per electron in the absence of a magnetic field and is given by

$$\varepsilon^{(0)} = \frac{2.2099}{r_s^2} - \frac{0.916}{r_s} + \varepsilon_c, \tag{6.2.22}$$

where ε_c is the correlation energy given by (4.2.18). The second term $\varepsilon^{(H)}$ represents the energy due to the magnetic field and is given by

$$\varepsilon^{(H)} = -\frac{5}{8}\frac{a^4}{k_{\mathrm{F}}^4}\frac{2.2099}{r_s^2}$$

$$\times\left\{\frac{g^2}{2} - \frac{2}{3} + \frac{4}{\pi}\left(\frac{4}{9\pi}\right)^{1/3}r_s\left(-0.06969 + \frac{g^2}{8} - \frac{1}{36}\ln r_s\right)\right.$$

$$+\left[\frac{4}{\pi}\left(\frac{4}{9\pi}\right)^{1/3}r_s\right]^2\left[-0.04096g^2 + 0.05616\right]$$

$$\left.+ (0.01563g^2 - 0.02228)\ln r_s + \ldots\right\} \qquad (6.2.23)$$

This expression is complicated. For the particular case of $r_s = 1$ and $g = 2$, we find that the energy change $\Delta\varepsilon$ due to the magnetic field is given by

$$\frac{\Delta\varepsilon}{\varepsilon} = -0.9819\frac{a^4}{k_{\mathrm{F}}^4}. \qquad (6.2.24)$$

Hence, the energy is decreased approximately by the square of the ratio of the field energy to the ideal Fermi energy. Note that $a^4/k_{\mathrm{F}}^4 = (\mu_{\mathrm{B}}H/\varepsilon_{\mathrm{F}})^2$.

The para- and diamagnetic susceptibilities can be expressed in powers of r_s. The ideal gas results are given by (6.2.9). Starting from the ideal gas expressions, which we denote as χ_p^0 and χ_d^0 the para- and diamagnetic susceptibilities have been obtained to order e^4 as follows [6.8]:

$$10^6\chi_p = 2.5886/r_s(m*/m) + 0.4293 - 0.0711r_s + 0.0356r_s\ln r_s + \ldots; \qquad (6.2.25)$$

$$10^6\chi_d = -(0.8629/r_s)(m*/m) - 0.05984 - 0.02385\ln r_s$$
$$+ 0.02459r_s - 0.01269r_s\ln r_s + \ldots, \qquad (6.2.26)$$

where χ_p corresponds to the case $g = 2$.

Each term in these series has been evaluated precisely except that we have estimated the contribution from the second order exchange graphs to be $0.06727r_s$. This choice results in $\chi_p = 1 \times 10^{-6}$ for Na with $m* = m$. An effective mass $m*$ has been introduced in the ideal gas term in order to compensate the neglected effects such as higher-order electron correlation, electron–phonon coupling, and lattice structure. The effective mass depends on the method of determination and a variety of values has been given for a given metal.

Table 6.1 lists the experimental effective masses of alkali metals determined by *Martin* [6.9a] from the electronic specific heat based on

$$m^*/m = \gamma_{\mathrm{exp}}/\gamma_{\mathrm{free}}. \qquad (6.2.27)$$

The specific heat effective mass differs from those determined by some other methods. For comparison, *Ham*'s values [6.9b] based on a distorted lattice model are also listed.

Table 6.1. Effective mass and other electron parameters of alkali metals

	Li	Na	K	Rb	Cs
ε_F [eV]	4.7	3.1	2.1	2.8	1.5
γ_{free}[μcal/K mole]	179.9	272.7	402.6	469.7	563.7
γ_{exp}^a [μcal/K mole]	390 ± 4	330 ± 5	497 ± 20	576^{+70}_{-40}	764 ± 250
m^*/m	2.168	1.210	1.234	1.226	1.355
γ_{dist}^b	1.66	1.00	1.09	1.21	1.76

[a] From [6.9a]
[b] From [6.9b]

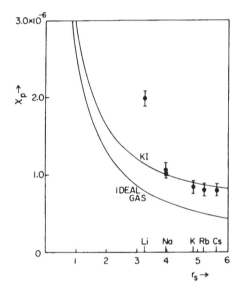

Fig. 6.4. The r_s dependence of the paramagnetic susceptibility of alkali metals (KI: Kojima and Isihara) [6.8]

Figure 6.4 illustrates the paramagnetic susceptibility as a function of r_s without effective mass corrections. The case of Na is distinguished by a square to indicate our choice for the second-order exchange contribution. The dots and bars represent respectively the data of several authors which are arithmetically averaged and the ranges. Such averaged values are used here because experimental values are scattered, and a comparison with a particular set of data is not especially meaningful. The variation in data points is inevitable because of the differences in experimental temperatures or in separating out the susceptibility components from the total susceptibilities. The theoretical χ_p is above the ideal gas curve, and decreases fast for small r_s, but due to the constant and higher- order terms in (6.2.25) it decreases more slowly at larger r_s. The theoretical solid curve is reasonably close to the data points except for Li, which is somewhat special. However, we must remember

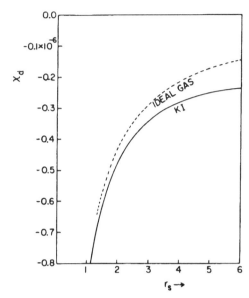

Fig. 6.5. The r_s variation of the diamagnetic susceptibility (KI: Kojima and Isihara) [6.8]

that the above r_s series becomes invalid for large r_s, even though the use of the effective mass correction is helpful in this respect.

Figure 6.5 shows the r_s variation of the diamagnetic susceptibility without effective mass corrections. The theoretical solid curve obtained by *Kojima* and *Isihara* [6.8] is slightly below the ideal gas curve for small r_s. That is, the electron correlation increases the diamagnetic susceptibility. The theoretical χ_d crosses χ_d^0 at a small r_s and then decreases its magnitude more slowly due to electron correlations for larger r_s. The difference from the ideal gas curve increases for large r_s due to electron correlations.

However, the theoretical results based on perturbation must eventually be renormalized.

Experimentally the above susceptibilities are determined by separating the contribution from the ions out of the total susceptibility. The total susceptibility χ_{tot}, is given by

$$\chi_{\text{tot}} = \chi_p + \chi_d + \chi_i. \tag{6.2.28}$$

Table 6.2 lists the theoretical χ_{tot} obtained by *Kojima* and *Isihara* [6.8] based on the above relations in comparison with the experimental values of *Collings* [6.10]. In this Table, *Ham*'s effective mass in Table 6.1 and χ_i determined by *Wilson* [6.9c] have been used.

Apart from the above approach, the paramagnetic susceptibility has been evaluated by a method in which a polarization parameter P is used [6.11]. This parameter distinguishes the population of electrons in spin-up and spin-down states. The use of this parameter is convenient because it determines the Fermi energy on which the correlation energy is dependent. That is, the correlation energy in the absence of a magnetic field is modified as a function of P, which is then determined by the condition of minimum energy. Even though the diamagnetic susceptibility cannot be determined in the same way it has been studied by a few authors [6.12]. Detailed analyses of various theories are given in [6.1.8].

Table 6.2. Susceptibility of alkali metals ($\times 10^6$ cgs volume units)

	Li	Na	K	Rb	Cs
r_s	3.25	3.93	4.86	5.28	5.63
χ_i [6.9c]	−0.0539	−0.2320	−0.2969	−0.3942	−0.4846
χ_p [6.8]	1.66	1.00	0.94	0.97	1.18
χ_d [6.8]	−0.22	−0.28	−0.24	−0.22	−0.17
χ_{tot} [6.8]	1.39	0.49	0.40	0.36	0.52
Exp[6.10]	1.789	0.583	0.402	0.321	0.398

On the experimental side, the paramagnetic susceptibility of alkali metals including those obtained by several other researchers has been tabulated by *Kushida* et al. [6.13], who also plotted many theoretical curves. In the case of Li, the χ_p values listed range from 1.96×10^{-6} at 300 K to 2.18×10^{-6} in cgs volume units. These values are not very far from the theoretical value of 1.96×10^{-6}, but a strong deviation from the ideal gas volume dependence has been observed. In the case of Na, a low value of 0.89×10^{-6} and a high value of 1.77×10^{-6} have been reported. The value found by *Kushida* et al. is 1.07×10^{-6}.

In comparing theory and experiment, particularly for the volume dependences of the susceptibilities, the ion contribution cannot be ignored. This requires the evaluation of the nuclear contact density, which is expected to depend on the crystal potential. Therefore, for detailed theoretical analyses of actual metals, it is desirable to take the band structure into consideration [6.13]. Even within the OCP model, many-electron theory applicable to low densities should be developed. Note that formally high density theory is limited to small r_s. The r_s value 3.25 of Li is already high, while that of Rb or Cs is even higher. Nevertheless, it is encouraging to note in Table 6.2 that the effective mass corrections resulted in reasonable theoretical values. Therefore, it is also desirable to evaluate the effective mass and its density variation.

6.2.5 Local Field Correction

The paramagnetic susceptibility χ_p is related to the polarization function χ introduced in Chap. 2 because it represents a linear response of electronic spins to an applied magnetic field. Therefore, a local field correction can be introduced as in that chapter. Apart from a numerical factor which includes the square of the Bohr magneton, a dimensionless susceptibility may be introduced in the form

$$\chi_s = \frac{\chi(q,\omega)}{1 + u(q)\overline{G}(q,\omega)\chi(q,\omega)}. \tag{6.2.29}$$

Since the susceptibility is due to the difference in the population of spin-up and spin-down states, the appropriate local field correction is expressed by

$$\overline{G}(q) = \frac{1}{2}[G_{\uparrow\uparrow}(q) - G_{\uparrow\downarrow}(q)]. \tag{6.2.30}$$

Here, $\chi(q,\omega)$ is the polarization function, and

$$G_{\sigma,\sigma'}(q) = -\frac{1}{n(2\pi)^3} \int \frac{\mathbf{q}\cdot\mathbf{k}}{k^2}[S_{\sigma,\sigma'}(|\mathbf{q}-\mathbf{k}|) - \delta_{\sigma\sigma'}]d\mathbf{k},$$

$$S_{\sigma,\sigma'}(q) = n\int (g_{\sigma,\sigma'}(r) - \frac{1}{2})e^{i\mathbf{q}\cdot\mathbf{r}}d\mathbf{r}. \tag{6.2.31}$$

$g_{\sigma,\sigma'}(r)$ is the spin–spin correlation function. The evaluation of the susceptibility is then dependent on numerical calculations as in the case of the dielectric function. The function $\overline{G}(q,\omega)$ is unknown, but its limiting expression has been given by *Zhu* and *Overhauser* [6.14] as

$$\lim_{q\to\infty} \overline{G}(q,\omega) = \frac{1}{3}[4g(0) - 1]. \tag{6.2.32}$$

This expression is analogous to (2.2.21) for the dielectric function. Here $g(0)$ is the radial distribution function at the origin.

6.3 Effective g-Factor in Si Inversion Layers

In 1968, *Fang* and *Stiles* [6.15] observed the 2D character of the electrons in an n-type Si inversion layer through the temperature dependence of the amplitude of the Shubnikov–de Haas oscillations, and determined the density dependence of the effective g-factor by placing the sample in a tilted magnetic field. Their finding that the g-factor was strongly enhanced toward low densities indicated increasing electron correlations and opened the door for the investigation of strongly correlated 2D electron systems. Their idea for determining the g-factor is as follows.

If a magnetic field is at an angle θ from the direction perpendicular to the 2D system, the orbital energy depends only on the vertical component, $H\cos\theta$, while the spin–magnetic field coupling depends on the total field. The energy levels are given by

$$\varepsilon_n = \left(n + \frac{1}{2}\right)\hbar\omega_c \cos\theta \pm \frac{g}{2}\mu_B H, \tag{6.3.1}$$

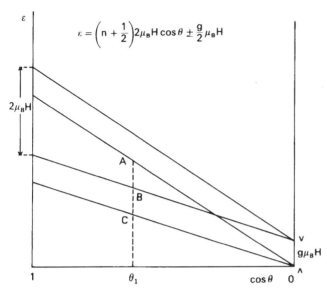

Fig. 6.6. Landau levels in tilted magnetic field as a function of $\cos\theta$, θ being the tilt angle. The upper parallel lines correspond to $n = 1$ and the lower to $n = 0$. θ_1 is the point where $AB = BC$

where μ_B is the Bohr magneton. These levels are characterized by a pair of parallel lines corresponding to the two spin directions for a given level index n when plotted against $\cos\theta$ as in Fig. 6.6. The energy difference BC due to spin splitting for $n = 0$ becomes equal to the difference AB between the lower $n = 1$ Landau level and the upper $n = 0$ level at a particular angle θ_1. Below this angle the orbital energy difference is larger than the spin splitting, and above it is less. At this particular angle, the phase of the Shubnikov–de Haas oscillations is expected to change. *Fang* and *Stiles* used this phase change, working on a [100] Si inversion layer, and determined the effective g factor through the relation

$$\hbar\omega_c \cos\theta_1 = 2g\mu_B H.$$

On the left side the cyclotron frequency ω_c associated with orbital motion is represented by the effective mass m, while on the right side the Bohr magneton μ_B originating from electron spin is expressed in terms of the bare electron mass. Hence, the effective g-factor was determined by

$$g = (m_0/m)\cos\theta_1, \tag{6.3.2}$$

where m is the effective mass in the plane [100].

However, a later experiment of *Smith* and *Stiles* [6.15b] revealed that the effective mass depends on electron density n such that

$$m/m^* = 1 - 0.071r_s^*. \tag{6.3.3}$$

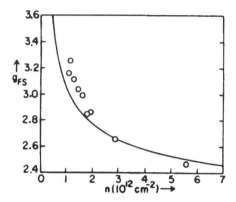

Fig. 6.7. Effective g-factor of electrons in a Si[100] inversion layer. *Points*: from [6.15a]. *Solid curve*: from [6.16]

Here, r_s^* is the effective parameter defined by

$$r_s^* = me^2\hbar^2\kappa_{sc}^{-1}(\pi n)^{1/2}.$$

In this expression, $\kappa_{sc} = 12$ and $m^* = 0.195m_0$ are appropriate values for the inversion layer.

The effective g-factor determined by *Fang* and *Stiles* varies with electron density n strongly, as illustrated in Fig. 6.7. The solid curve is a theoretical curve due to *Isihara* and *Ioriatti* [6.16] with the above Smith–Stiles correction. In this theory, the effective g-factor was determined from the field-dependent energy of the system in the following manner.

We note that the g-factor is associated with electron spin. Hence, we introduce a polarization parameter P that represents the difference in the fractions of spin-up and spin-down electrons. If p_F is the Fermi momentum of the unpolarized state, the Fermi momenta of the two spin states are given by

$$p_F^+ = (1 + P)^{1/2}p_F,$$
$$p_F^- = (1 - P)^{1/2}p_F. \tag{6.3.4}$$

To second order in P the shift in the ground-state energy of the system due to the polarization can be expressed as

$$E_g(r_s, P) - E_g(r_s, 0) = \frac{Ne^2m}{2\kappa^2a_0m_0}A(r_s)P^2 - N\mu_B HP, \tag{6.3.5}$$

where r_s is given by

$$\pi r_s^2 \left(\frac{m_0a_0\kappa}{m}\right)^2 = \frac{1}{n}. \tag{6.3.6}$$

κ is the average dielectric constant and a_0 is the Bohr radius. The factor $A(r_s)$ is a function of r_s and the valley degeneracy g_v. It can be evaluated from three contributions such that

$$A(r_s) = A^0(r_s) + A^x(r_s) + A^c(r_s). \tag{6.3.7}$$

The right side terms represent respectively the kinetic, exchange and correlation contributions. The first two are given by

$$A^0(r_s) = \frac{1}{g_v r_s^2}, \qquad A^x(r_s) = -\frac{2^{1/2}}{\pi g_v^{1/2} r_s}. \tag{6.3.8}$$

The correlation contribution will be discussed shortly.

The polarization parameter can be determined from the condition that the energy is minimum. This results in

$$P = \mu_B H a_0 \frac{m_0}{m} \frac{\kappa^2}{e^2 A(r_s)}. \tag{6.3.9}$$

In terms of the magnetization given by

$$M = n \mu_B H,$$

the susceptibility is determined from

$$\chi = (\partial M / \partial H)_n.$$

In terms of the ratio of the susceptibility to the ideal susceptibility for 2D defined by

$$\chi_0 = \frac{g_v m}{\pi \hbar^2} \mu_B^2, \tag{6.3.10}$$

we obtain

$$\chi_0/\chi = 1 - \frac{(2g_v)^{1/2}}{\pi} r_s + g_v r_s^2 A(r_s). \tag{6.3.11}$$

An effective g-factor can be introduced such that χ assumes the form of χ_0.

Let us now discuss the correlation contribution $A^c(r_s)$. In the correlation contributions, the second-order exchange energy is constant in two dimension and is independent of P. Hence, the correlation contribution may be obtained from the ring diagram contribution to the energy. Fortunately, this contribution has been obtained for all densities as discussed in Sect. 4.3. Since it is a function of the Fermi momentum, it is straightforward to obtain the energy difference between the polarized and unpolarized cases. We arrive at exactly

$$E_r(r_s, P) - E_r(r_s, 0) = -\frac{N e^2 m g_v P^2}{2\pi \kappa^2 m_0 a_0}$$

$$\times \int_0^\infty dz \int_0^\infty d\rho \left[4\pi^2 \lambda \frac{\partial}{\partial \rho} \left(\frac{\partial \lambda}{\rho d\rho} \right) \left(1 + \frac{2\pi r_s g_v^{3/2} \rho \lambda}{2^{1/2}} \right)^{-1} \right], \tag{6.3.12}$$

where $\rho = 2p_F/q$, $z = 2\pi j/\beta q^2$ and

$$\lambda(\rho, z) = \frac{\rho}{2\pi} \int_0^\infty dx \frac{\sin x}{x} e^{-|z|x} J_1(\rho x). \tag{6.3.13}$$

Hence, the correlation contribution can be expressed by

$$
A^c(r_s) = \frac{2^{1/2}}{\pi r_s g_v^{1/2}}
$$
$$
+ \frac{2g_v}{\pi r_s g_v^{3/2}} \int_0^\infty \int_0^\infty \frac{dz d\rho}{\rho} \frac{\partial}{\partial\rho} \left(\frac{\partial\lambda}{\rho\partial\rho} \right) \frac{2\pi}{1 + \frac{2\pi r_s g_v^{3/2}\rho\lambda}{2^{1/2}}}. \tag{6.3.14}
$$

Comparing this result with the exchange contribution given by (6.3.8) we find that there will be no exchange effect. Therefore, the correlation contribution is very important. We remark also that this contribution for an arbitrary g_v can be obtained from the case with $g_v = 1$ by using the relation

$$
A^c(r_s, g_v) = g_v A^c(g_v^{3/2} r_s, 1). \tag{6.3.15}
$$

The solid theoretical curve in Fig. 6.7 has been obtained in consideration of the density dependence of the effective mass. If g_{FS} represents the data of *Fang* and *Stiles*, it needs a correction due to the density dependence of the effective mass. The corrected g-factor which is denoted as g^* may be given by

$$
g^* = (m/m^*)g_{FS}. \tag{6.3.16}
$$

Since the data of *Fang* and *Stiles* fall into a smooth curve, the above weak field treatment is effective, and the agreement between theory and experiment is good. However, we remark that in strong magnetic fields the effective g-factor is expected to oscillate. Such an oscillating g-factor has been given theoretically by [6.17]

$$
g^* = 2 + \frac{\pi}{\alpha\beta n} \sum (-)^s \frac{\cos(gs/2)}{\sinh(\pi^2 s/\alpha)} \coth(\pi^2 s/\alpha) \cos(\pi s/\gamma_0). \tag{6.3.17}
$$

where $\alpha = \mu_B H/k_B T$, $\beta = 1/k_B T$ and $\gamma_0 = \mu_B H/\varepsilon_F$, ε_F being the ideal Fermi energy. Note that the g-factor appears in the sign factor, the amplitude of the oscillations is the same as that used by *Fang* and *Stiles* and the period is determined by the Fermi energy, which is proportional to the density of the electrons.

We have evaluated the effective g-factor analytically. We remark that the effective g-factor determined from the spin susceptibility may depend on the field strength because of the nonlinearity of the susceptibility. We remark also that short range electron correlations affect the enhancement of the g-factor at low densities. While these correlations are dominantly determined by the contribution from the ladder diagrams and can be treated analytically, an entirely numerical approach to the effective g-factor is also possible based on χ_s in (6.2.29). Although the local field correction $\overline{G}(q, \omega)$ is unknown, its limiting value for two dimensions has been obtained as [6.18]

$$
\lim_{q\to\infty} \overline{G}(q, \omega) = g(0), \tag{6.3.18}
$$

where $g(0)$ is the limiting value of the radial distribution function at $r = 0$. The corresponding local field correction for the 2D polarization function is given by $1 - g(0)$. The 3D limiting expression $\overline{G}(\infty, \omega)$ for χ_s has been given by (6.2.32).

6.4 Coulomb Effects on 2D de Haas–van Alphen Oscillations

Two-dimensional electron systems show interesting and unusual properties especially in a strong perpendicular magnetic field. Unlike in three dimensions, these electrons cannot move in the field direction and are confined to Landau levels. The electrons fill Landau levels at low temperatures, one by one from the bottom up to the level determined by the Fermi energy. When the electron density is increased and one level is filled, the Fermi energy will jump to the next level. Such a jump is expected to cause sharp changes in the electronic properties. The above picture is general but is based more or less on an ideal gas. It is then interesting to find the interplay between a magnetic field and electron correlations since the correlation effects are also strong in two dimensions. In this section we shall discuss correlation effects on magnetic susceptibilities in a systematic way to parallel the 3D case in Sect. 6.2.

6.4.1 Chemical Potential

The ideal Fermi energy depends only on electron density. Hence it can be considered as a density parameter. At finite temperatures or in the presence of Coulomb interaction the chemical potential μ plays the role of the actual Fermi energy. This Fermi energy can be determined by the golden rule of grand ensemble theory. Let us first investigate how it varies with electron density in the case without Coulomb interaction. The grand partition function of an ideal electron gas is given by [6.19]

$$\ln \Xi_0 = \frac{A\beta p_F^4}{4\pi} \left\{ 1 + \frac{\pi^2}{3\eta^2} - \left[\left(\frac{1}{2g} \right)^2 - \frac{1}{3} \right] \gamma^2 \right.$$
$$\left. + \frac{4\alpha}{\eta^2} \sum_{s=1}^{\infty} (-)^{s+1} \frac{\cos(\pi s/\gamma) \cos(gs\pi/2)}{s \sinh(\pi^2 s/\alpha)} \right\}, \qquad (6.4.1)$$

where A is the surface area, $\beta = 1/k_B T$, and

$$\eta = \beta p_F^2, \quad \alpha = \beta a^2, \quad \gamma = a^2/p_F^2. \qquad (6.4.2)$$

$a^2 = \mu_B H$ is the field energy, p_F is the actual Fermi momentum, and g is the Landé g-factor. The parameter γ is the ratio of the field energy to the actual Fermi energy. The partition function given by (6.4.1) is correct in the neglect

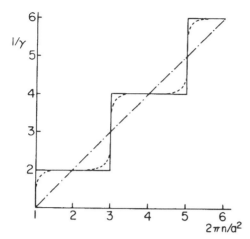

Fig. 6.8. Variation of the chemical potential μ as a function of filling factor ν of Landau levels. $1/\gamma = \mu/a^2$, $a^2 = \mu_B H$. In units in which $\hbar = 1$, and $2m = 1$, $2\pi n$ represents the ideal Fermi energy, and the ratio $2\pi n/a^2 = \nu$

of terms of order $\exp(-\eta)$ and higher. That is, it is essentially exact near absolute zero. In the zero temperature limit of the case $g = 2$, the number density is given in a dimensionless form by

$$\frac{1}{\gamma_0} = \frac{1}{\gamma} + \frac{2}{\pi} \sum_{s=1}^{\infty} \frac{1}{s} \sin \frac{s\pi}{\gamma}. \tag{6.4.3}$$

The quantity γ_0 is defined by

$$\gamma_0 = \frac{a^2}{2\pi n}, \tag{6.4.4}$$

that is, it is the ratio of the field energy to the ideal Fermi energy.

At absolute zero, (6.4.3) can be simplified to [6.20]

$$\frac{1}{\gamma} = \begin{cases} 2j, & 2j-1 < 1/\gamma_0 < 2j+1, \\ 2j+1, & 1/\gamma_0 = 2j+1, \end{cases} \tag{6.4.5}$$

where j is an integer. The variation of $1/\gamma$ as a function of $1/\gamma_0$ is illustrated in Fig. 6.8. The function $1/\gamma$ is constant about the points $1/\gamma_0 = 2j$, and jumps from $2j$ to $2(j+1)$ at $1/\gamma_0 = 2j+1$. $\gamma = \gamma_0$ when $1/\gamma_0$ is an even integer. Hence, $1/\gamma$ makes zigzag changes about the line $1/\gamma = 1/\gamma_0$. These abrupt changes are of course characteristic of absolute zero and are eased at low but finite temperatures. The dashed curve illustrates the case of $0.1\,\mathrm{K}$ and $H = 10^4 G$.

The internal energy is given by differentiating $\ln \Xi$ with respect to β. For absolute zero, it is given by

$$\frac{U/N}{p_0^2/2} = \begin{cases} 4j\gamma_0 - 4j^2\gamma_0^2, & 2j-1 < 1/\gamma_0 < 2j+1, \\ 4j(j+1)/(2j+1)^2, & 1/\gamma_0 = 2j+1. \end{cases} \tag{6.4.6}$$

The left side is the internal energy per electron in units of the average kinetic energy $p_0^2/2$.

The magnetization and susceptibility are obtained for the interval $2j-1 < 1/\gamma_0 < 2j+1$ as follows:

$$M/2N\mu_B = 2j^2\gamma_0 - j, \tag{6.4.7}$$

$$\chi/\chi_0 = 2j(2j - 1/\gamma_0), \tag{6.4.8}$$

$$\chi_0 = (1/2\pi)(e^2/c^2). \tag{6.4.9}$$

χ_0 is the ideal susceptibility which corresponds to the Pauli susceptibility in 3D. Different from the 3D case, the 2D susceptibility has a dimension. In the ordinary units,

$$\chi_0 = (4\pi m/h^2)\mu_B^2, \tag{6.4.10}$$

where m is the effective mass.

6.4.2 Exchange Effect

In order to find the effects of exchange and correlation on the magnetic properties, it is convenient to make use of the eigenvalues λ_j of the electron propagator which are given in the present case by [6.20]

$$\lambda_j(q) = \frac{1}{2\pi} \sum_{s=1}^{\infty} (-)^{s+1} z^s \frac{\cosh(g\alpha s/2)}{\sinh(s\alpha)}$$
$$\times \int_0^{\infty} dx \exp(-\pi|j|x/\alpha) \sin(Q \sin x)$$
$$\times \exp[-Q(1 - \cos x)\coth \alpha s], \tag{6.4.11}$$

where

$$Q = q^2/\omega_c, \quad \omega_c = 2eH/c, \quad z = e^\eta, \quad \eta = \beta p_F^2 = \beta\mu. \tag{6.4.12}$$

The above eigenvalue expression is exact but requires further calculations for practical applications. We note that the dummy index s appears not only with z but also with the spin factor g. Hence, it is convenient to split the spin-up and spin-down contributions,

$$\lambda_j(q) = \frac{1}{2}\left[\lambda_j^*\left(q, \mu + \frac{g}{4}\right) + \lambda_j^*\left(q, \mu - \frac{g}{4}\right)\right]. \tag{6.4.13}$$

From $\lambda_j^*(q)$ we can obtain $\lambda_j(q)$ by simply replacing μ by $\mu \pm g/4$.

In order to evaluate λ_j^* we make use of the following mathematical formula:

$$\frac{1}{1-z}\exp\left(-\frac{zx}{1-z}\right) = \sum_{k=0}^{\infty} z^k L_k(x),$$

where $L_k(x)$ is the Laguerre polynomial of the zeroth order. We find then

$$\lambda_j^*(q) = \frac{1}{2\pi} \int_0^\infty \exp(-\pi|j|x/\alpha)\sin(Q\sin x)$$

$$\times \sum_{k=0}^\infty 2\Theta(k)\exp[-Q(1-\cos x)]L_k[2Q(1-\cos x)]dx, \quad (6.4.14)$$

where

$$\Theta(k) = \frac{1}{\exp[2\alpha(k+1/2)-\eta]+1}. \quad (6.4.15)$$

At absolute zero, $\Theta(k)$ is a step function. The k sum in (6.4.14) can be expressed by a contour integral, in particular for the interval $2j-1 < 1/\gamma < 2j+1$, in the following form

$$\sum_{k=0}^\infty 2\Theta(k)\exp(-y/2)L_k(y) = \frac{1}{2\pi i}\int_c dt\exp(-yt/2)g(t) = F_j(y),$$

where the path c encircles 1, $y = 2Q(1-\cos x)$, and j is an arbitrary integer. In the domain $2j-1 < 1/\gamma < 2j+1$ and for $0\,\mathrm{K}$, the function $g(t)$ is given by

$$g(t) = -1 + \left(\frac{t+1}{t-1}\right)^j. \quad (6.4.16)$$

Hence, we obtain

$$F_j(y) = 4j\frac{J_1(X)}{X} + \frac{X^2}{24j}J_2(X) + \ldots, \quad 2j-1 < 1/\gamma < 2j+1 \quad (6.4.17)$$

with $X = 2(jy)^{1/2}$. The integrand of (6.4.14) is large for small x so that the eigenvalues are given by

$$\lambda_j^*(q) = \frac{1}{2\pi}\int_0^\infty dx\exp(-\pi|j|x/\alpha)\sin(Qx)F_k(c_kx)$$

$$= \frac{1}{2\pi}(1-r)+\ldots, \quad 2k-1 < 1/\gamma < 2k+1, \quad (6.4.18)$$

where k is an arbitrary integer, $c_k = (4kQ)^{1/2}$, and r is the solution of

$$s^4 = \frac{8k\gamma s^2}{1-r^2} - \frac{J^2}{r^2}, \quad (6.4.19)$$

where $s = q/p_\mathrm{F}$, $J = 2\pi j/\eta$.

Once the eigenvalues are obtained, the exchange contribution to the grand partition function can easily be calculated. Perhaps the simplest method is to employ a sum rule of the eigenvalues. According to this rule, and actually as one can easily verify

$$\ln\Xi_x = -\frac{\beta A}{2(2\pi)^2}\int d\mathbf{q}[\Lambda(q)-\Lambda(0)]u(q), \quad (6.4.20)$$

where $u(q) = 2\pi e^2/q$, and

$$\Lambda(q) = \lim_{\beta \to \infty} \sum_j \lambda_j(q). \tag{6.4.21}$$

By using $\lambda_j^*(q)$, the exchange contribution is given as

$$\ln \Xi_x = \frac{2\beta Ae^2}{3\pi^2} \frac{(p_F\kappa_j)^3 + (p_F\kappa_{j-1})^3}{2}, \quad 2(j-1) < 1/\gamma < 2j, \tag{6.4.22}$$

where $\kappa_j = (2j\gamma)^{1/2}$.

Once the grand partition function is given, the thermodynamic functions can be obtained in a straightforward way. In combination with the ideal gas contribution, the internal energy is given by

$$\frac{U/A}{p_0^4/4\pi} = 4u_1 j\gamma_0 - 4u_2 j^2\gamma_0^2, \quad 2j - 1 < 1/\gamma_0 < 2j + 1, \tag{6.4.23}$$

where u_1 and u_2 depend on r_s and the integer j. In the high density limit $u_1 = u_2 = 1$, while for large j

$$u_1 = u_2 = 1 - \frac{8\sqrt{2}}{3\pi} r_s. \tag{6.4.24}$$

In general, u_1 and u_2 are different from each other.

The magnetization and susceptibility are given in the interval $2j - 1 < 1/\gamma_0 < 2j + 1$ by

$$\frac{M}{2N\mu_B} = 2u_2 j^2 \gamma_0 - u_1 j, \tag{6.4.25}$$

$$\frac{\chi}{2\chi_0} = 2u_2 j^2 - u_1 \frac{j}{\gamma_0}. \tag{6.4.26}$$

6.4.3 Correlation Contribution

The correlation contribution to the grand partition function can be expressed in terms of the eigenvalues $\lambda_j(q)$. We note that in the interval $2(k-1) < 1/\gamma < 2k$, the $\lambda_j(q)$ are given by

$$\lambda_j(q) = \frac{1}{2}[\lambda_j(q, k-1) + \lambda_j(q, k)]$$
$$\sim \lambda_j(q, k - 1/2)$$

provided that the integer k is not small. We can make use of

$$\lambda_j(q, k) = \lambda_j^*(q)$$

with (6.4.18) for the right side. We introduce the variables

$$z = \pi j/Q\alpha, \quad \rho_k = 2p_F\kappa_k/q,$$

and then transform these variables into (ξ_k, ϕ_k) given by

$$\rho_k = \cosh \xi_k \sin \phi_k, \quad |z| = \sinh \xi_k \cos \phi_k.$$

These are the variables chosen in Sect. 4.3 for establishing an electrostatic analog of the eigenvalues. In terms of these new variables, the eigenvalues are greatly simplified,

$$\lambda_j(q, k) = \frac{1}{2\pi}(1 - \cos \phi_k). \tag{6.4.27}$$

Accordingly, the grand partition function is also expressed by a simple angle integral as in Chap. 4. Thus, the results given in that chapter can be used with minor modifications. Therefore, omitting further details, we give the final outcome [6.20].

The internal energy including the kinetic, exchange and correlation contributions is given for the interval $2j - 1 < 1/\gamma_0 < 2j + 1$ by

$$\frac{U}{(p_0^2/2)N} = 4[1 + f(r_s)]j\gamma_0 - 4\left(1 + \frac{(2j - 1)(2j + 1)}{(2j)^2}f(r_s)\right)j^2\gamma_0^2, \tag{6.4.28}$$

where the function $f(x)$ is

$$f(x)/x^2$$

$$= \begin{cases} -\dfrac{1.2004}{x} - 0.6137 - 0.1726x \ln x + 0.8653x + \cdots, & x < 1, \quad (6.4.29a) \\ -\dfrac{1.2935}{x^{2/3}} - \dfrac{0.14018}{x^{4/3}} + \cdots, & x > 1. \quad (6.4.29b) \end{cases}$$

Equations (6.4.29a and b) correspond respectively to high and low density expressions. Note that, in the low density expression, the first-order exchange contribution has been cancelled out by a term from the correlation contribution.

6.4.4 de Haas–van Alphen Oscillations

In general, in strong magnetic fields, the susceptibility is defined by

$$\frac{\chi}{\chi_0} = \frac{M}{\gamma_0 N \mu_B}, \tag{6.4.30}$$

where M is the magnetization, and χ_0 is the ideal susceptibility. In the present case in which the exchange and correlation contributions are taken into consideration, the magnetization is given by

$$\frac{M}{2N\mu_B} = 2\left(1 + \frac{(2j - 1)(2j + 1)}{(2j)^2}r_s^2\varepsilon(r_s)\right)j^2\gamma_0 - [1 + r_s^2\varepsilon(r_s)]j. \tag{6.4.31}$$

Here

$$\varepsilon(r_s) = -\frac{8\sqrt{2}}{3\pi}\frac{1}{r_s} + \frac{8}{\pi}\Phi, \tag{6.4.32}$$

and Φ is given by

$$\Phi = \int_0^{\pi/2} d\phi (\cos\phi - \cos 2\phi)(1 - \cos\phi)\sin\phi \left\{ \left(\frac{\pi}{2} - xF(x)\right) I_5 \right.$$

$$\left. - \left(\frac{\pi}{4} - x + \frac{\pi}{2}x^2 - x^3 F(x)\right) I_3 \right\}. \tag{6.4.33}$$

where $x = r_s(1 - \cos\phi)\sin\phi$ and

$$F(x) = \int_0^\infty dt \frac{1}{1 + x\cos t}, \tag{6.4.34}$$

$$I_3 = -\frac{\cos\phi}{2\sin^2\phi} + \frac{1}{2}\ln\left(\tan\frac{\phi}{2}\right),$$

$$I_5 = -\frac{\cos\phi}{4\sin^4\phi} + \frac{3}{8}\left[\ln\left(\tan\frac{\phi}{2}\right) - \frac{\cos\phi}{\sin^2\phi}\right].$$

These integrals appear from the double integral for ε_r in (4.3.14).

The internal energy is parabolic in intervals of $1/\gamma_0 = \varepsilon_F/\mu_B H$ as shown in Fig. 6.9. In this graph the ordinate represents the energy in units of the average kinetic energy for the case $g = 2$. The left (right) side set of curves corresponds to relatively small (large) values of $1/\gamma_0$ which is proportional to electron density or to H^{-1}. The right side curves show stronger density variations than the left side because the magnetic field is larger. As we go down in Fig. 6.9, r_s increases from 0.6 to 1.4757. Note that the curves are at first convex-up, gradually flatten out and then change their curvature. This curvature change takes place at a point where the ground state energy in the absence of a magnetic field becomes negative. If this negativeness is associated with a liquid state, the concavity is its characteristic.

The magnetization is plotted in Fig. 6.10 as a function of $1/\gamma_0 = \nu$ for the case $g = 2$. The left set of graphs corresponds to smaller values of the abscissa. The sharp saw-tooth oscillations are characteristic of absolute zero. For small r_s, that is, when the system is in a gaseous phase, the amplitude is reduced as r_s increases. However the phase of oscillations changes at a point between $r_s = 0.6$ and 0.7 where the ground state energy in the absence of a magnetic field becomes negative. The magnetic response of the system changes at this particular point. Beyond this point the amplitude increases with r_s.

In general it is difficult to observe dHvA oscillations in small 2D systems. Nevertheless, successful experiments have been performed. For instance, *Eisenstein* et al. [6.21] used modulation-doped GaAs/GaAlAs heterostructures on GaAs [100] substrates. In order to enhance the signal, many, well-separated layers of GaAs/GaAlAs were stacked upon one another. *Eisenstein* et al. determined the relative amplitude M/M_0 of the oscillations as a function of the magnetic field for two samples as shown in Fig. 6.11. Here the white circles correspond to sample 2 with mobility $28.5\,\text{m}^2/\text{V s}$ with electron density $n = 5.4 \times 10^{11}\,\text{cm}^{-2}$, and the black circles represent sample 1 with

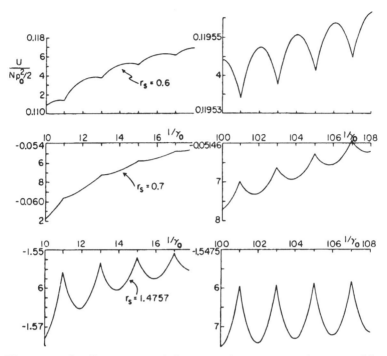

Fig. 6.9. Oscillating part of the internal energy as a function of Landau level filling factor $1/\gamma_0 = \nu$ [6.20]

mobility $8\,\mathrm{m^2/V\,s}$ and $n = 5.4 \times 10^{11}\,\mathrm{cm^{-2}}$. They tried to interpret this variation in terms of a one-electron theory with a density of states corresponding to a short range impurity potential. In this case the density of states turns out to be proportional to $\sqrt{H}$. Although the amplitude varied roughly as $\sqrt{H}$, the theoretical density of states was found to be four times larger than the experimentally determined one. Therefore, their analyses based on one-electron theory is not convincing. On the other hand, the relative amplitude is proportional to $\gamma_0 = \mu_B H/\varepsilon_F$ according to (6.4.31). Hence it increases with H and decreases with electron density n. Such a variation is apparent in Fig. 6.11. Indeed sample 1 has a lower slope than sample 2 even though the latter has a higher electron density. That is, the relative amplitude increases for the larger r_s sample. This is what we expect for strongly interacting cases. Note also that the relative amplitude is less than 1. Thus many-body effects appear to be strong in these systems.

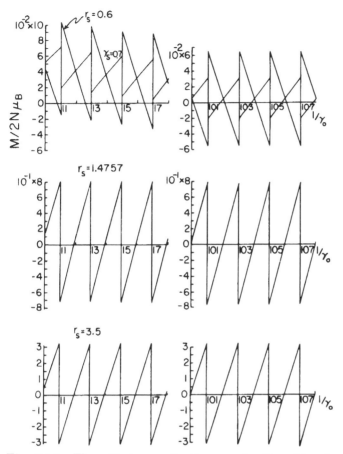

Fig. 6.10. Theoretical magnetization as a function of Landau level filling factor $1/\gamma_0 = \nu$ [6.20]

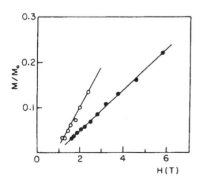

Fig. 6.11. Normalized magnetization as a function of magnetic field. Data from [6.21]

6.5 Field Induced SDW States of 1D Conductors

An unusual phenomenon has been revealed recently in 1D Bechgaard salts $(TMTSF)_2X$, where X is an inorganic anion such as PF_6^- AsF_6^-, ReO_4^- or ClO_4^-. These salts are one-dimensional in the sense that the conductivity is high along the stacking direction (a axis) in which the π orbitals of selenium in the TMTSF molecules overlap most strongly because the interlayer distance is 3.6 Å in comparison with the lattice constants 7.7 Å in the b direction and 13.5 Å in the c direction. The structure of these salts is illustrated schematically in Fig. 6.12, where black circles represent Se atoms in TMTSF. The conductivity ratio in the a, b and c directions is $1:10^{-2}:10^{-5}$. They are superconductors below around 1 K and in a magnetic field which is lower than about $0.1T$ in the direction of the weakest conductivity. The superconducting transition temperature decreases when small amounts of nonmagnetic impurities are added. Such a decrease is not expected under the usual mechanism of electron pairing via a phonon. The salts are metallic between 0.1 and $4T$.

It was found in 1981 [6.22] that in this metallic state a temperature- and pressure-sensitive threshold magnetic field H_{th} exists above which the magnetoresistance shows sharp Shubnikov–de Haas (SdH) type oscillations. Subsequent NMR, Hall effect and specific heat measurements revealed that the threshold field corresponds to a second order phase transition into a spin density wave (SDW) state with a reduced number of carriers [6.22]. The metallic phase is unstable above this threshold field. When the magnetic field is changed, a surprising new phenomenon is observed: a series of weak first-order transitions from one SDW state to another takes place under the same pressures that cause superconductivity. The phase transitions between these *field induced SDW states* occur with an approximate periodicity in H^{-1} as in the case of the SdH oscillations. Only the magnetic field in the c direction is relevant to these transitions. At the same time, the Hall effect shows a series of plateaus in resemblance to the quantized Hall effect. The superconducting state occurs in competition with the SDW states due probably to electron pairing via spin fluctuations instead of the ordinary s-type coupling. If so, the

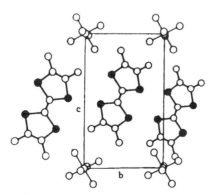

Fig. 6.12. Structure of the Bechgaard salts $(TMTSF)_2X$. Black circles represent Se atoms

unusual phenomenon in these salts may not be related to electron–phonon coupling. It is important to note that the transitions occur in the presence of a magnetic field in the c direction.

In the absence of a magnetic field and at low temperatures, the electrons can be considered on the Fermi circle in the ab plane. When a magnetic field is switched on in the c direction, the electrons move on orbits which are open in the a direction but are restricted in the b direction. The width of motion in this direction is inversely proportional to the magnetic field. Hence, as the field strength is increased the system becomes more one–dimensional.

Let us consider a system of parallel stacks at a distance b in a single (x, y) plane. The electrons most easily move in the x direction and hop in the b direction. The ideal Hamiltonian $\mathcal{H}_0$ may be chosen to represent the motion in the x direction and hopping in the b direction between the stacks with an overlap integral t_b. Let us assume on-site interaction with strength U for the interaction Hamiltonian $\mathcal{H}_1$ [6.23],

$$\mathcal{H}_0 = \sum_{n,\alpha=\pm 1} \int dx \left(-i\alpha v_F \psi^\dagger(x,n,\alpha)\frac{\partial}{\partial x}\psi(x,n,\alpha) \right.$$

$$\left. +t_b \sum_{m=\pm 1} \psi^\dagger(x,n,\alpha)\psi(x,n+m,\alpha)\exp(-imqx) \right), \tag{6.5.1}$$

$$\mathcal{H}_1 = U\sum_n \int \psi^\dagger(x,n,+)\psi^\dagger(x,n,-)\psi(x,n,-)\psi(x,n,+)dx. \tag{6.5.2}$$

Here, v_F is the Fermi velocity, U is the strength of electron–electron interaction, $q = ebH/c$, x is the coordinate along the stacks which are numbered by n, the index $\alpha = \pm$ corresponds to electrons with momenta $\pm p_F$, and the exponential factor in $\mathcal{H}_0$ corresponds to the gauge in which $A_y = Hz$, $A_x = A_z = 0$. For simplicity, let us neglect spin. The eigenvalue equation for $\mathcal{H}_0$ is

$$[-i\alpha v_F d/dx + 2t_b \cos(k_y n - qx)]\psi_n(x) = \varepsilon\psi_n(x), \tag{6.5.3}$$

$$\psi_n(x) = \exp\{i[-\varepsilon t + k_x x + k_y n - \alpha\lambda\sin(k_y n - qx)]\}, \tag{6.5.4}$$

where

$$\varepsilon = \alpha v_F k_x, \quad \lambda = 2t_b/(v_F q). \tag{6.5.5}$$

Note that ε is proportional to αk_x independent of k_y. This enables us to Fourier transform ψ in k_y, resulting in a transformed eigenfunction

$$\tilde{\psi}_n(x,m) = \exp\{i[-\varepsilon t + k_x x + (n-m)qx]\}J_{n-m}(-\alpha\lambda), \tag{6.5.6}$$

where $J_n(z)$ is the Bessel function, and m is an integer. This wave function is localized at the nth chain.

It is convenient to introduce creation and annihilation operators $a^\dagger(k,m,\alpha)$ and $a(k,m,\alpha)$ in the states represented by (6.5.6). In this representation, the Hamiltonians are given by

$$\tilde{\mathcal{H}}_0 = \sum_{m,k_x,\alpha} \alpha k_x v_F a^\dagger(k_x, m, \alpha) a(k_x, m, \alpha), \qquad (6.5.7)$$

$$\tilde{\mathcal{H}}_1 = \sum_{k_i m_i} g(m_1, m_2, m_3, m_4) \delta(k_1 + k_2 - k_3 - k_4 - (m_1 + m_2 - m_3 - m_4)q)$$
$$\times a^\dagger(k_1, m_1, +) a^\dagger(k_2, m_2, -) a(k_3, m_3, -) a(k_4, m_4, +), \qquad (6.5.8)$$

where

$$g(m, n, p, q) = U k_F \sum_s J_{s-m}(-\lambda) J_{s-n}(\lambda) J_{s-p}(\lambda) J_{s-q}(-\lambda). \qquad (6.5.9)$$

The new Hamiltonian $\tilde{\mathcal{H}}_1$ represents nonlocal interactions between electrons on four chains. This type of Hamiltonian can be treated by the so-called parquet approximation with logarithmically divergent diagrams. Summing over these diagrams is equivalent to a renormalizing process for the amplitude $g(m_1, m_2, m_3, m_4; \omega)$ which represents an effective interaction of electrons with energy close to ω and momentum k_x of order ω/v_F in four chains. Although the amplitudes include a set of four integers designating the stacks with electrons, those in which $(m + n) \neq (p + q)$ can be neglected for small energy because they do not conserve momentum. Hence, a set of three integers characterizes the interaction parameter instead of four. One can parameterize these three integers such that

$$g(n, m, l; \omega) = g(n + m, n + l, n + m + l, n; \omega).$$

where a new notation has been introduced on the right side. This notation is convenient because if the momenta k_i are Fourier transformed one can rewrite the interaction Hamiltonian in the form

$$\tilde{\mathcal{H}}_1 = \sum_{n,m,l} g(n, m, l; \omega) b^\dagger(x, n + m, l) b(x, n, l), \qquad (6.5.10)$$

where b is the conjugate of

$$b^\dagger(x, n, l) = \hat{a}^\dagger(x, n, +) \hat{a}(-x, n + l, -), \qquad (6.5.11)$$

and where $\hat{a}(x, n, \alpha)$ is the Fourier transform in k_y of $a(k_y, n, \alpha)$.

The new operator $b^\dagger$ defined above creates an electron–hole pair with separation l. The new interaction Hamiltonian indicates that such a pair hops a distance m. The parquet approximation amounts to taking into consideration the two types of logarithmically divergent diagrams which are shown in Fig. 6.13 and results in a renormalization group equation

$$dg(n, l, \xi)/d\xi = (2\pi U k_F)^{-1} \sum_m [g(m, l, \xi) g(n - m, l, \xi)$$
$$- g(m, n + l - m, \xi) g(n - m, l - m, \xi)], \qquad (6.5.12)$$

where $\xi = \ln(\Omega/\omega)$, and $\Omega = ebH v_F/c$ is a characteristic field energy which is assumed to be larger than the electron's energy ω. This assumption is

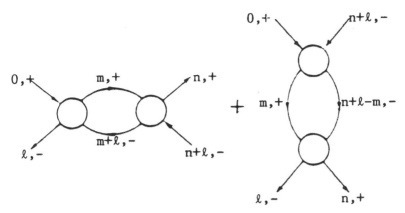

Fig. 6.13. Logarithmically divergent diagrams for the parquet approximation

valid for low temperatures since $\omega \sim k_\mathrm{B}T$. The initial condition for the above differential equation is

$$g(n,l,0) = U k_\mathrm{F} \sum_m J_{m-n}(-\lambda) J_{m-n-l}(\lambda) J_{m-l}(\lambda) J_m(-\lambda). \qquad (6.5.13)$$

Equation (6.5.12) is nonlinear in U, but in the neglect of the second term a convolution method can be adopted for the m sum. Under this approximation and using a Fourier transformation in m, we arrive at a simpler differential equation for the Fourier transform $\hat{g}(k_y, l, \xi)$ of $g(n, l, \xi)$,

$$\mathrm{d}\hat{g}(k_y,l,\xi)/\mathrm{d}\xi = \hat{g}^2(k_y,l,\xi)/U k_\mathrm{F}. \qquad (6.5.14)$$

The solution of this equation has the form

$$\hat{g}(k_y,l,\xi) = \frac{U k_\mathrm{F}}{c(k_y,l) - \xi} \qquad (6.5.15)$$

with an integration constant $c(k_y, l)$ which can depend on k_y and l. As the magnetic field H increases, ξ increases. When ξ reaches a point at which $c(k_y, l)$ is a minimum, the above solution diverges. This divergence may be associated with a critical point. In the vicinity of this point which we designate as (k_0, l_0), (6.5.15) may be approximated by

$$\hat{g}(k_y,l,\xi) = A\delta(l,l_0)[\xi_c - \xi + \beta(k_y - k_0)^2]^{-1}, \qquad (6.5.16)$$

where A can be of order 1 and β is inversely proportional to the magnitude $|U|$. Substituting the Fourier transform of (6.5.16) back into (6.5.12), we find that the first term is of order $|\xi_c - \xi|^{-2}$ while the second term is of order $|U||\xi_c - \xi|^{-1}$. Hence, the above approximation is valid for $|\xi_c - \xi| \ll |U|^{-1}$. The solution is characterized by a pole which is not constant but moves with the variable k_y. Since the existence of an electron hole pair depends on $b(x, n, l)$, its average over n is characteristic of this state with a moving pole.

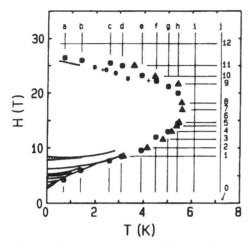

Fig. 6.14. Phase diagram of $(TMTSF)_2ClO_4$ due to *Naughton* et al. [6.25]. Light vertical and horizontal lines: Paths taken on different sweeps. Circles: H sweeps. Triangles: T sweeps. Lines below $10T$: Previous data. Crosses, open circles and thick solid line: Data from other authors

This average may be used as an order parameter. Taking a Fourier transform, we find that the average has the form

$$\langle b(x,n,l) \rangle = Y(k_0, l_0, x, n)\delta(l, l_0)\exp(ik_0 n). \tag{6.5.17}$$

It has a magnitude Y which depends on the distance l_0 between the electron and hole in the direction perpendicular to the stack, fluctuates with x, and is periodic in n with the center-of-mass momentum k_0.

Yakovenko [6.23] has determined the CDW wave which depends on the above order parameter and found that its longitudinal momentum is quantized and its perpendicular momentum is equal to k_0. Furthermore, he has numerically found that the behavior of the system depends strongly on the sign of the interaction U. For $U > 0$, and $H = 0$, a CDW (or SDW, if spins are included) state occurs. For $U < 0$, the system exhibits superconductivity for $H = 0$. As H increases, a series of phase transitions occurs between different CDW (SDW) states which depend on the wave vector with the above-mentioned character.

There are, of course, some other theories [6.24], but let us now discuss a new experimental finding. Figure 6.14 illustrates the phase diagram of $(TMTSF)_2ClO_4$ obtained by *Naughton* et al. [6.25] based on magnetization and magnetotransport measurements for $0.5 < T < 10K$ and $H < 30T$. The graph includes the data of some other authors. The field induced spin density wave states appear only in a limited regime because, at high magnetic fields, reentrance into the metal phase takes place. The black data and the solid lines in the graph are those taken from the three sets of other data. The solid lines are previously determined phase boundaries between metal–FISDW and FIFSW–FIDSW states. No reentrance has been found in $(TMTSF)_2PF_6$.

It is appropriate to give comments on the theories [6.26] which are strictly two-dimensional. Although the two-dimensionality has been demonstrated, for example in the observation by *Yan* et al. [6.27] of a cosine dependence of the threshold field for the appearance of the SDW states, small hopping in the c direction and an anion gap may play important roles in bringing about the complex properties. The former is expected to add fast magnetic oscillations. Similar fast oscillations have been predicted to appear in the dHvA oscillations of a strongly interacting 3D electron gas in superposition to the slow ones [6.1].

With an anion gap, the electron spectrum may be expressed as [6.28]

$$\varepsilon_\pm = \pm v_F(k_x \mp k_F) \pm [4t_b^2 \cos^2(k_y b) + \Delta^2]^{1/2} + 2t_c \cos(k_z c), \qquad (6.5.18)$$

where Δ is the anion gap. The hopping terms in the b and c directions are characterized by t_b and t_c respectively. The former is of order 100–200 K, while the latter is small, 5–10 K. The first term represents the free-electron motion along the chain in the right $(+)$ and left $(-)$ sheets of the Fermi surface. These two sheets are related to each other through a displacement $(2k_F, \pi/b, \pi/c)$ so that a SDW instability may be expected. It has been shown [6.28] that the gap Δ causes neighboring chains not to be equivalent to each other. In particular in the presence of strong magnetic fields, hopping in the b direction is suppressed, electron orbits in the two inequivalent chains are localized, and reentrance to the metallic state can take place.

7. Electronic Specific Heat

The specific heat of electron systems may be enhanced strongly by correlations.

7.1 Correlation Effects

We have seen how Coulomb interaction affects the specific heat of a classical OCP (Fig. 3.3). The parameter Γ plays the role of temperature in that case, and the specific heat can be obtained once the internal energy is given as a function of Γ. Although it is difficult to evaluate the internal energy for large Γ, Monte Carlo simulations have shown how the specific heat curve looks like. Therefore, we shall be concerned in what follows in this chapter with low temperatures, where new problems concerning the electronic specific heat evaluation arise.

Sommerfeld showed a long time ago that the electronic contribution to the specific heat of simple metals is significant at low temperatures because it varies linearly with temperature. How does Coulomb interaction affect such a variation? As long as the electrons retain free motion a linear specific heat can still be expected, although the electronic mass may have to be replaced by an effective mass as Landau's Fermi liquid theory suggests. However, the evaluation of the internal energy and specific heat as functions of temperature in the presence of Coulomb interaction involves difficulties. Moreover, there are unusual cases to be investigated.

7.1.1 Lidiard Approximation

In order to obtain the electronic specific heat for low but finite temperatures in the presence of Coulomb interaction the internal energy including the correlation energy must be evaluated as a function of temperature. Since this evaluation is difficult, a simple and approximate formula due to *Lidiard* [7.1] has often been adopted. Let us first discuss this formula.

For this purpose we assume that the energy per unit volume of a many electron system can be expressed by

$$U(T) = \int_0^\infty \varepsilon f(\varepsilon, T) g(\varepsilon, T) d\varepsilon, \qquad (7.1.1)$$

where $f(\varepsilon, T)$ is the Fermi distribution and $g(\varepsilon, T)$ is the density of states per unit volume, which satisfies

$$n = \int_0^\infty f(\varepsilon, T) g(\varepsilon, T) \mathrm{d}\varepsilon. \tag{7.1.2}$$

Since electron correlations depend on density as well as temperature, the density of states can depend on temperature as so indicated.

Upon differentiation, (7.1.1) yields

$$c_v = \int_0^\infty \mathrm{d}\varepsilon \left(g \frac{\mathrm{d}f}{\mathrm{d}T} + f \frac{\mathrm{d}g}{\mathrm{d}T} \right). \tag{7.1.3}$$

We assume near absolute zero a δ-function-type variation of $\mathrm{d}f/\mathrm{d}T$ at the Fermi energy and use the relation $\mathrm{d}n/\mathrm{d}T = 0$ to obtain

$$c_v = \frac{\pi^2}{3} k_B^2 T g(\varepsilon_F) + \int_0^{\varepsilon_F} \mathrm{d}\varepsilon f(\varepsilon - \varepsilon_F) \frac{\mathrm{d}g(\varepsilon)}{\mathrm{d}T}, \tag{7.1.4}$$

where $g(\varepsilon_F)$ represents the density of states at the Fermi energy at 0 K, and where a zero temperature limit may be taken for $\mathrm{d}g/\mathrm{d}T$ after differentiation.

We have neglected higher-order terms in the low temperature expansion of the Fermi distribution function.

When the energy is expressed as in (7.1.1) the density of states $g(\varepsilon, T)$ is expected to include many-body effects. Hence, it can be a complicated function of ε, r_s and temperature. If this temperature dependence is neglected, the specific heat is simply given by

$$c_v = \frac{1}{3} \pi^2 k_B^2 T g(\varepsilon_F). \tag{7.1.5}$$

Moreover, we avoid the evaluation of $g(\varepsilon)$ and adopt the free-electron energy $\varepsilon = q^2$ and density of states, q being the wave number. With a spin factor 2, the density of states depends only on q or ε and is given by

$$g(\varepsilon) = \frac{q^2}{\pi^2} \left(\frac{\mathrm{d}\varepsilon}{\mathrm{d}q} \right)^{-1}. \tag{7.1.6}$$

Therefore,

$$c_v = \frac{k_F^2 k_B^2}{3} T \left(\frac{\mathrm{d}\varepsilon}{\mathrm{d}q} \right)^{-1}_{k_F}. \tag{7.1.7}$$

According to this formula the electronic specific heat can be obtained in terms of the derivative of the ground state energy at k_F, making the evaluation of the energy for finite temperatures unnecessary. Therefore, immediately after the evaluation of the correlation energy for absolute zero, the above formula was applied to the specific heat [7.2]. However, it is not exact in the presence of electron correlations [7.3].

7.1.2 Exchange and Correlation Effects

The specific heat can be evaluated for low but finite temperatures if all the energies discussed in Chap. 4.2 are given as functions of temperature. Therefore, let us evaluate low temperature corrections to the grand partition functions given in that chapter. These corrections start with terms of order $1/\eta^2$, where $\eta = \beta\mu$ and $\mu = p_F^2$ is the chemical potential which is generally different from the ideal Fermi energy k_F^2. The temperature-dependent corrections to the grand partition function will be denoted by $\Delta \ln \Xi(T)$ with appropriate indices to distinguish various contributions.

The free-electron correction is given by

$$\Delta \ln \Xi_0(T) = \frac{2\eta^{5/2}V}{15\pi^2\beta^{3/2}} \left(\frac{5}{8}\frac{\pi^2}{\eta^2} \right), \tag{7.1.8}$$

where the factor outside the brackets is $\ln \Xi_0$ for 0 K.

The first-order exchange contribution given by the first equality of (4.2.5) can be rewritten as [7.4]

$$\ln \Xi_x = \frac{Ve^2}{8\pi^3\beta}J, \tag{7.1.9}$$

where

$$J = \pi \int_{-\infty}^{\eta} d\eta [F_{-1/2}(\eta)]^2,$$

$$F_{-1/2}(\eta) = \frac{1}{\Gamma(-1/2+1)} \int_0^{\infty} dx \frac{x^{-1/2}}{e^{x-\eta}+1}. \tag{7.1.10}$$

Note that the integrand of J includes two Fermi distribution functions. This makes an immediate application of the Sommerfeld method difficult. Hence, we first decompose J into two parts

$$J = J_0 + J_1, \tag{7.1.11}$$

where

$$J_0 = 4\int_0^{\infty} \varepsilon f(\varepsilon)d\varepsilon, \quad J_1 = \int_0^{\infty} g(\varepsilon)f(\varepsilon)d\varepsilon,$$

$$g(\varepsilon) = 2\int_0^{\infty} t^{-1/2}(1-t)^{-1}\ln(1-e^{-\varepsilon|1-t|})dt. \tag{7.1.12}$$

Note that both J_0 and J, include only one Fermi distribution function.

The first integral can be easily obtained as

$$J_0 = 2\eta^2 + (2\pi^2/3). \tag{7.1.13}$$

It can also be evaluated from (4.2.5) by considering the common volume of the two Fermi spheres.

For J_1, the domain of the t integration for $g(\varepsilon)$ can be split into two domains, $[0, 1]$ and $[1, \infty]$. In the former domain, we change the variable from

t to $(1 - t)$, expand $(1 - t)^{-1/2}$ in powers of t, and then integrate term by term. In the latter, we change the variable such that $x - 1 \to x$ and then use the Sommerfeld method. For small $e^{-\eta}$, we arrive at

$$\int_0^\eta g(x)\mathrm{d}x = A - \frac{\pi^2}{3} \ln \eta + \sum_{j=2,4,\ldots} \frac{(2j+1)!!}{j2^{j-1}(j+1)} \left(\sum_{n=1}^\infty n^{-j-2} \right) \eta^{-j}$$
$$- \frac{2\pi^2}{9\eta^{3/2}} \left(1 - \frac{3}{10\eta} - \frac{3}{8\eta^2} - \cdots \right), \tag{7.1.14}$$

where

$$A = - \frac{\pi^2 \gamma}{3} - 2 \sum_{n=1} n^{-2} \ln n - \frac{\pi^2}{6} \sum_{j=2,4,\ldots} \frac{(2j+1)!!}{j(j+1)2^{j-1}}$$
$$+ \frac{\pi^2}{3} \left(2^{1/2} - \ln \left| \frac{2^{1/2}+1}{2^{1/2}-1} \right| \right)$$
$$= - 4.035.$$

γ is the Euler constant.

The temperature-dependent part of the exchange contribution to the grand partition function is given by

$$J_1 = A - \frac{\pi^2}{3} \ln \eta + \frac{10\pi^4}{2.12^2 \eta^4} - \frac{\pi^2}{9\eta^{7/2}} \left(1 - \frac{3}{10\eta} - \cdots \right) + \cdots . \tag{7.1.15}$$

The electronic specific heat is obtained by combining the ideal and exchange contributions to the grand partition function. Note that p_F in η must be determined as a function of density, which is represented by its ideal value k_F, r_s and T from the number density relation in accordance with grand ensemble theory. To order e^2, the electronic specific heat is found as follows:

$$c_v = c_v^0 \left\{ 1 + \left(\frac{4}{9\pi^4} \right)^{1/3} r_s \left[\frac{5}{2} + \frac{3A}{\pi^2} - \ln \left(\frac{k_F}{k_B T} \right)^2 \right] + \cdots \right\}, \tag{7.1.16}$$

where

$$c_v^0 = \frac{k_B^2 T}{6} k_F, \quad \text{(per volume)} \tag{7.1.17}$$

is the ideal gas specific heat per unit volume. In (7.1.16), the second term, which is proportional to r_s, includes a logarithmic term. This term becomes increasingly large towards absolute zero, even though c_v itself does not diverge. Generally, the appearance of such a logarithmic term is characteristic of certain perturbations from the Fermi distribution function of ideal electrons. In the present case, perturbation is caused by temperature, and in the case of ε_c by Coulomb interaction. The presence of such a logarithmic term reduces the domain of validity of a perturbation series.

In order to study the convergence of the above result for c_v, let us examine the ring and second-order exchange contributions [7.3]. The temperature-

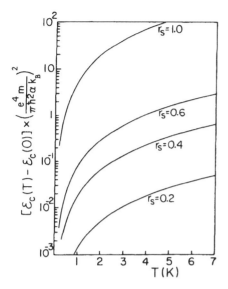

Fig. 7.1. Temperature variation of the correlation energy [7.3b]

dependent part of the ring diagram grand partition function, $\Delta \ln \Xi_r(T)$, is obtained by expanding the eigenvalues $\lambda_j(T)$ in $1/\beta$, which starts with $1/\eta^2$. By a calculation similar to the case for absolute zero discussed in Sect. 4.2 and with $\Lambda = 2e^2/\pi p_F$, we arrive at

$$\Delta \ln \Xi_r(T) =$$
$$-\frac{e^4\eta^{3/2}V}{2\pi^4\beta^{1/2}} \left[-\frac{\pi}{12\eta^2} \left(0.492 + \frac{\pi}{4}(1 - \ln 4)\ln\Lambda \right) + \cdots \right]. \tag{7.1.18}$$

Concerning the second-order exchange graphs, only the contribution from the anomalous diagrams has been analytically obtained as

$$\Delta \ln \Xi_{a2x}(T) =$$
$$\frac{e^4\eta^{3/2}V}{4\pi^4\beta^{1/2}} \left[\frac{\pi^2}{12\eta^2} \left(-1.273 + 4.026\ln\eta + (\ln\eta)^2 \right) + \cdots \right]. \tag{7.1.19}$$

The regular second-order exchange contribution has not been evaluated as yet but presumably it has a similar form. Omitting this contribution, we can write down the temperature-dependent correlation energy

$$\Delta\varepsilon_c(T) =$$
$$-\frac{\pi^2}{12\eta_0^2}[1.427 - 0.152\ln r_s - 0.616\ln\eta_0 - 0.304(\ln\eta_0)^2] + \cdots . \tag{7.1.20}$$

Here, the right side is now expressed as a function of $\eta_0 = \beta k_F^2$. That is, $\eta = \beta p_F^2$ has been renormalized in terms of η_0. The deviation of the correlation energy per electron from its value at absolute zero is given in the ordinary units by

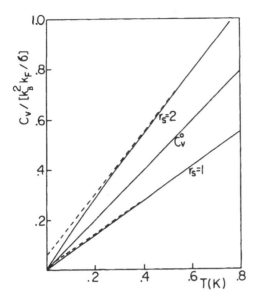

Fig. 7.2. Temperature variation of the electronic specific heat [7.3a]

$$\Delta\varepsilon_c(T) = \left(\frac{e^4 m}{\hbar^2 \alpha k_B}\right)^{-2} T^2 r_s^4 \times$$
$$\times [5.458 - 0.786\ln(r_s^2 T) + 0.076\ln(r_s^2 T)^2 + 0.0138\ln r_s], \qquad (7.1.21)$$

where T is in kelvin and where $\hbar$ and $2m$ have been restored. The parameter $\alpha = (4/9\pi)^{1/3} = 0.52106$. The above energy difference increases parabolically as a function of r_s such that the higher the temperature the stronger the increase. It increases with temperature parabolically also for a given r_s. Figure 7.1 illustrates this variation.

Once the temperature dependence of the internal energy is found, the specific heat is obtained easily. To order r_s^2 and in the neglect of a small term, it is given by

$$c_c/c_v^0 = 1 + 0.162 r_s - 0.166 r_s \ln \eta_0 - 0.157 r_s^2 + 0.0138 r_s^2 \ln r_s$$
$$+ r_s^2 [0.0282 \ln \eta_0 + 0.0275(\ln \eta_0)^2] + \cdots. \qquad (7.1.22)$$

We remark that the last logarithmic term is exactly the square of the first logarithmic term. Hence, quite possibly they are a part of a geometric series. In the lack of higher-order terms, no other similar functional relation has been found.

Figure 7.2 illustrates the temperature variation of the specific heat c_v in units of $k_B^2 k_F/6$. The specific heat varies almost linearly with T as indicated by the dashed straight lines. Note that the curve for $r_s = 1$ is below the ideal specific heat curve while it is above it for $r_s = 2$. As a function of r_s the ratio c_v/c_v^0 is less than 1 below around $r_s = 0.6$ and is larger the higher the temperature. It is larger than 1 above this r_s in such a way that it is larger

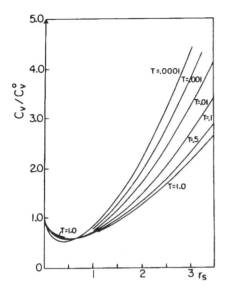

Fig. 7.3. Dependenee or the electronic specific heat on r_s [7.3a]

the lower the temperature. That is, a complete reversal in the r_s dependence takes place at around $r_s = 0.6$. Hence, in the range $r_s = 2-3$, which includes Cu, the ratio is larger than 1 with its value increasing towards absolute zero as in Fig. 7.3.

7.1.3 Specific Heat of 2D Electrons

The specific heat of interacting electrons in 2D can be calculated in powers of r_s for high densities. The ideal gas contribution to the grand partition function is given for low temperatures by

$$\ln \Xi_0 = \frac{A\beta p_F^4}{4\pi}\left[1 + \frac{\pi^2}{3\eta^2} + O\left(\frac{1}{\eta^4}\right)\right], \qquad (7.1.23)$$

where $\eta = \beta\mu$, μ being the chemical potential p_F^2 and A is the surface area.
The first-order exchange contribution is given by

$$\ln \Xi_x = \frac{\beta A}{(2\pi)^4}\int u(q)d\mathbf{q}\int f(p)f(\mathbf{p}+\mathbf{q})d\mathbf{p}. \qquad (7.1.24)$$

The right side integral is difficult to obtain, but it has been evaluated rigorously via a hypergeometric function [7.5] or based on a sum rule [7.6]. The result yields the energy per electron in the form

$$\varepsilon(T) = \frac{1}{r_s^2}\left[1 + \frac{\pi^2}{3\eta_0^2} - \frac{8\sqrt{2}}{3\pi}r_s - \left(\frac{1}{6}\ln \eta_0 + 4C - \frac{1}{6}\right)\frac{\pi\sqrt{2}}{\eta_0^2}r_s\right], \qquad (7.1.25)$$

where $\eta_0 = \beta k_F^2 = \beta(2\pi n)$ and $C = 0.02553$. Hence, to order r_s the specific heat is given by

$$c_v = c_v^0 \left[1 + r_s \left(\frac{3\sqrt{2}}{4\pi}(1 - 16C) - \frac{1}{(2\pi)^{1/2}} \ln \eta_0 \right) \right], \qquad (7.1.26)$$

where

$$c_v^0 = \frac{\pi^2 k_B^2 T}{3 k_F^2}, \quad \text{(per electron)}. \qquad (7.1.27)$$

More explicitly, the specific heat is given by

$$c_v/c_v^0 = 1 + r_s(0.1997 - 0.2251 \ln \eta_0). \qquad (7.1.28)$$

In comparison with the corresponding expression for the 3D case given by (7.1.22), the 2D case is larger due to relatively more effective Coulomb interaction.

The correlation energy $\varepsilon_c(T)$ has also been evaluated for finite temperatures as follows [7.7],

$$\varepsilon_c(0) = -0.3946 + 0.86 r_s - 0.17 r_s \ln r_s,$$

$$\varepsilon_c(T) = \frac{4\pi}{\eta_0^2} \left[-0.1824 - 0.02968 \ln \eta_0 + \frac{1}{24\pi}(\ln \eta_0)^2 \right] + O(r_s). \qquad (7.1.29)$$

The term with $(\ln \eta_0)^2$ is exactly the square of the logarithmic term in the first exchange energy, and thus these two may represent a part of a geometric series. In combination, they lead to a temperature dependent part of the internal energy in the form

$$\Delta \varepsilon(T) = \frac{\pi^2}{3 r_s^2} \frac{1}{\eta_0^2} \left(1 + 0.0871 r_s + \frac{1}{1 + (r_s/\sqrt{2\pi}) \ln \eta_0} \right) + \cdots. \qquad (7.1.30)$$

It is unclear at present whether other logarithmic terms can be summed in similar forms or not. If these terms are neglected, one can force $\varepsilon(T)$ to assume the ideal gas form with an effective mass m^*. This limiting specific heat effective mass m^* is given approximately by

$$m^* = m(1 + 0.0436 r_s). \qquad (7.1.31)$$

For Si inversion layers with $m = 0.2 m_0$ and a dielectric constant 7.8, the effective mass is

$$m^* = m(1 + 0.0168 \times 10^6 n^{-1/2}). \qquad (7.1.32)$$

Since the density n is of order 10^{12} cm^{-2}, m^* is slightly increased due to the exchange interaction in inverse proportion to the square of n.

7.2 Magnetothermal Effect in Two Dimensions

Two-dimensional electrons are primarily in Landau levels in a perpendicular magnetic field. When the field is turned off adiabatically their kinetic energy is converted into thermal energy, where the conversion can be total in distinction from the case of 3D. Hence, adiabatic changes of strong magnetic field can cause significant temperature oscillations as was shown first by *Shiwa* and *Isihara* in 1983 [7.8c].

The temperature variation under adiabatic changes of the magnetic field can be given by

$$dT = \left(\frac{\partial T}{\partial H}\right)_{S,n} dH$$

$$= -\left(\frac{\partial S}{\partial H}\right)_{T,n} \left[\left(\frac{\partial S}{\partial T}\right)_{H,n}\right]^{-1} dH. \tag{7.2.1}$$

Here, n is the density, and S is the entropy, which can be evaluated from the grand partition function through

$$S = \frac{\partial}{\partial T}(\beta^{-1}\ln \Xi)_{H,A,\mu}, \tag{7.2.2}$$

where A is the surface area and μ is the chemical potential.

The Landau levels are expressed as

$$\varepsilon_i = (2i+1)\mu_B H \pm (g/2)\mu_B^0 H, \tag{7.2.3}$$

where $\mu_B = e\hbar/2mc$ is the effective Bohr magneton with the effective mass m, μ_B^0 is the real Bohr magneton with the bare mass m_0 and g is the effective Landé factor of a given system. We can write the grand potential as

$$\Omega(\mu) = \frac{1}{2}\Omega_0(\mu + \mu_B H) + \frac{1}{2}\Omega_0(\mu - \mu_B H), \tag{7.2.4}$$

where

$$\Omega_0(\mu) = -k_B T \sum_i \ln\{1 + \exp[\beta(\mu - \varepsilon_i^0)]\} \tag{7.2.5}$$

with

$$\varepsilon_i^0 = (2i+1)\mu_B H. \tag{7.2.6}$$

The sum over Landau states i in (7.2.5) can be replaced by integration in terms of the density of states given in the natural unit by

$$g(\varepsilon) = \mu_B H \frac{A}{\pi}\sum_i \delta(\varepsilon - \varepsilon_i^0). \tag{7.2.7}$$

We obtain

$$\Omega_0(\mu) = -k_B T \int d\varepsilon g(\varepsilon)\ln\{1 + \exp[\beta(\mu - \varepsilon)]\}. \tag{7.2.8}$$

Although the ideal case can be treated, it is more realistic to consider the case in which the Landau levels are broadened. As a typical example, let us examine the case of Lorentzian broadening. Replacing each δ function in $g(\varepsilon)$ by a corresponding Lorentzian function with a half-width $1/\tau$, we obtain

$$g_0(\varepsilon) = \frac{A\alpha}{\pi\beta} \sum_{i=0}^{\infty} \frac{1}{\pi} \frac{1/\tau}{(\varepsilon - \varepsilon_i^0)^2 + (1/\tau)^2}, \tag{7.2.9}$$

where $\alpha = \beta\mu_\mathrm{B} H$ with $\beta = 1/k_\mathrm{B}T$. The grand potential is then given by

$$\Omega(\mu) = -\frac{A\eta^2}{4\pi\beta^2} \left\{ 1 + \frac{\pi^2}{3\eta^2} + \left[\left(\frac{g}{2}\right)^2 - \frac{D(\Gamma)}{3} \right] \frac{\alpha^2}{\eta^2} \right.$$
$$\left. + \frac{4\alpha}{\eta^2} \sum_{l=1}^{\infty} (-)^{l+1} W^l(\Gamma) \frac{\cos(\pi\gamma)\cos(g\pi l/2)}{l\sinh(\pi^2 l/\alpha)} \right\}, \tag{7.2.10}$$

here

$$\eta = \beta\mu, \quad \gamma = \alpha/\eta$$

are dimensionless and

$$W(\Gamma) = \exp(-\pi/\tau\alpha^2), \tag{7.2.11}$$

$$D(\Gamma) = \frac{12}{\pi^2} \sum_{k=1}^{\infty} \frac{(-)^{k+1}}{k^2} \exp(-2\pi\Gamma k), \tag{7.2.12}$$

$$\Gamma = 1/(2\tau a^2), \quad a^2 = \mu_\mathrm{B} H. \tag{7.2.13}$$

Note that $W(\Gamma)$ is simply the Fourier transform of the Lorentzian function. When each Landau level is broadened in the Lorentzian form in the energy space, the contribution from the level to the grand potential is reduced progressively by the factor $W(\Gamma)$ as its index increases. Note also that level broadening affects the non-oscillating part of the grand potential through the function $D(\Gamma)$ with $D(0) = 1$.

Equation (7.2.1) involves electron density n. Hence, we need an equation which relates the density with the chemical potential μ. This relation is given by

$$\frac{1}{\gamma_0} = \frac{1}{\gamma} - \frac{2\pi}{\alpha} \sum_{t=1}^{\infty} (-)^{l+1} W^l(\Gamma) \frac{\sin(\pi l/\gamma)\cos(g\pi l/2)}{\sin(\pi^2 l/\alpha)}, \tag{7.2.14}$$

where

$$1/\gamma_0 = 2\pi n/a^2, \quad 1/\gamma = \mu/a^2. \tag{7.2.15}$$

In (7.2.14), the actual chemical potential μ is given implicitly as a function of electron density n. This relation is very important, and will be illustrated shortly.

For $g = 2$, the entropy is found to be given by

$$\frac{4\pi S}{A k_\mathrm{B}} = \frac{2\pi^2}{3\beta} + \frac{4\pi^2}{\beta} \sum_{l} W^l L\left(\frac{\pi^2 l}{\alpha}\right) \frac{\cos(\pi l/\gamma)}{\sin(\pi^2 l/\alpha)}, \tag{7.2.16}$$

where

$$L(x) = \coth x - 1/x$$

is the Langevin function.

Once the entropy is obtained, the temperature change dT is obtained without much difficulty. We note that its dimensionless derivative defined by

$$S_T = (\pi/Ak_Ba^2)T(\partial S/\partial T)_{H,n} \qquad (7.2.17)$$

is related to the electronic specific heat. In fact, the specific heat per area is given by

$$c_{A,H} = (k_Ba^2/\pi)S_T, \quad \text{(per area).} \qquad (7.2.18)$$

Its analytical expression near absolute zero is

$$c_{A,H} = \frac{\pi k_B^2 T}{6} \frac{1 + 2W[\cos(\pi/\gamma) - W]}{1 - 2W\cos(\pi/\gamma) + W^2}. \qquad (7.2.19)$$

Near absolute zero and if $W \neq 1$ and $1/\gamma = m$ is an integer, the following simple formulas can be derived,

$$S_T = \frac{\pi^2}{3\alpha} \left(\frac{1}{2} + (-)^m \frac{W}{1 - (-)^m W} \right), \qquad (7.2.20)$$

and

$$c_{A,H} = \frac{\pi k^2 T}{6\alpha} \left(1 + (-)^m \frac{2W}{1 - (-)^m W} \right), \quad \text{(per area),} \qquad (7.2.21)$$

$$\frac{dT}{T} = \pm \frac{2W\tilde{\Gamma}}{1 - W^2} \frac{dH}{H}, \qquad (7.2.22)$$

where $\tilde{\Gamma} = 2\pi\Gamma$.

These expressions indicate that the temperature fluctuations are very strong when W approaches 1. This corresponds to the case without broadening. Such large fluctuations can be traced to the behavior of the electronic specific heat. One can show that in the absence of broadening the linear term in the electronic specific heat is canceled and the specific heat varies exponentially as a function of temperature. This disappearance of the linear specific heat can be seen by putting $W = 1$, $m = 2j + 1$ in (7.2.21). Obviously, the same formula cannot be used for $W = 1$ and $m = 2j$ because it has been based on $W \neq 1$.

A similar calculation can be performed for the elliptically broadened case in which

$$g_0(\varepsilon) = \frac{A\alpha}{\pi\beta} \sum_{i=0}^{\infty} \frac{2\tilde{\tau}}{\pi} [1 - \tilde{\tau}^2(\varepsilon - \varepsilon_i^0)^2]^{1/2}, \qquad (7.2.23)$$

where a new relaxation time $\tilde{\tau}$ has been used.

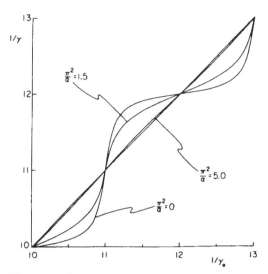

Fig. 7.4. Relation between the chemical potential μ and electron density n for $\tilde{\Gamma} = 0.3$. The ordinate $1/\gamma$ is equal to $\mu/\mu_B H$ and the abscissa $1/\gamma_0$ is equal to the Landau level filling factor ν [7.8b]

The relaxation time which has been introduced phenomenologically may be estimated from the scattering time $\tau_s = m\mu/e$, where ξ is the mobility. Restoring $\hbar$ we can use for Si inversion layers

$$\hbar/2\tau_s = 0.3047(0.19m_0/m)(10^4/\mu\,[\mathrm{cm^2V^{-1}s^{-1}}])\,\mathrm{meV}.$$

Using the cyclotron energy given by

$$\hbar\omega_c = 0.6093H(0.19m_0/m)\,\mathrm{meV},$$

we find

$$\tilde{\Gamma} = (\pi/H)(10^4/\mu\,[\mathrm{cm^2V^{-1}s^{-1}}]).$$

From this expression, we find a typical value of $\tilde{\Gamma} = 0.3$. This corresponds to $W = 0.7408$.

Figure 7.4 shows the relation between the actual chemical potential and electron density n for three different temperatures at fixed $\tilde{\Gamma} = 0.3$. The ordinate is the chemical potential in units of $\mu_B H$, and the abscissa is the filling factor $\nu = 1/\gamma_0$ of Landau levels. The graph corresponds to the case in which $g = 2$. If there is no broadening the chemical potential jumps from one Landau level to another. Such sudden changes are eased by broadening particularly at finite temperatures. The curve for $\pi^2/\alpha = 0$ corresponds to absoute zero, but due to broadening it is smooth. The two curves $\pi^2/\alpha = 1.5$ and 5.0 correspond to finite temperatures. The former corresponds roughly to $1\,\mathrm{K}$ at $2T$ in the region of electron densities of order $10^{12}\,\mathrm{cm^{-2}}$. The latter case is very close to the classical limit represented by the straight line. Note that

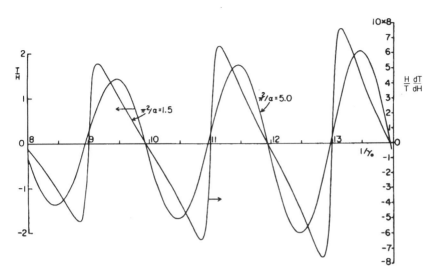

Fig. 7.5. Magnetothermal oscillations in 2D [7.8b]

all the curves pass through the same points at which $1/\gamma = 1/\gamma_0 = 2j + 1$, where j is an integer or 0. The curves are sharp about these points at finite temperatures. Such sharp changes are responsible to strong oscillations in a magnetic field.

Figure 7.5 illustrates normalized temperature oscillations plotted against filling factor $1/\gamma_0 = \nu$ for $\tilde{\Gamma} = 0.3$. Let curve A correspond to $\pi^2/\alpha = 5.0$ and B to $\pi^2/\alpha = 1.5$. Note that the period of oscillation is constant and is given by

$$\Delta(1/\gamma_0) = \Delta(\varepsilon_{\rm F}/\mu_{\rm B}H) = 2. \tag{7.2.24}$$

The amplitude of A is given by the left-hand ordinate, while that of B by the right side. In general, the amplitude increases toward the right, which is the direction of increasing density, and it is larger in the lower temperature case of B. The corresponding 3D case is far smaller. In fact even for high metallic densities and even in the absence of broadening the 3D amplitude corresponding to the case B is several times smaller.

Magnetothermal oscillations are associated with oscillations in sound velocity. Figure 7.6 represents $(v^2 - v_0^2)/2v_0^2$, where v and v_0 are sound velocities in the presence and absence of a magnetic field. The solid curve illustrates the case $\pi^2/\alpha = 1.5$. The corresponding ordinate is on the left side. The dotted curve corresponds to $\pi^2/\alpha = 5.0$ with the right ordinate. The broadening parameter is the same as in Fig. 7.5 in both cases. The two oscillating curves are asymmetric about the abscissa, but the period of oscillations is constant as in the case of the magnetothermal effect given by (7.2.24). Note that in three dimensions the magnitude of such oscillations is

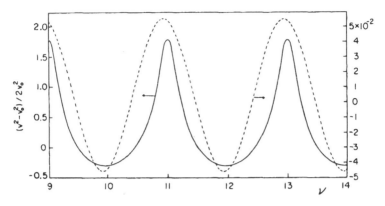

Fig. 7.6. Magnetoacoustic oscillation in 2D [7.8e]

$$(v^2 - v_0^2)/2v_0^2 \sim 10^{-5} - 10^{-4}. \tag{7.2.25}$$

The relatively large amplitudes in Fig. 7.6 are characteristic of two dimensions.

7.3 Heavy Fermions

In the past ten years, very unusual low temperature properties have been found in systems that contain rare earth (such as Ce) or actinide (such as U) ions. A linear specific heat is observed below a certain characteristic temperature T^* in such systems but the proportionality constant γ is unusually high. In fact, γ can be of order 1 J/mole K^2 in contrast to mJ/mole K^2 in ordinary metals. Table 7.1 lists γ and the Pauli susceptibility at 0 K of such systems [7.9]. The specific heat type and a constant R will be discussed shortly.

Table 7.1. Specific heat constant and Pauli susceptibility of heavy fermions

	γ[mJ/mol K^2]	Type	χ_P [10^{-3} emu/mol]	R
UPt$_3$	450	2	7	0.52–0.66
U$_2$Zn$_{17}$	535	1	12.5	0.79
UCd$_{11}$	840	1	38	1.55–1.82
CeAl$_3$	1600	3	36	0.75
Ce$_2$Cu$_2$Si$_2$	1100	3	~ 7	0.17–0.44
UBe$_{13}$	1100	3	15	0.46

These systems have a nearly constant spin susceptibility at low temperatures, indicating that they are Fermi liquids. Their large γ suggests that the effective mass is several hundred times the free-electron mass. Hence, their electrons are called *heavy Fermions*. Above T^*, they show a Curie–Weiss susceptibility which originates from the localized f electrons ($4f$ in Ce and $5f$ in U). The conduction electrons are "light" and interact with these localized electrons. Below T^* the f electrons are delocalized and form a Fermi liquid together with the conduction electrons. The above description of heavy Fermion systems is basic and general. In detail, even the specific heat behavior is somewhat complicated due to sample dependence and system variation. In fact, there are at least three types of temperature variation of the specific heat. First, there is type 1 (U_2Zn_{17}, UCd_{11}), which follows the ordinary temperature variation except for a high γ. In type 2 (UPt_3), the specific heat c varies as

$$c/T = \gamma + \beta T^2 + \delta T^2 \ln T. \tag{7.3.1}$$

Here, the first two terms represent the normal temperature variation, although γ is large. The appearance of the last term is characteristic of type 2. This term appears to be due to spin fluctuations in antiferromagnetic configurations. A similar term was first observed in liquid ^{3}He, with which the present systems have a certain similarity. In type 3, c/T changes as $T^{-1.6}$ ($CeCu_2Si_2$, UBe_{13}) or as $T^{-0.88} \sim T^{-1.8}$ ($CeCu_6$, $CeAl_3$).

In analyzing heavy Fermion systems, the *Sommerfeld–Wilson ratio* R may be used. If the degeneracy of the f level is labeled by the magnetic quantum number J such that $-J, -J+1, \ldots, J$, it is defined by

$$R = \frac{\pi^2 k_B^2 \chi_p}{3g^2 \mu_B^2 J(J+1)\gamma}. \tag{7.3.2}$$

Here,

$$g\mu_B[J(J+1)]^{1/2} = \mu_{\text{eff}}$$

represents an effective magnetic moment.

The ratio R is 2 in the spin one-half *Kondo problem*. As we shall discuss shortly, the heavy Fermion and Kondo problems have a certain similarity. In actual systems, the determination of μ_{eff} involves uncertainty. Nevertheless, R is an indicator for correlations. In $CeCu_2Si_2$, it varies from around 0.17 in superconducting samples to 0.44 in nonsuperconducting samples. In UBe_{13} and UPt_3, R is around 0.5, while in UCd_{11}, R is of order 1.7.

Since an important mechanism for the occurrence of heavy Fermions can be seen through analyses of the Kondo effect, let us briefly discuss this effect and then its relationship with heavy Fermions. The Kondo effect is manifested by a logarithmic increase of the resistivity of metals with a small amount of magnetic impurities when the temperature is reduced. It is caused by spin flip scatterings of conduction electrons by the magnetic impurities [7.10]. These impurities are isolated and independently interact with the conduction electrons. Due to the localized magnetic moments of these impurities, the magnetic susceptibility increases towards absolute zero following the Curie–Weiss law. However, below a Kondo temperature T_K, the magnetic moments

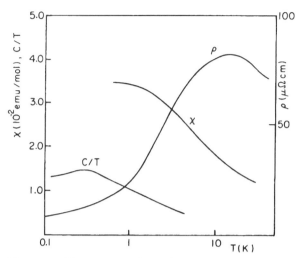

Fig. 7.7. Temperature dependences of the resistivity ρ, susceptibility χ and specific heat c [J/mole] divided by T [K] of CeCu$_6$ [7.11]

may be sealed by conduction electrons so that the susceptibility becomes constant.

When the concentration of magnetic impurities increases two distinctive cases occur. First, in the case of transition elements with a relatively wide d orbital, overlapping of d orbitals can take place when the concentration is increased. As a consequence, the Kondo effect will disappear and will be replaced by a magnetically ordered state or by a spin glass state.

Second, rare earth atoms such as Ce or Yb can be used as impurities. These atoms have characteristic f orbitals which are strongly localized near the atomic core. Hence, even if the concentration of these atoms is increased, the overlapping of the f orbitals may not take place easily. When this happens, rare earth impurities form a *Kondo lattice* in which the Kondo effect occurs at each impurity site as in the dilute case.

On the other hand, it may be advantageous energetically if some of the f electrons move out of their shell. These electrons will then mix with the conduction electron to form a Fermi liquid. As a result *valence fluctuation* will occur. If this change takes place at a low temperature the susceptibility, which increases in inverse proportion to T, can be very large. Heavy Fermion properties may then be expected.

Figure 7.7 illustrates the susceptibility χ, specific heat c divided by T and resistivity ρ of CeCu$_6$ with T_K of around $3K$ [7.11]. These quantities increase as the temperature is decreased, but their temperature variations change below T_K. Note especially that the resistivity shows a marked decrease at around $10K$. Similar decreases have been observed in Ce$_x$La$_{1-x}$Cu$_6$, when the concentration of Ce atoms is above 80%. Apparently, the La atoms play the role of reducing antiferromagnetic spin correlations among the Ce atoms.

While actual heavy Fermion systems show differences in their properties, the disappearance of a magnetic moment is common. Theoretically, the transition from a magnetic to a nonmagnetic state can be expected to be due to energy gain. Therefore, let us try to find whether or not such an energy gain is possible based on the simplest model in which a single magnetic ion is placed in a Fermi liquid of conduction electrons. A similar model has been used for the Kondo effect.

Let us treat such a single f ion based on the following Anderson Hamiltonian [7.12–14],

$$\mathcal{H} = \sum_{k\sigma} \varepsilon(\mathbf{k}) c_{k\sigma}^\dagger c_{k\sigma} + \varepsilon_f \sum_m n_m^f + \frac{U}{2} \sum_{m \neq m'} n_m^f n_{m'}^f$$
$$+ \sum_{km\sigma} [V_{m\sigma}(\mathbf{k}) f_m^\dagger c_{k\sigma} + c.c]. \tag{7.3.3}$$

The first term represents the conduction electrons. ε_f is the position of the f level, which is presumably below the Fermi level. The notation

$$n_m^f = f_m^\dagger f_m$$

represents the number operator of the f electrons in the mth state with m varying from 1 to

$$\nu_f = 2J + 1,$$

which is the degeneracy of the f orbital. U represents the interaction among the f electrons and $V_{m\sigma}(\mathbf{k})$ is the hybridization matrix element between the f and the conduction electrons.

From the Hamiltonian, we separate out the energy associated with only the conduction electrons. Furthermore, we assume that the coupling between the f and the conduction electrons depends only on the magnitude $k = |\mathbf{k}|$ with a constant interaction parameter V so that the Hamiltonian is simplified such that

$$\mathcal{H} = \sum_{km} \varepsilon(k) c_{km}^\dagger c_{km} + \varepsilon_f \sum_m n_m^f + \frac{U}{2} \sum_{m \neq m'} n_m^f n_{m'}^f$$
$$+ V \sum_{km} (f_m^\dagger c_{km} + c_{km}^\dagger f_m) + \mathcal{H}_0. \tag{7.3.4}$$

The interaction parameter V can be given at the Fermi surface so that $V = V(k_F)$. The index m appears in the first term due to the rewriting of the original Anderson Hamiltonian. $\mathcal{H}_0$ is due to the conduction electrons only and may be forgotten in the following argument. We measure $\varepsilon(k)$ from ε_F.

As a trial wave function let us introduce a linear combination of state $|\Phi_0\rangle$, which is completely filled, and states $f_m^\dagger c_{km}|\Phi_0\rangle$ in the form

$$|\psi\rangle = (1 - n_f)^{1/2} \left[1 + \sum_{km} b(k) f_m^\dagger c_{km}\right] |\Phi_0\rangle. \tag{7.3.5}$$

This represents a spin zero singlet state which is normalized so as to yield $1 - n_f$ for $\langle\psi|\psi\rangle$, n_f being the occupancy of the f orbital. $b(k)$ is a hybridization parameter. The magnetic multiplet state will be represented by $|\psi_m\rangle$:

$$|\psi_m\rangle = f_m^\dagger |\Phi_0\rangle. \tag{7.3.6}$$

This state has an additional electron. However, we shall measure all energies from the Fermi energy so that the difference in the numbers of electrons becomes irrelevant. The energy of the magnetic state is expressed as

$$\varepsilon_m = \varepsilon_0 + \varepsilon_f, \tag{7.3.7}$$

where ε_0 is the energy of the filled state. The energy of the singlet state is

$$\varepsilon_s = \langle\psi|H|\psi\rangle / \langle\psi|\psi\rangle. \tag{7.3.8}$$

If this energy is associated with ε_m such that

$$\varepsilon_s = \varepsilon_m + \varepsilon \tag{7.3.9}$$

the sign of ε is crucial to the disappearance of the magnetic moment.

The energy ε_s is determined through variations with respect to n_f and also to the hybrid parameter $b(k)$. The stationary conditions are

$$\varepsilon = -\varepsilon_f + V\sqrt{\nu_f} \sum_k b(k);$$

$$\varepsilon b(k) = V\sqrt{\nu_f} - \varepsilon(k)b(k).$$

Elimination of $b(k)$ between these equations yields

$$\varepsilon = -\varepsilon_f + \nu_f V^2 \sum_k \frac{1}{\varepsilon + \varepsilon(k)}. \tag{7.3.10}$$

An examination of this equation leads to the conclusion that there is a solution $\varepsilon < 0$ for $\varepsilon(k) < 0$. That is, the system gains energy by entering the singlet nonmagnetic state. For $|\varepsilon_f| \gg |\varepsilon|$, this solution is given approximately by

$$\varepsilon = -D\exp[-|\varepsilon_f|/\nu_f g(\varepsilon_F)V^2]$$
$$= -D\exp(-\pi|\varepsilon_f|/\nu_f \Gamma), \tag{7.3.11}$$

where D is a cutoff energy and $\Gamma = \pi g(\varepsilon_F)V^2$, $g(\varepsilon_F)$ being the density of conduction electrons per spin on the Fermi surface. The condition for this solution is that V is small in such a way that $\pi\nu_f g(\varepsilon_F)V^2 \ll |\varepsilon_f|$.

In the presence of a magnetic field, the degeneracy of the f level will be lifted such that ε_f becomes $\varepsilon_f - g_J\mu_B mH$ where $-J \leqq m \leqq J$. We arrive at a variational equation similar to (7.3.10) with the replacement

$$\nu_f \sum_k \to \sum_{km}.$$

The magnetic susceptibility is obtained by differentiating ε twice with respect to H. The result is

$$\chi = (g_J \mu_B)^2 \frac{J(J+1)}{3} \frac{1}{\nu_f \Gamma} \frac{n_f^2}{1-n_f}. \tag{7.3.12}$$

The specific heat constant can be shown to be inversely proportional to $|\varepsilon|$. Reexpressing this quantity, we find that γ is given by

$$\gamma = \frac{1}{3}\pi^2 k_B^2 \frac{1}{\nu_f \Gamma} \frac{n_f^2}{1-n_f}. \tag{7.3.13}$$

Therefore, χ and γ increase as n_f approaches 1.

Although the above model has interesting features, it is definitely an over-simplification. In actual heavy Fermion systems, the ions form a crystal, and a certain consideration of their magnetic interaction is necessary. Such a consideration has been made based on a model with two ions, a mean field approximation, perturbation, etc. Although the details of all these attempts cannot be delivered in a short space, let us outline a mean field approach.

When the ions form a crystal, their site index i, locations $\mathbf{R}_i$ and a band index n of the conduction electrons must be introduced in (7.3.3). For instance, the second term is replaced by $\varepsilon_{fm} f_m^\dagger(i) f_m(i)$ and the last term by $N^{-1/2} V_{m\sigma}(\mathbf{k},n) c_{kn\sigma}^\dagger f_m(i) e^{-i\mathbf{k}\cdot\mathbf{R}_i}$, where N is the total number of f ions. Since this so-called Anderson lattice Hamiltonian is complicated, let us consider the limiting case $U \to \infty$. In this limit, each lattice site has at most one f electron due to their high mutual repulsion. Hence, the probability that the f orbital at site i is empty can be represented by the operator [7.12–14]

$$b^\dagger(i)b(i) = 1 - \sum_m f_m^\dagger(i) f_m(i). \tag{7.3.14}$$

Note that the second term is either 1 (one f electron), or 0 (no electron). It turns out that the U term in the Hamiltonian can be eliminated in terms of the operators $b(i)$ obeying Bose statistics. The Hamiltonian becomes

$$\mathcal{H} = \mathcal{H}_b + \sum_{mi} \varepsilon_{fm} f_m^\dagger(i) f_m(i)$$
$$+\frac{1}{\sqrt{N}} \sum_{imkn\sigma} [V_{m\sigma}(\mathbf{k},n) b^\dagger(i) c_{kn\sigma}^\dagger f_m(i) e^{-i\mathbf{k}\cdot\mathbf{R}_i} + c.c.], \tag{7.3.15}$$

where $\mathcal{H}_b$ is the band Hamiltonian of the conduction electrons. Note the appearance of the b operators in the mixing term. The requirement that $b^\dagger(i)b(i)$ is either 1 or 0 can be accommodated by introducing Lagrange multipliers $\Lambda(i)$ into (7.3.14) and adding the resulting expression to the Hamiltonian.

In a mean field approximation, $b^\dagger(i)$ may be replaced by an average

$$\langle b^\dagger(i) \rangle = r.$$

Correspondingly, a site-independent Lagrange multiplier Λ is introduced into the Hamiltonian to ensure the above subsidiary condition. This Hamiltonian is

$$\mathcal{H} = \mathcal{H}_b + \sum_{mk}(\varepsilon_{fm} + \Lambda)f^\dagger_{km}f_{km} + \sum_{nmk\sigma} r[V_{m\sigma}(\mathbf{k},n)c^\dagger_{kn\sigma}f_{km} + c.c.]$$
$$+\Lambda N(r^2 - 1). \tag{7.3.16}$$

Note that f-electron energies ε_{fm} are shifted by Λ and that the hybridization matrix element $V_{m\sigma}$ is renormalized in terms of the mean field parameter r. The last term in the Hamiltonian is a c number. It is due to the subsidiary condition

$$r^2 = 1 - n_f. \tag{7.3.17}$$

Here,

$$n_f = \sum_{km}\langle f^\dagger_{km}f_{km}\rangle$$

is the f-orbital occupation with $<\dots>$ representing a mean field average. Taking such a mean field average of the Hamiltonian and minimizing the result with respect to r, we arrive at

$$\sum_{nkm\sigma} V_{m\sigma}(\mathbf{k},n)\langle c^\dagger_{kn\sigma}f_{km}\rangle + N\Lambda r = 0. \tag{7.3.18}$$

Equations (7.3.17) and (7.3.18) constitute self-consistent equations. We may multiply (7.3.18) by r and use an effective coupling parameter $V^*_{m\sigma} = rV_{m\sigma}$. Since the result is quadratic in f_{km}, one can treat the equations by expanding f_{km} and $c_{kn\sigma}$ into a new set of operators.

Let us assume for explicit results that each of the ν_f orbitals hybridizes with one of the conduction bands. Hence, the total number of conduction bands is also ν_f instead of $2\nu_f$. If the hybridization takes place in the same way, one can drop the band index n and orbital index m. The problem is then reduced to hybridization of two bands. Without further details the specific heat is given approximately by

$$c = \frac{\pi^2 k_B^2 T}{3}\frac{n_f}{\Delta\varepsilon_f}(2J + 1), \tag{7.3.19}$$

where $\Delta\varepsilon_f$ is the renormalized ε_f measured from the Fermi energy. The magnetic

$$\chi_p = \mu^2_{eff}\frac{1}{\nu_f}\frac{n_f}{\Delta\varepsilon_f}\frac{(2J+1)}{3}. \tag{7.3.20}$$

Note that $\Delta\varepsilon_f$ appears in the denominator in these expressions and that, for a small energy difference, γ and χ_p can be large. Note also in this respect that (7.3.12) and (7.3.13) can also be expressed with $|\varepsilon|$ in the denominator.

The above treatment neglects valence fluctuations. A consideration of these fluctuations has been shown to reduce the specific heat by

$$\Delta c = -(1/3)\pi^2 k_B^2 T n_f / \Delta \varepsilon_f.$$

Therefore, with this correction, the Sommerfeld–Wilson ratio is

$$R = \frac{\nu_f}{\nu_f - 1}. \qquad (7.3.21)$$

Hence, in the limit $\nu_f \to \infty$, $R = 1$, as should be the case in the classical limit.

We now turn our attention to electron–electron interactions, and note that in liquid ^{3}He, spin-triplet pairing takes place. That is, the strong short-range repulsion is overcome by forming such pairs. The strong local repulsion (large U) among f electrons and the existence of spin fluctuations, as is indicated by the appearance of a $T^3 \ln T$ term in the specific heat of UPt$_3$, are similar to the case of liquid ^{3}He. Hence, a similar formation of electron pairs may be possible in heavy Fermion systems [7.11]. In fact, UPt$_3$ is a superconductor below 0.5 K, just as liquid ^{3}He shows superfluidity. It is perhaps due to p-wave pairing rather than s-wave pairing in the ordinary superconductors. Also, the boson field b which has been assumed to be constant in the mean field theory may actually be fluctuating $b(k)$. Such fluctuations may mediate pairing. Interestingly, there are other heavy Fermion systems such as UBe$_{13}$ and CeCu$_2$Si$_2$ which also become superconducting. Because there are differences in the specific heat behaviors of these heavy Fermions the mechanism for their superconductivity may not be the same [7.12].

8. Magnetoconductivity in Two Dimensions

The magnetoconductivity and cyclotron resonance of 2D electron systems show peculiar density and frequency variations.

8.1 Magnetoconductivity of 2D Electrons

It has been observed that the magnetoconductance of 2D systems shows strong density and magnetic field variations. The conductance as a function of electron density increases slowly at first, then rapidly until finally it reaches an asymptotic value. A magnetic field tends to suppress the conductance such that the higher the field the lower the asymptotic value. When a magnetic field reaches a certain strength, the conductance shows significant Shubnikov–de Haas oscillations in the density range where the conductivity increases rapidly. That is, these oscillations are not very visible on either side of this density range.

The theory of conductivity in 2D has been developed since around 1971, after the development of linear response theory for 3D. In this section we shall describe a more recent theory [8.1] and discuss the above experimental variations. We shall explicitly write $\hbar$ and the electron mass m in the equations for conductivity-related problems.

The conductivity is determined by a current–current correlation function according to linear response theory. In 2D it is a tensor with the longitudinal component σ_{xx} or σ_{yy}, and the transverse component σ_{xy} or σ_{yx}. In the isotropic case one can adopt

$$\sigma_{xx} = \sigma_{yy}, \quad \sigma_{xy} = -\sigma_{yx}.$$

For absolute zero, these components can be expressed as

$$\sigma_{xx} = \frac{i\hbar e^2}{A} \langle \mathrm{Tr}\{v_x \delta(\varepsilon - \mathcal{H})v_x \delta(\varepsilon - \mathcal{H})\}\rangle; \tag{8.1.1}$$

$$\sigma_{xy} = \frac{ec\partial N(\varepsilon, 0)}{A\partial H}$$

$$+ \frac{i\hbar e^2}{2A} \langle \mathrm{Tr}\{v_x G^+(\varepsilon)v_y \delta(\varepsilon - \mathcal{H}) - v_x \delta(\varepsilon - \mathcal{H})v_y G^-(\varepsilon)\}\rangle, \tag{8.1.2}$$

where A is the surface area, $v_i (i = x, y)$ are the velocity components, and $\mathcal{H}$ is the Hamiltonian. At low temperatures, the electron energy ε can be replaced by the Fermi energy ε_{F}, $N(\varepsilon, 0)$ is the number of electrons with energies less than ε,

$$N(\varepsilon, 0) = \int_{-\infty}^{\varepsilon} \langle \delta(\varepsilon' - \mathcal{H}) \rangle \mathrm{d}\varepsilon', \tag{8.1.3}$$

and $G^{\pm}$ are the resolvents, often called the Green's functions, defined by

$$G^{\pm}(z) = (z - \mathcal{H} \pm \mathrm{i}0)^{-1}. \tag{8.1.4}$$

The conductivity at finite temperatures is obtained from the above formulas as

$$\sigma_{ij}(\varepsilon_{\mathrm{F}}, T) = -\int_{-\infty}^{\infty} \frac{\partial f}{\partial \varepsilon} \sigma_{ij}(\varepsilon, 0) \mathrm{d}\varepsilon, \tag{8.1.5}$$

where $f(\varepsilon)$ is the Fermi distribution function. The resolvents depend on the Hamiltonian $\mathcal{H}$, which may be assumed to consist of the ideal Hamiltonian $\mathcal{H}_0$ and impurity potential $\mathcal{H}_1$.

In the coherent potential approximation [8.2] (abbreviated as CPA), new resolvents are introduced by averaging $G^{\pm}$ over the impurity configurations. The concept behind this averaging process comes from the expectation that the averaged resolvents, to be denoted by $\langle G^{\pm}(z) \rangle$, have the same structure as the resolvent of the ideal case without impurities, except that the energy is slightly modified. The ideal resolvent is of the form

$$G_0 = (z - \mathcal{H}_0)^{-1}. \tag{8.1.6}$$

The energy modification is expressed by a self-energy Σ such that

$$\langle G(z) \rangle = (z - \mathcal{H}_0 - \Sigma)^{-1}. \tag{8.1.7}$$

In general the self-energy is expected to be complex

$$\Sigma = \Delta - \mathrm{i}\Gamma. \tag{8.1.8}$$

The real part Δ represents an energy shift. The imaginary part Γ causes level broadening and is related to the mean lifetime τ by

$$\tau = \hbar/2\Gamma. \tag{8.1.9}$$

Let us now discuss how the self-energy can be determined. In multiple scattering theory, a T matrix is introduced such that

$$G = G_0 + G_0 T G_0. \tag{8.1.10}$$

When averaged over impurity configurations, this equation becomes

$$\langle G \rangle = G_0 + G_0 \langle T \rangle G_0 \tag{8.1.11}$$
$$= G_0 + G_0 \Sigma \langle G \rangle. \tag{8.1.12}$$

Equation (8.1.12) follows from (8.1.7) and the definition of G_0. We make use of the identity

$$G = G_0 + G_0(\mathcal{H} - \mathcal{H}_0)G. \tag{8.1.13}$$

Combining this with (8.1.10), we arrive at

$$T = \frac{\mathcal{H} - \mathcal{H}_0}{1 - G_0(\mathcal{H} - \mathcal{H}_0)}. \tag{8.1.14}$$

On the other hand we obtain from (8.1.11) and (8.1.12)

$$\Sigma = \langle T\rangle(1 + G_0\langle T\rangle)^{-1}. \tag{8.1.15}$$

Hence a combination of these two equations can in principle determine Σ.

For a short-range impurity potential characterized by a strength V_0 and for a small impurity fraction c, a convenient self-consistent equation has been given as [8.2]

$$\Sigma = cV_0 + \frac{c(1-c)V_0^2 K(z-\Sigma)}{1 + [\Sigma + (c-1)V_0]K(z-\Sigma)}, \tag{8.1.16}$$

where

$$K(z - \Sigma) = \mathrm{Tr}\{\langle G(z)\rangle\}. \tag{8.1.17}$$

Equations (8.1.11, 16 and 17) represent a set of equations to determine Σ.

Let us consider the particular case of a 2D electron system in a magnetic field which is so strong that the Landau level mixing can be ignored and the resolvents can be defined per each Landau level. Neglecting the energy shift given by the first term in (8.1.16) and all other Landau levels but a particular one with index n, we may express the self-energy for this level by

$$\Sigma_n = \tfrac{1}{4}\Gamma^2\langle G_n\rangle, \tag{8.1.18}$$

where the numerical factor $1/4$ and a parameter Γ have been introduced for a later convenience. Equation (8.1.12) becomes

$$\langle G_n^2\rangle - \frac{4(\varepsilon - \varepsilon_n)}{\Gamma^2}\langle G_n\rangle + \frac{4}{\Gamma^2} = 0.$$

The self-energy for the nth level is then given by

$$\Sigma_n = \Im\langle G_n\rangle$$

$$= \frac{2}{\Gamma}\left[1 - \left(\frac{\varepsilon - \varepsilon_n}{\Gamma}\right)^2\right]^{1/2}. \tag{8.1.19}$$

Thus the parameter Γ represents the width of the nth level. It is expected that Γ is proportional to the strength of a short range potential and Γ^2 to the concentration of the impurities. The above form agrees with that obtained by Ando et al. [8.3].

Let us now consider the case of a strong magnetic field in a somewhat general way [8.1]. The function K can be expressed by a sum of the contributions from the Landau levels,

$$K(z - \Sigma) = \frac{m\omega_c \Omega}{2\pi\hbar} \sum_{n=0}^{n_M} \frac{1}{z - \hbar\omega_c(n + 1/2) - \Sigma}, \tag{8.1.20}$$

where $\Omega = 2\pi\hbar^2/m\varepsilon_M$, determined by the maximum energy ε_M is the area per single impurity site. The energy ε_M is related to n_M by

$$n_M = \frac{\varepsilon_M}{\hbar\omega_c} - \frac{1}{2}.$$

n_M is a cut-off parameter which must be introduced in order to avoid a divergence which appears in taking the trace for K. For large values of ε_M, the last term $1/2$ on the right side of this equation can be neglected.

The function $K(z - \Sigma)$ can be split into a smooth part K_0 and an oscillating part ΔK,

$$K(z - \Sigma) = K_0(z - \Sigma) + \Delta K(z - \Sigma), \quad \Sigma = \Delta - \mathrm{i}\Gamma. \tag{8.1.21}$$

Under the condition that $0 \ll |z - \Sigma| \ll \varepsilon_M$, and for $z = \varepsilon + \mathrm{i}0$, we obtain

$$K_0(z) = R_0(\varepsilon) + \mathrm{i}F_0(\varepsilon); \tag{8.1.22}$$

$$R_0(\varepsilon) = \frac{1}{\varepsilon_M}(\ln|\varepsilon| - \ln|\varepsilon - \varepsilon_M|); \tag{8.1.23}$$

$$F_0(\varepsilon) = -\frac{\pi}{\varepsilon_M}[\theta(\varepsilon) - \theta(\varepsilon - \varepsilon_M)], \tag{8.1.24}$$

where θ is a step function. The functions R_0 and F_0 are plotted against $\varepsilon/\varepsilon_M$ in Fig. 8.1. Note that the former shows a logarithmic variation, while the latter is constant in the interval $0 < \varepsilon < \varepsilon_M$.

The oscillating part of the auxiliary function K is found to be

$$\Delta K = -\frac{2\pi\mathrm{i}}{\varepsilon_M} \sum_{s=1}^{\infty} \exp\left(\frac{2\pi\mathrm{i}s}{\hbar\omega_c}(z - \Sigma) - \mathrm{i}\pi s\right). \tag{8.1.25}$$

The density of states, given by the imaginary part of K, consists of the following smooth and oscillating parts:

$$g_0(\varepsilon) = -\frac{1}{\pi}\Im\{K_0\}; \tag{8.1.26}$$

$$\Delta g(\varepsilon) = \frac{2}{\varepsilon_M} \sum_{s=1}^{\infty} \exp(-\pi s/\omega_c\tau)\cos\left(\frac{2\pi s}{\hbar\omega_c}[\varepsilon - \Delta(\varepsilon)] - \pi s\right). \tag{8.1.27}$$

When $\omega_c\tau$ is of order 1, only the first few terms of the s-sum are important. Generally, the oscillating parts can be considered to be small in comparison with the corresponding non-oscillating parts.

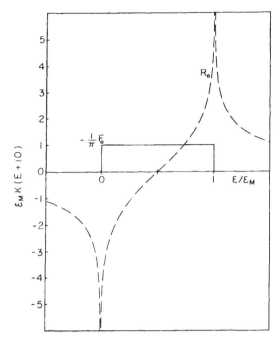

Fig. 8.1. Real and imaginary parts of the smooth part of the function given by (8.1.20)

Equation (8.1.16) can be considerably simplified when $c^2 \ll 1$ and $|\Sigma_0 + cV_0| \ll V_0$. We obtain for weak fields

$$\Delta_0 = -\frac{c}{R_0}\left(1 + \frac{1}{V_0 R_0} + \cdots\right), \tag{8.1.28}$$

$$\Gamma_0 = -\frac{cF_0}{R_0^2}\left(1 + \frac{2}{V_0 R_0} + \cdots\right). \tag{8.1.29}$$

Under the condition that $\Delta g \ll g_0$, the conductivity components can be given by [8.1]

$$\sigma_{xx} = \frac{\sigma_0}{1 + \omega_c^2\tau_0^2}\left(1 + \frac{2\omega_c^2\tau_0^2}{1 + \omega_c^2\tau_0^2}\frac{\Delta g}{g_0}\right), \tag{8.1.30}$$

$$\sigma_{xy} = -\frac{\sigma_0\omega_c^2\tau_0^2}{1 + \omega_c^2\tau_0^2}\left(1 - \frac{3\omega_c^2\tau_0^2 + 1}{\omega_c^2\tau_0^2(\omega_c^2\tau_0^2 + 1)}\frac{\Delta g}{g_0}\right). \tag{8.1.31}$$

Here

$$\sigma_0 = \frac{e^2}{m}N_0\tau_0 \tag{8.1.32}$$

is the conductivity in the absence of a magnetic field, and the electron mass m is displayed explicitly. For convenience, we shall do the same for most of the final results concerning magnetoconductivity. N_0 is the number of carriers with energies below ε in the absence of a magnetic field and is given by

$$N_0(\varepsilon) = \frac{1}{\Omega}(\varepsilon - \Delta_0)g_0(\varepsilon), \qquad (8.1.33)$$

where

$$g_0(\varepsilon) = -\frac{1}{\pi}F_0(\varepsilon). \qquad (8.1.34)$$

The relaxation time τ_0 is obtained from (8.1.9) through the smooth part of the imaginary part of the self-energy.

The magnetoconductivity depends on $\Delta g/g_0$ which oscillates with the period $\hbar e/[mc(\varepsilon - \Delta_0)]$, which depends on ε. Therefore, it is useful to write

$$(\varepsilon - \Delta_0)/\hbar\omega_c = \varepsilon/\hbar\omega_c^*,$$

where ω_c^* is an effective cyclotron frequency defined by

$$\omega_c^* = eH/m^*c.$$

The effective mass m^* is given by

$$m^* = m[1 - \Delta_0(\varepsilon)/\varepsilon]. \qquad (8.1.35)$$

At low temperatures, ε here may be replaced by ε_F. This replacement leads to a finite temperature expression

$$\frac{\Delta g}{g_0} = 2\sum_{s=1}^{\infty} \exp(-\pi s/\omega_c\tau_0)$$

$$\times \frac{2\pi^2 s k_B T/\hbar\omega_c^*}{\sinh(2\pi^2 s k_B T/\hbar\omega_c^*)} \cos\left(\frac{2\pi s\varepsilon_F}{\hbar\omega_c^*} - \pi s\right). \qquad (8.1.36)$$

In order to take into consideration the effectiveness of impurity scattering, which is expected to vary with the electron's energy, we introduce a model in which the impurities cause shallow localized states with energies just below the conduction threshold. The actual wave function describing the conduction states can be considered to be orthogonal to these impurity states. For the impurity effect, only the most effective localized state $|b\rangle$ with energy ε_b may be used, its pseudopotential being given by [8.4]

$$V_i = v + |b\rangle(\varepsilon - \varepsilon_b)\langle b| + |\gamma b\rangle\frac{1}{\varepsilon - \varepsilon_b}\langle b|, \qquad (8.1.37)$$

where v is the impurity potential, $|b\rangle$ denotes the localized state, and γ is a parameter which hybridizes the localized and conduction states. The form given by (8.1.37) for a pseudopotential is somewhat general. The matrix element of the pseudopotential represents an effective strength of the impurity effect. It is given by

$$V_0 = \langle 0|V_i|0\rangle. \qquad (8.1.38)$$

On the other hand, in the first approximation (8.1.28) and (8.1.29) yield

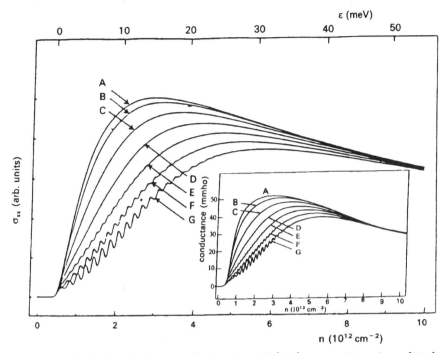

Fig. 8.2. Shubnikov–de Haas oscillations in a Si [100] accumulation layer [8.1a]. Magnetic fields are: A: 0, B: 0.36, C: 0.72, D: 1.08, E: 1.44, F: 1.80, G: 2.16 T. *Inset*: from [8.5a]

$$\Gamma_0 = \pi c V_0^2 / \varepsilon_F. \tag{8.1.39}$$

Using (8.1.9, 38 and 39), we can show that τ_0 is given by

$$\tau_0 = a \left(\frac{\varepsilon_\sigma \varepsilon}{\varepsilon_\sigma \varepsilon^2 + (\varepsilon_\sigma^2 - 3\varepsilon_\tau^2)\varepsilon + \varepsilon_\sigma \varepsilon_\tau^2} \right)^2, \tag{8.1.40}$$

where a is a constant inversely proportional to the impurity concentration which appears together with two other constants, ε_τ and ε_σ, in correspondence to the three constants in (8.1.37). These parameters may be obtained from experimental data.

The density and magnetic field variations of the magnetoconductance in Si[100] accumulation layers have been determined experimentally by *Fang* et al. [8.5]. Their data on a Si [100] accumulation layer at 1.9 K is illustrated in Fig. 8.2 (inset). The larger curves represent theoretical results based on (8.1.30, 36, 40). Both the theoretical and experimental curves show the density dependence of the magnetoconductivity in increasing order in magnetic fields from A to G. The field strengths in tesla are

$$A:0; \quad B = 0.36; \quad C = 0.72; \quad D = 1.08; \quad E = 1.44; \quad F = 1.80; \quad G = 2.16.$$

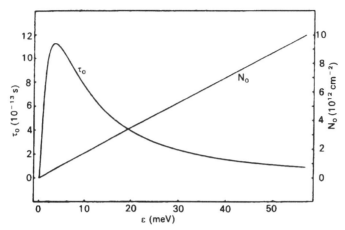

Fig. 8.3. Dependences of the relaxation time τ_0 and effective number of carriers N_0 on the Fermi energy [8.1a]

The temperature is $1.9\,\mathrm{K}$ and the effective mass is $m = 0.21m_0$ in both cases. Note that the oscillatory behavior appears for the field above $1.08T$. Note also that the position of the conductance maximum shifts to a larger electron density as the magnetic field increases.

The overall variation of these curves depends on the zero field case. The zero-field conductivity σ_0 is determined by a product of N_0 and τ_0, which are energy dependent. Since the shift Δ_0 is small, N_0 is proportional to energy as seen from (8.1.33). On the other hand, the relaxation time reaches a maximum at $\varepsilon = \varepsilon_\tau = 3.5\,\mathrm{meV}$ independently of the values of ε_σ and a. The product of τ_0 with N_0 exhibits a maximum which is determined by ε_σ independent of the impurity concentration. These properties facilitate the determination of the three parameters.

The energy dependences of τ_0 and N_0 are illustrated in Fig. 8.3. The abscissa ε represents the Fermi energy. This graph helps to interpret the cause of the maximum in the conductance. It is due to the balance between the increase in the number of carriers and the decrease in their mobility. The scattering time increases with ε at low energies since electrons move more freely as their energy becomes higher than the localization energy at the bottom of the conduction band. The decrease in τ_0 is fast at high energies causing the decrease in the conductivity. At high energies electrons can get closer to the scattering centers and cause an increase in the scattering rate.

8.2 Memory Function Formalism

According to linear response theory the conductivity tensor is given by the current–current correlation function as in (8.1.1). The essential part of this correlation function can be expressed as

$$\sigma_{\mu\nu}(\omega) = \frac{1}{A} \int_0^\infty dt e^{i\omega t} [J_\nu^\dagger, J_\mu(t)], \tag{8.2.1}$$

where A is the size of the system, and

$$[J_\mu^\dagger, J_\nu] = \int_0^\infty d\tau \langle e^{\tau H} J_\mu^\dagger e^{-\tau H} J_\nu \rangle. \tag{8.2.2}$$

$\langle \ldots \rangle$ represents a thermal average and $\dagger$ denotes the Hermitian conjugation; the Greek letters such as μ and ν are used to specify the components of vectors. In the low-frequency limit the conductivity can be evaluated by using the Green's function. However, the static limit of the result may not be automatically taken for a finite frequency because of the appearance of a singularity. In the opposite high frequency limit a perturbation method may be used.

Apart from these two limiting cases, it is generally difficult to derive the conductivity explicitly for the entire frequency range. Therefore, in 1972 *Götze* and *Wölfle* [8.6] adopted a form for the conductivity that yields a reasonable approximation in the complete frequency range. The form is

$$\sigma(\omega) = \frac{i}{4\pi} \frac{\omega_p^2}{[\omega + M(\omega)]}. \tag{8.2.3}$$

$M(\omega)$ is called the *memory function*, and $\omega_p = (4\pi n e^2/m)^{1/2}$ is the plasma frequency. They then derived $M(\omega)$ explicitly in first order in the impurity concentration and the electron-phonon coupling. In this case, $M(\omega)$ becomes pure imaginary.

By identifying the inverse of such an imaginary memory function as a relaxation time τ, the dynamical conductivity in the absence of a magnetic field can be expressed in the real Drude form given by

$$\sigma(\omega) = \frac{ne^2\tau/m}{1 + \omega^2\tau^2}. \tag{8.2.4}$$

This form has been tested for electrons in Si inversion layers by *Allen* et al.[8.5b]. They used four carrier densities ranging from a high of 15.3×10^{11} cm^{-2} to a low of 1.8×10^{11}. At densities above 4.5×10^{11}, the Drude form was found to reproduce the data well. At lower densities, the formula reproduced the data points only for $\omega > E_A$, where E_A is the activation energy. Such an activation energy was not observed in the metallic regime. The Drude form was found to show significant deviations at microwave frequencies. The appearance of the activated conductivity at low densities may be due to excitation of electrons across a gap to conducting states. The appearance of such a gap may be due to electron–electron interaction.

What happens to the dynamical conductivity if a magnetic field is applied? *Ting* et al. [8.7] showed in 1977 that the dynamical magnetoconductivity of 2D electrons can be given approximately for a small impurity concentration by

$$\sigma_\pm(\omega) = \frac{ine^2}{m(\omega \mp \omega_c)}\left(1 - \frac{M(\omega)}{\omega \mp \omega_c}\right), \tag{8.2.5}$$

where $\omega_c = eH/mc$ is the cyclotron frequency, and

$$\sigma_\pm(\omega) = \sigma_{xx}(\omega) \pm i\sigma_{xy}(\omega). \tag{8.2.6}$$

This equation exhibits cyclotron resonance explicitly and can be considered as a generalization of (8.2.3). However, the second term shows a stronger cyclotron divergence than the first free-electron term. This problem has been removed by the exact theory developed by *Shiwa* and *Isihara* [8.8] in 1983. The outline of their approach is as follows.

The Hamiltonian of a system in a uniform magnetic field $\mathbf{H}$ is given by

$$\mathcal{H} = \mathcal{H}_0 + \mathcal{H}_1, \tag{8.2.7}$$

where

$$\mathcal{H}_0 = \sum_i \frac{\pi_i^2}{2m} + \sum_{i>j} u(\mathbf{r}_i - \mathbf{r}_j), \tag{8.2.8}$$

and where $\pi_i = \mathbf{p}_j + (e/c)\mathbf{A}_i$ is the momentum of the ith electron with the vectorpotential $\mathbf{A}_i$, and $u(r)$ is the Coulomb interaction. The perturbation part of the Hamiltonian is given by the impurity potential

$$\mathcal{H}_1 = \sum_{i,j} v(\mathbf{r}_i - \mathbf{R}_j), \tag{8.2.9}$$

where the $\mathbf{R}_j$ are the impurity positions. The current vector is given by

$$\dot{J}_\mu = -(e/mc)\varepsilon_{\mu\nu\lambda}J_\nu H_\lambda - F_\mu, \tag{8.2.10}$$

where $\varepsilon_{\mu\nu\lambda}$ is the unit axial tensor, H_λ is the λ-component of the magnetic field, and it is understood that a summation is taken over any repeated Greek subscript. $\mathbf{F}$ is the force due to the impurities defined by

$$\mathbf{F} = \sum_{i,j} \frac{\partial}{\partial \mathbf{R}_j} v(\mathbf{r}_i - \mathbf{R}_j). \tag{8.2.11}$$

The conductivity components can be written formally as

$$\sigma_{\mu\nu}(\omega) = A^{-1}[J_\nu^\dagger, J_\lambda] \cdot \Xi_{\lambda\mu}(\omega), \tag{8.2.12}$$

where

$$\Xi(\omega) = \int_0^\infty dt e^{i\omega t}[\mathbf{J}^\dagger, \mathbf{J}]^{-1} \cdot [\mathbf{J}^\dagger, \mathbf{J}(t)], \tag{8.2.13}$$

where the dot denotes a matrix multiplication, $(\mathbf{J}^\dagger, \mathbf{J})$ being a 3×3 square matrix formed by the column and row vectors $\mathbf{J}^\dagger$ and $\mathbf{J}$ as indicated in (8.2.2). From the equation of motion which is related to $\dot{J}$ in (8.2.10), it can be shown for 2D that

$$\Xi(\omega) = [-i\omega\mathbf{1} + i\Omega + \Gamma(\omega)]^{-1}, \tag{8.2.14}$$

1 is a unit matrix, and

$$i\Omega_{\mu\nu} = -(e/mc)\varepsilon_{\mu\nu\lambda}H_\lambda, \tag{8.2.15}$$

$$i\Gamma_\pm(\omega) = \frac{M(\omega)}{1 - [M(\omega)/(\omega \mp \omega_c)]}, \tag{8.2.16}$$

where

$$M(\omega) = \frac{i}{Nm} \int_0^\infty dt e^{i\omega t} [F_x^\dagger(t), F_x]. \tag{8.2.17}$$

That is, the memory function is given in terms of the force–force correlation function.

The final conductivity expression is

$$\sigma_\pm(\omega) = \frac{ine^2/m}{\omega \mp \omega_c + M(\omega)/[1 - M(\omega)/(\omega \mp \omega_c)]}. \tag{8.2.18}$$

This formula is exact within linear response theory. Note that it is reduced to the approximate form given by (8.2.5) for small $M(\omega)$. In particular, to first order inimpurity concentration the memory function is given by

$$M(\omega) = \frac{n_i}{2\omega m N} \sum_q q^2 |v(q)|^2 |[\chi_q(0) - \chi_q(\omega)], \tag{8.2.19}$$

where n_i is the impurity concentration, and

$$\chi_q(\omega) = i \int_0^\infty dt e^{i\omega t} \langle [\rho_q^\dagger(t), \rho_q(0)] \rangle \tag{8.2.20}$$

is the density–density correlation function. In terms of this correlation function, the dielectric function of the system can be expressed as

$$\varepsilon^{-1}(q,\omega) = 1 - u(q)\chi_q(\omega). \tag{8.2.21}$$

Thus, the memory function can be written as

$$M(\omega) = \frac{n_i}{2\omega m N} \sum_q q^2 |v(q)|^2 u(q)^{-1} [\varepsilon^{-1}(q,\omega) - \varepsilon^{-1}(q,0)]. \tag{8.2.22}$$

For small impurity effects, the memory function is small and can be neglected in comparison with ω_c. In the first approximation and in the static limit,(8.2.18) yields

$$\sigma_{xx}(0) = \frac{ne^2}{m} \frac{M''(0)}{\omega_c^2}. \tag{8.2.23}$$

Here, $M''(0)$ is the imaginary part of the memory function in the static limit. More explicitly

$$\sigma_{xx}(0) = \frac{n_i c^2}{2e^2 H^2} \int_0^\infty dq q^4 |v(q)|^2 \Im \left(\frac{\partial}{\partial\omega} \frac{1}{\varepsilon(q,\omega)} \right)_{\omega=0}, \tag{8.2.24}$$

H is the magnetic field and c is the speed of light. This expression can be extended to a finite temperature by applying (8.1.5) [8.9].

Let us now derive another useful formula for $\sigma_{xx}(0)$. For this purpose, it is convenient to make use of the expression for the imaginary part $M''(\omega)$ of the memory function which is due to *Götze* and *Hajdu* [8.10]:

$$M''(\omega) = \frac{n_i}{2\pi nm} \int dq q^2 |v(q)|^2 \sum_{n,n'} |\rho_{nm'}|^2$$

$$\times \int d\varepsilon \frac{f(\varepsilon - \omega) - f(\varepsilon)}{\omega} \Im\{\langle G_n(\varepsilon)\rangle\}\Im\{\langle G_{n'}(\varepsilon - \omega)\rangle\}, \quad (8.2.25)$$

where

$$\rho_{nn'}(q) = \langle n|e^{i\mathbf{q}\cdot\mathbf{r}}|n'\rangle. \quad (8.2.26)$$

The matrix element on the right side can be taken in terms of the known harmonic oscillator eigenfunctions. Note that

$$|\rho_{n,n+s}(q)|^2 = \frac{n!}{(n+s)!}\left(\frac{lq}{\sqrt{2}}\right)^{2s} \exp(-l^2 q^2/2)[L_n^s(l^2 q^2/2)]^2,$$

where l is the magnetic length. To first order in a short range impurity potential,(8.1.20) can be used for $\Im\{\langle G_n\rangle\}$. The parameter Γ for this case is

$$\Gamma^2 = 2n_i V_0^2/\pi l^2, \quad (8.2.27)$$

where l is the magnetic length and V_0 is the strength of the impurity potential. Note that $1/2\pi l^2 = eH/ch$ is the natural number of states per unit area. In taking the matrix element, the following integral formula is useful,

$$\frac{1}{4\pi} \int_0^\infty dq q^3 \exp(-q^2 l^2/2)[L_n(q^2 l^2/2)]^2 = \frac{1}{\pi l^4}\left(n + \frac{1}{2}\right).$$

Using the static limit $M''(0)$ and the ordinary units we arrive at the following formula for static conductivity $\sigma_{xx}(0)$ which is correct to first order in the impurity effect:

$$\sigma_{xx}(0) = \frac{e^2}{\pi^2\hbar}\sum_n \left(n + \frac{1}{2}\right)\int d\varepsilon \left(-\frac{\partial f(\varepsilon)}{\partial\varepsilon}\right)\left[1 - \left(\frac{\varepsilon - \varepsilon_n}{\Gamma}\right)^2\right]. \quad (8.2.28)$$

The right side shows that each Landau level contributes to the static magnetoconductivity in proportion to its level index. Such a proportionality is clear in the works of *Bastin* et al. and others [8.11]. The above form was first obtained by *Ando* and *Uemura* [8.12].

8.3 Anomalous Cyclotron Resonance

Since around 1974, cyclotron resonance has been used extensively to probe electrons in Si inversion layers and some other 2D systems. The density of electrons in these system can be in the range from around 10^{10} to $10^{13}\,\mathrm{cm}^{-2}$. This wide range helps to study the density dependence of cyclotron resonance. The density dependence can be due to electron correlations and becomes anomalous at low densities.

In principle, cyclotron resonance provides information concerning the two important parameters, effective mass and relaxation time, which characterize the dynamical conductivity. For small impurity effects, the conductivity can be expressed by

$$\sigma(\omega) = \frac{ine^2/m}{\omega \mp \omega_c + M(\omega)}.$$

(8.3.1)

In general, the memory function $M(\omega)$ consists of real and imaginary parts,

$$M(\omega) = M'(\omega) + iM''(\omega).$$

(8.3.2)

The real part M' can be derived from the imaginary part M'' by using the Kramers–Kronig relation. From (8.3.1), the cyclotron effective mass m^* and relaxation time τ are given formally by

$$m^* = m[1 + M'(\omega)/\omega],$$

(8.3.3)

$$\tau = M''^{-1}[1 + M'(\omega)/\omega].$$

(8.3.4)

The frequency and density variations of M' and M'' have been determined for Si inversion layers in which the density of electrons can be varied easily. It has been found that the memory function shows different behaviors in accordance with $\nu \gtrless 1$, where ν is the filling factor defined by

$$\nu = n\left(\frac{\hbar c}{eH}\right).$$

(8.3.5)

In the region $\nu < 1$, only the lowest Landau level is partially filled. For such low densities, Coulomb effects are expected to be strong.

Shiwa and *Isihara* [8.13] developed a theory which goes beyond the RPA for the ν dependences of M' and M''. They adopted *Götze*'s method [8.14] to feedback density fluctuations into the density–density correlation function which characterizes the memory function. The memory function is given to first order in the impurity concentration by (8.2.19) which can be expressed as

$$M(\omega) = \frac{n_i}{4\pi nm} \int d\mathbf{q}q^2 |v(q)|^2 C(q, \omega),$$

(8.3.6)

where

$$C(q, \omega) = \frac{\chi(q, \omega) - \chi(q, 0)}{\omega},$$

(8.3.7)

$$\chi(q,\omega) = i \int_0^\infty dt e^{i\omega t} \langle [\rho_q^\dagger(t), \rho_q] \rangle. \tag{8.3.8}$$

The correlation function may be given in the first approximation based on the RPA. However, such an approximation not only leads to a cyclotron divergence but also is poor for low densities where electron correlations are strong. One can go beyond the RPA by introducing a feedback of density fluctuations into the response function.

In order to discuss such a feedback mechanism, let us introduce a correlation function $C_{ij}(\omega)$ in a somewhat general form by

$$C_{ij}(\omega) = i \int_0^\infty dt e^{i\omega t} (A_i(t), A_j). \tag{8.3.9}$$

For the present case, the A's are

$$A_0 = (\rho_q^\dagger, \rho_{-q})^{-1/2} \rho_q,$$
$$A_1 = (j_q^\dagger, j_{-q})^{-1/2} j_q = (m/ne^2)^{1/2} j_q, \tag{8.3.10}$$

where j_q is the Fourier transform of the current density. Note that

$$(A_i, A_j) = \delta_{ij}; \quad \langle A_i \rangle = 0. \tag{8.3.11}$$

The correlation function C satisfies a matrix equation,

$$[\omega 1 - \Omega + M(\omega)] \cdot C(\omega) = -1, \tag{8.3.12}$$

where

$$\Omega_{ij} = i(\dot{A}_i, A_j); \quad M(\omega) = 1 \cdot i\Gamma.$$

Let us denote

$$M = i\Gamma,$$
$$\Gamma_{00} = \Gamma_{i0} = \Gamma_{0j} = 0.$$

The first approximation to the matrix equation (8.3.12) is given by

$$(\omega 1 - \Omega)C^{(0)}(\omega) = -1. \tag{8.3.13}$$

In order to obtain the second approximation, we first note

$$\Gamma_{\alpha\beta} C_{\beta\gamma}^{(1)} = \Gamma_{11}(\delta_{\alpha\beta} C_{\beta\gamma}^{(1)} - \delta_{\alpha 0} C_{0\gamma}^{(1)}),$$

and hence

$$(\omega \delta_{\alpha\beta} - \Omega_{\alpha\beta} + i\Gamma_{11}\delta_{\alpha\beta})C_{\beta\gamma}^{(1)} = -\delta_{\alpha\gamma} + i\Gamma_{11}\delta_{\alpha 0}C_{0\gamma}^{(1)}. \tag{8.3.14}$$

One can use $C^{(0)}$ of (8.3.13) in the left side of the above equation to obtain

$$\{[C^{(0)}(\omega + i\Gamma_{11})]^{-1}C^{(1)}\}_{\alpha\beta} = -\delta_{\alpha\beta} + i\Gamma_{11}\delta_{\alpha 0}C_{0\beta}^{(1)}. \tag{8.3.15}$$

This results in

$$C_{00}^{(1)} = \frac{C_{00}^{(0)}(\omega + i\Gamma_{11})}{1 + i\Gamma_{11}C_{00}^{(0)}(\omega + i\Gamma_{11})}. \tag{8.3.16}$$

That is, an improved expression for the memory function can be obtained from its first approximation by using

$$M(\omega) = \frac{n_i}{4\pi nm} \int_0^\infty \mathrm{d}\mathbf{q}\, q^3 |v(q)|^2 \frac{C_0(q, \omega + M(\omega))}{1 + M(\omega)C_0(q, \omega + M(\omega))/\chi_0(q)}. \tag{8.3.17}$$

Here,

$$C_0(q, \omega) = \chi(q)C_{00}^{(0)}(\omega). \tag{8.3.18}$$

Equation (8.3.17) yields the memory function which goes beyond the first approximation which is based on the RPA.

The RPA response function $\chi_0(q, \omega)$ is given by

$$\chi_0(q, \omega) = \frac{\lambda(q, \omega)}{1 + u(q)\lambda(q, \omega)}, \tag{8.3.19}$$

where $u(q)$ is the Coulomb potential, and $\lambda(q, \omega)$ is given by replacing $2\pi j/\beta$ by $-i\omega + \eta$ in (6.4.11), η being a small parameter to be brought to zero after integration. From the expression thus obtained, one can easily show that [8.15]

$$\lambda(q, \omega) = \frac{1}{2\pi} \sum_{n=1}^\infty (-)^{n+1} z^n \frac{\coth(gn\alpha/2)}{\sinh(n\alpha)} \phi_n(q, \omega). \tag{8.3.20}$$

Here

$$\phi_n(q, \omega) = \frac{\exp[-Q\coth(n\alpha)]}{2\sinh(\pi\alpha)} \int_0^\pi \mathrm{d}t \exp[-Q\coth(n\alpha)\cos t]$$
$$\times [\cos(Q\sin t - \Omega t) - \cos(Q\sin t + \Omega t)], \tag{8.3.21}$$

where $Q = q^2/\omega_c$ and $\Omega = \omega/\omega_c$ are dimensionless.

For low temperatures $\coth(n\alpha)$ can be replaced by 1. The integral on the right side can be expressed by the incomplete gamma function of the first kind. Furthermore, following the arguments in Sect. 6.4 and using a contour integral we identify

$$\sum_{n=1}^\infty (-)^{n+1} z^n \frac{\cosh(gn\alpha)}{\sinh(n\alpha)} = \frac{1}{\gamma_0}, \tag{8.3.22}$$

where $1/\gamma_0 = \varepsilon_\mathrm{F}/\mu_\mathrm{B} H$ as in (6.4.4) and z is the absolute activity. Equation (8.3.20) becomes

$$\lambda(q, \omega) = \frac{e^{-Q}}{4\pi\gamma_0} [(-Q)^{-\Omega}\gamma(\Omega, -Q) + (-Q)^{\Omega}\gamma(-\Omega, -Q)]. \tag{8.3.23}$$

Here, $\gamma(\Omega, -Q)$ is the incomplete gamma function. More explicitly, $\lambda(q, \omega)$ can be expanded in powers of Q,

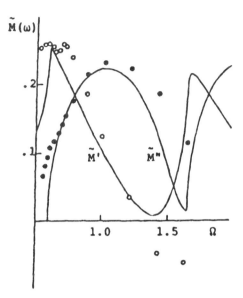

Fig. 8.4. Frequency dependences of the real and imaginary parts of the memory function $M(\omega)$ for a Si inversion layer. *Solid curve*: from [8.13]. *Points*: from [8.16a]. $H = 6.15\,T$ and $n = 2.3 \times 10^{11}\,\mathrm{cm}^{-2}$

$$\lambda(q,\omega) = \frac{e^{-Q}}{2\pi\gamma_0} \sum_{n=1}^{\infty} \frac{Q^n}{n!} \frac{n}{n^2 - \Omega^2}. \tag{8.3.24}$$

Using this in (8.3.19) and then solving (8.3.17) self-consistently, one can obtain the second approximation. From the real and imaginary parts, M' and M'', of this solution, the conductivity is given as

$$\sigma_{xx} = \frac{(ne^2/m)M''}{(\omega - \omega_c + M')^2 + M''^2}, \tag{8.3.25}$$

$$\sigma_{xy} = \frac{(ne^2/m)(\omega - \omega_c + M')}{(\omega - \omega_c + M')^2 + M''^2}. \tag{8.3.26}$$

The frequency dependences of M' and M'' have been evaluated by *Shiwa* and *Isihara* based on the above theory for a Si inversion layer with electron density $n = 2.3 \times 10^{11}$ cm^{-2} and with a Gaussian impurity potential given by

$$n_i |v(q)|^2 = 4\pi r_0^2 u^2 \exp[-(qr_0)^2], \tag{8.3.27}$$

where r_0 is a cutoff parameter and u is a strength parameter. The theoretical results with $Q_c = 2.0$, $\tilde{u} = u/\omega_c = 1.9$ and $g_v = 2$ are illustrated in Fig. 8.4 by solid curves. The data represent what was obtained by *Allen* et al. [8.16] on a Si inversion layer by a piecewise Kramers–Kronig analysis. The magnetic field is 6.15 T and the temperature is 1.3 K.

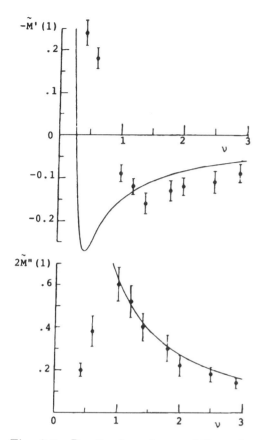

Fig. 8.5. Density dependences of the real and imaginary parts of the memory function. *Solid curve*: from [8.13]. *Points*: from [8.16b]. The data were obtained at 1.3 K and 6.15 T

The real part representing shifting shows a general decrease with frequency. This decrease is reproduced by the theoretical curve reasonably well. However, at low frequencies the data points do not change much, while the theoretical curve shows a sharp maximum. The imaginary part, representing broadening, is maximum at ω_c, around which the agreement between the theory and data is good. Let us use reduced variables $\tilde{M}' = M'/\omega_c$ and $\tilde{M}''/\omega_c$.

Figure 8.5 illustrates the density dependences of $\tilde{M}'$ and $\tilde{M}''$. The solid curves represent the theory of *Shiwa* and *Isihara* outlined above, and the data are taken from *Wilson* et al. [8.16b]. The theoretical parameters in this graph are slightly different from the above choices because Fig. 8.4 corresponds to a particular data point which is off from the rest of the data points in the density plot. The theory agrees with the data fairly well for $\nu > 1$ but only poorly for $\nu < 1$. This is understandable because the theory is based on a high density approach.

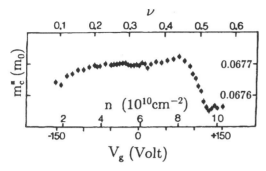

Fig. 8.6. Cyclotron effective mass in GaAs/GaAlAs obtained by *Chou* et al. [8.17]

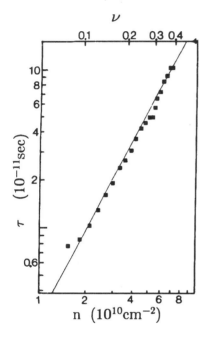

Fig. 8.7. Cyclotron relaxation time in GaAs/GaAlAs obtained by *Chou* et al. [8.17]

More complicated resonance behavior has been reported for GaAs/GaAlAs. However, a recent study by *Chou* et al. [8.17] resolved some experimental discrepancies. They measured the cyclotron resonance of high mobility 2D electrons in GaAs/GaAlAs heterostructures at low densities in the range between $n = 1.55 \times 10^{10}$ cm^{-2} and 1.03×10^{11} cm^{-2}. The range is from around $\nu = 0.6$ to 0.08 in terms of filling factor. The effective mass was found to decrease slightly with n due to the nonparabolic conduction band of GaAs. The corrected effective mass m_c^* at the band edge had an uncertainty of order 1%, but its relative variation was far more accurate because of the sharpness of resonance. In the region $0.4 < \nu < 0.14$, m_c^* was found to be nearly constant ($0.0677m_0$) with accuracy better than 0.025%. In detail, it increased slightly from $\nu = 0.1$ to 0.4, and then decreased at around $\nu = 0.5$. This variation is shown in Fig. 8.6, where the gate voltage V_g and electron density n are given in the bottom frame and filling factor ν in the top frame. The cyclotron relaxation time τ was found to be 15 times the dc scattering time and to increase with electron density. This increase was expressed empirically in the above range of ν such that

$$\tau \propto n^{1.9 \pm 0.1}.$$

Figure 8.7 illustrates this variation. The straight line represents the above relation. According to *Chou* et al.'s analysis the increase is due to scattering by screened residual ionized impurities in GaAs.

9. Localization

Scaling laws have been discovered in the weak localization of electrons in random media.

9.1 Anderson Localization and Scaling

The theoretical study of electric conduction in solids was traditionally concerned with the case in which the concentration of impurities was small. In such a case the correlation between the impurities can be neglected and the conductivity is proportional to the concentration. Moreover, most theories were developed for infinite systems even though it was known for many years that the conductance of actual systems depends on their size. Attempts to go beyond these idealizing conditions have been limited and did not bring about any conceptual change in the general view of conductance. In fact a finite system with a high impurity concentration was not a good theoretical object for many years. All this has changed in the past ten years, although the origin of this new trend can be traced back to 1958. *Anderson* [9.1] pointed out in that year the possibility of localization of electrons in random systems. That is, impurities may completely change the conductive character of such systems. However, theoretical developments did not follow immediately.

In 1978, *Mott* [9.2] discussed the existence of a *mobility edge* in the metal–insulator transition in disordered systems. He argued that when the concentration of impurities increases, the localized electron waves around the impurities start overlapping with each other until a continuous impurity band is formed. However, the bottom of this band is still localized and there must be a sharp boundary between the localized and extended states. The mobility is finite in this case if the electron energy is above the mobility edge ε_c and is zero below. Hence, there can be a minimum conductivity above which the system is conductive.

Subsequently, in 1979, *Abrahams* et al. [9.3] reported a scaling theory which opened the door to the basic understanding of conduction in disordered systems. According to this theory the conductance of disordered systems varies continuously to zero without a minimum. The scaling theory has triggered strong theoretical and experimental developments in the hitherto neglected area.

Anderson localization, as it is called now, has become the central subject in the study of conduction in disordered systems.

In what follows it becomes desirable to use different notations for conductance and conductivity. Therefore, the former will be denoted by G and the latter by σ. The conductance of a system with length L in each direction is considered as a function of L in the scaling theory. If the length L is large enough the conductance to be denoted by $G_L = G(L)$ should be proportional to the cross section L^{d-1} and inversely proportional to the length L, where d is the dimension. Therefore, the *conductance* can be expressed as

$$G(L) = \sigma L^{d-2}, \tag{9.1.1}$$

where σ is the conductivity which is independent of L. If $d = 2$, $G(L) = \sigma$, and it appears as though there is no L dependence. However, let us keep d arbitrary first and consider particular dimensions later. This process may be considered to correspond to taking limiting cases from arbitrary dimensions.

Let us now introduce another system with length L' in each direction. The conductance $G(L')$ is assumed to depend on L' in accordance with

$$G(L') = f(G(L), L'/L), \tag{9.1.2}$$

where f is a universal function which is independent of the microscopic structure of the system. This is a scaling hypothesis. Its consequence can be seen by taking a logarithmic differential of this equation with respect to a variable $x = L'/L$. The derivative is

$$\frac{d \ln G}{d \ln L} = \beta(G), \tag{9.1.3}$$

where $\beta(G)$ is a universal function.

In the metallic limit (9.1.1) suggests

$$\beta(G) = d - 2. \tag{9.1.4}$$

In the opposite limit, the system is expected not to conduct. In this case the conductance may decrease exponentially with L. If

$$G_L \sim \exp(-\alpha L)$$

as $L \to \infty$, we expect

$$\beta(G) = \ln G. \tag{9.1.5}$$

Let us assume that $\beta(G)$ is a continuous function which connects the two limits given by (9.1.4) and (9.1.5). Such a function is schematically shown in Fig. 9.1, in which, $\beta(G)$ is plotted against $\ln G$. This type of variation of $\beta(G)$ agrees with a numerical result [9.4]. In three dimensions $\beta(G)$ is negative under a certain value of G and beyond this value increases towards the asymptotic value 1. The point at which $\beta(G) = 0$ is called a fixed point in the sense that $d \ln G/dL$ vanishes, i.e. there is no L dependence. However,

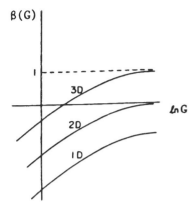

Fig. 9.1. Schematic variation of $\beta(G)$ as a function of $\ln G$

the point may depend on impurity or electron energy. Let the conductance at this point be G_c.

How does the conductivity σ in 3D vary when the conductance approaches the fixed point? In order to answer this question, let us be on the conductive side of the fixed point, i.e., $G(L) > G_c$. Near G_c, we may assume a linear variation of $\beta(G)$,

$$\beta(G) = a[G(L) - G_c].$$

Here, a is the derivative of $\beta(G)$ at G_c and is positive in accordance with Fig. 9.1. Since $\beta(G) \to 1$, as $G \to \infty$, one possible form for $\beta(G)$ which is correct in both limits is

$$\beta(G) = \frac{a[G(L) - G_c]}{1 + a[G(L) - G_c]}.$$

We can use this expression in (9.1.3) and integrate over G from $G(L_0)$ to $G(L)$ and over L from a sufficiently long length L_0 to L. Note that $G(L_0) > G(L)$ and $L_0 > L$ so that the integration is in the decreasing direction. We obtain

$$\left(\frac{G(L) - G_c}{G(L_0) - G_c} \frac{G(L_0)}{G(L)} \right)^{\nu} \frac{G(L)}{G(L_0)} = \frac{L}{L_0}.$$

Here

$$\nu = 1/aG_c. \tag{9.1.6}$$

Note that L_0 is long enough to assure the use of (9.1.1) for $G(L_0)$. That is, the conductivity is given by $\sigma = G(L_0)/L_0$. On the other hand, $G(L)$ can be sufficiently close to G_c. Since $G(L_0) \gg G_c$, we find

$$\sigma = \sigma_c(G - G_c)^{\nu}. \tag{9.1.7}$$

That is, σ vanishes as $G \to G_c$ because $\nu > 0$. Here, σ_c is a certain constant.

The conductivity given by (9.1.7) may depend on the system size L. In fact in the regime of weak localization, which will be defined shortly, such a dependence is conceivable. In the above derivation we have assumed a particular form of $\beta(G)$, but it is clear that (9.1.7) is a consequence of the behavior of β near G_c and is independent of the particularly chosen form. A form of $\beta(g)$ which connects (9.1.4) and (9.1.5) has been derived based on a self-consistent theory [9.5].

We may translate the above relation into energy space, because near the fixed point the conductance is expected to depend on electron energy. By assuming that $G(L) \sim \varepsilon$ and $G_c \sim \varepsilon_c$, we arrive at

$$\sigma = \sigma_c(\varepsilon - \varepsilon_c)^\nu. \tag{9.1.8}$$

The constant ε_c is the mobility edge. At this point the conductivity vanishes instead of approaching a minimum value.

The fixed point may be associated with a critical impurity concentration n_c. For a given impurity concentration n_i near n_c let us assume

$$G(L) - G_c \propto (n_i - n_c)$$

even though such a proportionality is not guaranteed. We obtain

$$\sigma = \sigma_c(n_i - n_c)^\nu. \tag{9.1.9}$$

Thus, the conductivity vanishes when the impurity concentration approaches a certain value n_c.

It is important conceptionally to recognize the appearance of a point in three dimensions where the conductivity vanishes. In contrast no such point appears in one and two dimensions since $\beta(G)$ stays below the abscissa as in Fig. 9.1. That is, according to the above consideration the electrons are localized at absolute zero in one and two dimensions regardless of the amount of disorder. However, a numerical calculation [9.6] indicated that an Anderson transition may occur in two dimensions.

One dimension is somewhat special in the sense that the only conduction channel is disrupted by a disorder at a site. Therefore, let us find how the conductance of a 2D system varies with its length L. For large G, Fig. 9.1 suggests that $\beta(G)$ approaches zero. Let us assume that this approach is simply proportional to $1/G$

$$\beta(G) = -\frac{G_a}{G}. \tag{9.1.10}$$

Such an asymptotic form has actually been justified to be correct to order $(1/G^3)$ based on a nonlinear σ model [9.7].

The dimension of conductivity is given in 2D by e^2/h. Hence, the constant G_a may be expressed by

$$G_a = \frac{e^2\lambda}{2\pi^2\hbar}, \tag{9.1.11}$$

where λ is a dimensionless constant. In combination with (9.1.3), (9.1.10) yields

$$G(L) = G(L_0) - G_a \ln(L/L_0),$$ (9.1.12)

where $G(L_0)$ and L_0 are appropriate constants.

Note that the form of (9.1.10) has been introduced for large G and the logarithmic term has been obtained approximately. Therefore, it should be smaller than the first term. We have found under this condition that a logarithmic decrease of the conductance takes place.

The above logarithmic variation has been derived for absolute zero. The temperature variation of conductance is logarithmic as we shall discuss below. According to the general conductivity formula, which is expressed by the current–current correlation function, phase coherence of electron waves is crucial to the conductivity. Electrons are scattered elastically and diffuse about until inelastic scattering takes place and destroys phase coherence. Hence, an elastic mean free path L_m is expected to be given by $(D\tau_i)^{1/2}$, where τ_i is the inelastic scattering time and D is a diffusion constant. The inelastic scattering time may be assumed to vary as

$$\tau_i \sim T^{-p},$$ (9.1.13)

because $\tau_i \to \infty$ at absolute zero, where p is of order 1. Hence, due to localization, a finite temperature correction, $\Delta\sigma(T)$, to the Drude-type conductivity, is expected in the form

$$\Delta\sigma(T) = \frac{e^2}{2\pi^2\hbar} p \ln T/T_0.$$ (9.1.14)

Below a certain temperature T_0, the conductance in two dimensions decreases with temperature. The prefactor

$$\frac{e^2}{2\pi^2\hbar} = 1.2 \times 10^{-5}\,\Omega^{-1}$$ (9.1.15)

is the natural conductivity unit.

Note that (9.1.14) is a correction. The logarithmic decrease can be considered as a precursor to a complete localization at absolute zero, and represents what is called weak localization. Such a logarithmic variation has been observed in films of Cu, Pt, AuPd, etc. since around 1979–1981. However, some of the early observations are now considered to be due to an interaction effect because the scattering time in noble and transition metals becomes independent of temperature when τ_i reaches the spin-flip scattering time. It has also been found that the proportionality constant of the logarithmic variation deviates from that in (9.1.14) due to Coulomb interaction. We shall discuss such an interaction effect shortly.

9.2 Weak Localization

9.2.1 Interference Effect

Weak localization has been studied extensively [9.8] since the advancement of the scaling theory. Among these studies is a microscopic derivation of $\beta(\sigma)$

which plays a primary role in the scaling theory. For this derivation, let us start with the dynamical conductivity $\sigma(\omega)$ which is given by the current–current correlation function in the form

$$\sigma(\omega) = \frac{1}{A\hbar\omega} \int_0^\infty e^{i\omega t - \eta t} \langle [j(t), j] \rangle \, dt, \tag{9.2.1}$$

where η is a parameter to be brought to zero after integration and $\langle \ldots \rangle$ represents a thermal average. The one electron current operator j can be replaced by $-e\mathbf{v}$ in terms of the velocity operator. After integrating over t, the conductivity is expressed by a product of the Green's functions. The corresponding conductivity expression for the static case has been given by (8.1.1), where the appearance of the Green's function can be foreseen from the relation

$$\delta(\varepsilon - \mathcal{H}) = \lim_{\eta \to 0} \frac{i}{2\pi} \left(\frac{1}{\varepsilon - \mathcal{H} + i\eta} - \frac{1}{\varepsilon - \mathcal{H} - i\eta} \right)$$

$$= -\frac{1}{2\pi i} [G^+(\varepsilon - H) - G^-(\varepsilon - \mathcal{H})],$$

where $G^\pm(x)$ denotes $G(x \pm i\eta)$.

The Green's function $G(\varepsilon - \mathcal{H})$ can be expressed in powers of the impurity potential V in terms of the ideal Green's function as

$$\frac{1}{\varepsilon - \mathcal{H}} = \sum_{n=0}^\infty \frac{1}{\varepsilon - \mathcal{H}_0} \left[V(r) \frac{1}{\varepsilon - \mathcal{H}_0} \right]^n.$$

In an approximation for a delta-function-type impurity potential $V = V_0 \sum_i \delta(\mathbf{r} - \mathbf{R}_i)$, such a series can be used to derive an averaged Green's function in the form given by (8.1.7). The averaged Green's function is characterized by a self-energy Σ.

The electron's energy is locked in the Fermi energy ε_F for low temperatures and the energy shift Δ may be neglected. The self-energy is then expressed as

$$\Sigma(\varepsilon_F) = -\frac{i\hbar}{2\tau} = -i\pi n_i V_0^2 g(\varepsilon_F), \tag{9.2.2}$$

where τ is a scattering time, n_i is the impurity concentration, which is assumed to be small, $g(\varepsilon_F)$ is the density of states at the Fermi energy, and V_0 is the strength of a delta-function-type impurity potential. The averaged Green's function can be expressed as

$$\langle G^\pm(\varepsilon) \rangle = \frac{1}{\varepsilon - \varepsilon(k) \pm i\hbar/2\tau}. \tag{9.2.3}$$

$\varepsilon(k)$ is the kinetic energy, and the real part of the self-energy representing an energy shift has been neglected.

The zeroth approximation to the conductivity is given by using these averaged Green's functions. If $f(\varepsilon)$ represents the Fermi distribution, the conductivity is given for two dimensions in the form

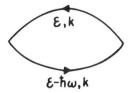

Fig. 9.2. Graphical illustration of the integrand for σ_0 in (9.2.4)

$$\sigma_0(\omega) = 2e^2 \int \frac{d\varepsilon}{2\pi} \frac{f(\varepsilon - \hbar\omega) - f(\varepsilon)}{\omega} \int \frac{d\mathbf{k}}{(2\pi)^2} v^2(k) G^+(\varepsilon - \hbar\omega, k) G^-(\varepsilon, k),$$

$$(9.2.4)$$

where the factor 2 is due to spin, $v(k) = d\varepsilon(k)/(\hbar dk)$, the average notation $\langle \ldots \rangle$ is dropped from the Green's functions, and only the contributing product of the Green's functions is retained. This contribution can be illustrated as in Fig. 9.2 where the lines represent the Green's functions. The integration over $\mathbf{k}$ can be converted into that in $\varepsilon(k)$. We note that

$$\int d\varepsilon(k) G^+(\varepsilon - \hbar\omega, k) G^-(\varepsilon, k) = \frac{2\pi\tau}{\hbar(1 - i\omega\tau)}.$$

As a result, we arrive at

$$\sigma_0(\omega) = \frac{ne^2\tau/m}{1 - i\omega\tau}. \tag{9.2.5}$$

The next approximation to the conductivity is obtained by taking the averages of the products of the Green's functions instead of replacing them by the products of their averages. That is, quantum interference of electron waves must be taken into consideration in order to go beyond (9.2.5).

It has been found that a significant contribution is given by the so-called maximally crossed diagrams, which are of the type shown in Fig. 9.3. The fan diagram on the left side in (a) represents all the graphs on the right side with one, two, three, ... dotted interaction lines with impurities which are represented by dots. One electron collides in sequence with impurities $1, 2, \ldots$, while the other follows the opposite sequence in a time-reversed path. The wave functions of the two electrons interfere with each other, resulting in a standing wave around the loop of impurities.

The scattering mechanism associated with the fan diagram is shown in (b). Here, an electron with momentum $\mathbf{k}$ is scattered to states $\mathbf{k}_1, \mathbf{k}_2$, etc. and reaches the state $-\mathbf{k}$ on the opposite side of the Fermi circle. The time-reversed process $\mathbf{k} \to -\mathbf{k}_2 \to -\mathbf{k}_1 \to -\mathbf{k}$ is complementary to the first process. Examining the momentum changes we find that the scattering amplitudes of these two pro- cesses are the same in the final state $-\mathbf{k}$. Therefore, the two processes interfere with each other constructively so that more current returns to the starting point. This means that less current flows through the sample, i.e. localization. The contribution from the fan diagram is treated mathematically as follows.

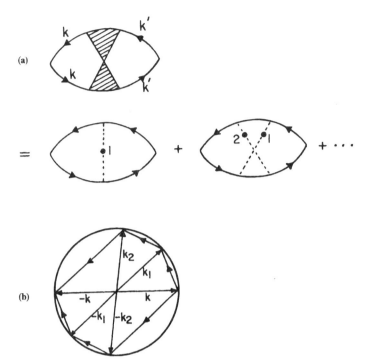

Fig. 9.3a,b. Maximally crossed diagram. (a) The right hand side gives all the fan diagrams. (b) A scattering mechanism of a fan diagram

In the static limit, these graphs result in the conductivity given by

$$\Delta\sigma = \frac{\hbar}{\pi}\left(\frac{e\hbar}{m}\right)^2 \int \frac{d\mathbf{k}d\mathbf{k}'}{(2\pi)^2} k_x k_x' G^+(k)G^-(k)G^+(k')G^-(k')\Lambda(\mathbf{k}+\mathbf{k}'),$$

$$(9.2.6)$$

where $\Lambda(q)$, which may be called a ladder function, is given by

$$\Lambda(q) = \frac{n_i V_0^2}{1 - n_i V_0^2 \lambda(q)}, \tag{9.2.7}$$

$$\lambda(q) = \int \frac{d\mathbf{k}}{(2\pi)^2} G^+(k)G^-(\mathbf{q}-\mathbf{k}). \tag{9.2.8}$$

The integral in (9.2.6) may be evaluated near $\mathbf{k}+\mathbf{k}'=0$, where a maximum contribution is expected as seen from Fig. 9.3b. Near $\mathbf{q} = \mathbf{k}+\mathbf{k}' = 0$, $\lambda(q)$ can be expanded in powers of q such that

$$\lambda(q) = \lambda(0)(1 - \tau D q^2), \tag{9.2.9}$$

where

$$D^{-1} = \tau(\hbar k_{\rm F}/m)^2/2. \tag{9.2.10}$$

τ is a certain relaxation time, and D is a diffusion constant. In fact it can be shown that the quadratic form of $\lambda(q)$ leads to electron propagation which represents diffusion.

The conductivity becomes

$$\sigma(L) = \sigma_0 - \frac{2e^2}{\pi\hbar}\frac{1}{(2\pi)^2}\int\frac{d q}{q^2}. \tag{9.2.11}$$

The q integral is divergent, but it is conceivable to cut-off the domain of integration such that $1/L < q < 1/L_0$. As a result we obtain

$$\sigma(L) = \sigma_0 - 2\left(\frac{e^2}{2\pi^2\hbar}\right)\ln\left(\frac{L}{L_0}\right) \tag{9.2.12}$$

where $\hbar$ has been restored. Differentiating this expression we obtain

$$\begin{aligned}
\beta(\sigma) &= \frac{d\ln\sigma(L)}{d\ln L} \\
&= -2\left(\frac{e^2}{2\pi^2\hbar}\right)\frac{1}{\sigma}. \tag{9.2.13}
\end{aligned}$$

Accordingly we find that the parameter λ in (9.1.11) is 2.

The above derivation is based on a particular set of graphs. However, it has been shown that to order $(1/\sigma^2)$ and $(1/\sigma^3)$ the above form still holds even if some other graphs are taken into consideration [9–7].

9.2.2 Magnetic Field Effects

In 1959, Aharonov and Bohm proposed a novel "Gedankenexperiment" to coherently split an electron beam by a thin solenoid into two paths, such that the two beams pass around the solenoid and then recombine at a detector located on the opposite side of the solenoid from the source. They predicted that these two waves will show interference which is determined by the magenetic flux and has a characteristic flux period of ch/e. This *Aharonov–Bohm effect* has been observed in vacuo.

Quantum interference of electron waves is sensitive to a magnetic flux even in condensed matter. As in the case of the Aharonov–Bohm effect, this interference is determined by the magnetic flux and takes place even if the waves are not directly exposed to the magnetic field. *Al'tshuler* et al. [9.9a] arrived at the surprising theoretical conclusion in 1981 that the magnetoresistance of a thin metal cylindrical shell, through which a magnetic flux is threaded, should show oscillations with a period of $ch/2e$ rather than ch/e because the phase shift between the two complementary waves going around the cylinder is $e\phi/c\hbar - (-e\phi/c\hbar) = 2e\phi/c\hbar$, where ϕ is the flux. This type of oscillation is due to coherent backscattering.

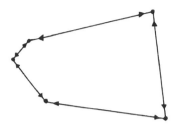

Fig. 9.4. A pair of time reversed paths

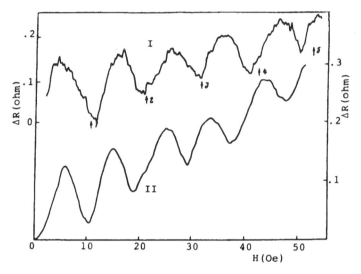

Fig. 9.5. Aharonov-Bohm oscillations obtained by Sharvin and Sharvin [9.9b]

Figure 9.4 illustrates a pair of time-reversed electron paths. The dots are impurities, and the arrows are the directions of propagation. The resulting interference of the wave functions is a standing wave around the loop. Due to phase coherence the current returning to the starting point is enhanced by a factor of two, resulting in large negative corrections to the Drude conductivity.

Figure 9.5 depicts the data of *Sharvin* and *Sharvin* [9.9b] which showed such oscillations in magnesium films I and II at 1.12 K. Sample I (left ordinate) showed more noise than sample II (right ordinate) due apparently to its composition. The arrows indicate the field values corresponding to integral multiples of flux ϕ. The conductivity per unit area of the film is related to the resistance by $\sigma = L/pR$, where p and L are the perimeter and length of the cylinder. The order of magnitude of the amplitude of oscillation of the conductivity is determined by the universal value

$$\Delta\sigma = (e^2/\pi^2\hbar) = 2.47 \times 10^{-5}\Omega^{-1}.$$

If a magnetic field is perpendicular to a disordered 2D system, spin–orbit coupling causes an increase of the magnetoconductivity. Even in the absence of this coupling, there is an important effect due to a perpendicular magnetic field because it confines an electron in a 2D system onto cyclotron orbits. This orbital effect is represented by a characteristic magnetic length $l = (c\hbar/eH)^{1/2}$. If this length is larger than the inelastic mean free path L_m, the conductance is determined by L_m. In the opposite case it must be l that characterizes the conductance. If $\Delta\sigma(H)$ denotes the change in the conductivity due to a magnetic field, we expect

$$\Delta\sigma(H) = \sigma(H) - \sigma(0)$$
$$= (e^2/2\pi^2\hbar)\ln H. \tag{9.2.14}$$

A more precise expression than the above has been given by *Hikami* et al. and *Al'tshuler* et al. [9.10] in consideration of spin orbit coupling and magnetic scattering due to impurity spin $\mathbf{S}$. Hence, they adopted the following form for the scattering amplitude:

$$f_{\alpha\beta} = V_0\delta_{\alpha\beta} + ib(\mathbf{p} \times \mathbf{p}')\sigma_{\alpha\beta} + c\mathbf{S} \cdot \sigma_{\alpha\beta}, \tag{9.2.15}$$

where α and β represent spin directions. The first term is the ordinary impurity scattering, the second term represents the spin–orbit coupling, and the third term is the magnetic scattering due to the impurity spin $\mathbf{S}$. In addition to the ordinary scattering time, two additional relaxation times appear in their ladder function $\Lambda(q)$. Moreover a cutoff parameter N_M is needed in the summation:

$$\sum_{n=0}^{N_M} \frac{1}{(n+1/2) + 1/a\tau_i}, \tag{9.2.16}$$

where N_M is a cutoff parameter which is expected to be of order $1/a\tau$. This is the ratio of a characteristic time a^{-1} due to a magnetic field and the elastic mean free time τ. The summation is approximately given by the digamma function for large N_M.

When the inelastic scattering time τ_i is much shorter than the other relaxation times $\Delta\sigma(H)$ is given by [9.10]

$$\Delta\sigma(H) = \alpha g_v \frac{e^2}{2\pi^2\hbar} \left[\Psi\left(\frac{1}{2} + \frac{1}{a\tau_i}\right) - \psi\left(\frac{1}{2} + \frac{1}{a\tau}\right) + \ln\frac{\tau_i}{\tau} \right]. \tag{9.2.17}$$

Here the parameter α is 1 if spin–orbit coupling and the magnetic scattering are weak. Hence, in the above expression α is not really needed, but it is convenient to indicate here that the prefactor changes in accordance with the scattering mechanism. The valley degeneracy factor g_v has also been introduced for application to Si inversion layers. ψ is the digamma function, $a^{-1} = (c\hbar/4DeH)$, and the diffusion constant

$$D = \varepsilon_{\mathrm{F}}\tau/m.$$

The digamma function varies logarithmically for large values of its argument. If $(a\tau)^{-1} > 1 > (a\tau_i)^{-1}$, (9.2.17) approaches (9.2.14). When $a\tau \ll 1$, it is proportional to H^2 due to cancellation of first-order terms. That is, for $g_v = 1$, the expressions for $\Delta\sigma(H)$ in these two limits are

$$\Delta\sigma(H) = \begin{cases} \dfrac{\alpha e^2}{48\pi^2\hbar}\left(\dfrac{4DeH\tau_i}{c\hbar}\right)^2, & (a\tau_i \ll 1), \\[3mm] \dfrac{\alpha e^2}{2\pi^2\hbar}\ln\left(\dfrac{4DeH\tau_i}{c\hbar}\right), & (a\tau_i \gg 1). \end{cases} \tag{9.2.18}$$

Note that when $n = 0$ the denominator of the series in (9.2.16) does not vanish, indicating that a magnetic field counteracts localization. The consequence of this counteraction appears as *negative magnetoresistance* or increased conductance. This effect originates from the finiteness of the denominator in contrast to $q = 0$ in (9.2.11) in the absence of a magnetic field.

A logarithmic variation of magnetoconductivity appears in the case of the Kondo effect. However, in that case the variation is independent of the field direction and the coefficient of $\ln H$ is not a universal constant. Therefore the logarithmic variation of negative magnetoresistance related to localization can be easily distinguished.

Figure 9.6 illustrates the field variation of 2D conductivity due to localization. The solid curves represent (9.2.17) in which the inelastic scattering time was chosen as an adjustable parameter. The data are due to *Kawaguchi* and *Kawaji* [9.11] on a Si inversion layer at electron density $1.9 \times 10^{12}\,\mathrm{cm}^{-2}$ between 1.9 K and 10.4 K. The agreement is excellent. They also confirmed the saturation of $\Delta\sigma(H)$ in accordance with (9.2.18). Note that in this expression one can write

$$DeH/c\hbar = 1/\tau.$$

Since $\tau_i \propto T^{-p}$ the saturation value depends on temperature,

$$\Delta\sigma(H) = \frac{\alpha e^2 p}{2\pi^2\hbar}\ln T, \quad (a\tau_i \gg 1).$$

Equation (9.2.17) has been derived for $\hbar\omega_c \ll \varepsilon_F \ll \hbar/\tau$, where ω_c is the cyclotron frequency, so that deviations take place as the magnetic field increases. Relaxing the condition such that $\hbar\omega_c \ll (\hbar/\tau, \varepsilon_F)$, *Kawabata* [9.12] treated the magnetic field effect in a more quantum mechanical way and obtained results which remove the discrepancies.

For three dimensions, negative magnetoconductance in doped semiconductors was found around 1955 [9.13] but its quantitative explanation based on localization was given some 30 years later. It is expected in the case of three dimensions that the n-sum in (9.2.16) over the Landau levels is modified such that

$$\sum_{n=0}^{N_M}\int_{-k_F}^{k_F} dq \frac{1}{(n+1/2) + q^2 l^2/4 + l^2/4D\tau}, \tag{9.2.19}$$

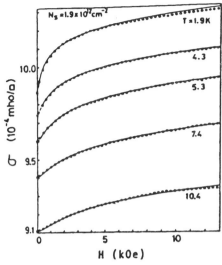

Fig. 9.6. Magnetic field dependence of the conductivity in a Si inversion layer determined by *Kawaguchi* and *Kawaji* for small fields. [9.11a]

where τ is a certain scattering time which may depend on phonons. It has been shown that the leading contribution to $\Delta\sigma(H)$ is given by [9.14]

$$\Delta\sigma(H) = \Delta\sigma_1 + \frac{e^2}{2\pi^2\hbar}(D\tau)^{-1/2}. \qquad (9.2.20)$$

Here

$$\Delta\sigma_1 = \frac{e^2 F(\delta)}{2\pi^2\hbar l}. \qquad (9.2.21)$$

$$F(\delta) = \sum_{n=0}^{\infty}[2(n+1+\delta)^{1/2} - (n+\delta)^{1/2} - (n+1/2+\delta)^{1/2}]. \qquad (9.2.22)$$

δ is a dimensionless parameter defined by

$$\delta = l^2/4D\tau. \qquad (9.2.23)$$

When the magnetic field is strong such that $\delta \ll 1$, a series expansion leads to

$$F(\delta) = 0.605 - 2\delta^{1/2} + 2.39\delta. \qquad (9.2.24)$$

Hence, $\Delta\sigma_1$ is proportional to $H^{1/2}$. In the opposite limit of weak fields $\delta \gg 1$. In this case $F(\delta)$ can be obtained by expansion in inverse powers of $(n+1/2+\delta)$, resulting in an approximation,

$$F(\delta) = \frac{\delta^{-3/2}}{48}. \qquad (9.2.25)$$

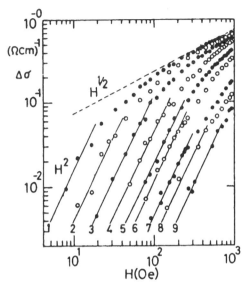

Fig. 9.7. Magnetic field dependence of the magnetoconductance in n-GaAs measured by *Morita* et al. [9.15]

Hence, $\Delta\sigma_1$ is proportional to H^2. In combination of these two limiting cases, the resistivity may be expressed as

$$\Delta\sigma = aH^{1/2} + bH^2, \tag{9.2.26}$$

where a and b are the limiting proportionality constants.

Figure 9.7 illustrates the magnetic field dependence of the magnetoconductivity of n-GaAs at electron density $7.8 \times 10^{16}\,\mathrm{cm}^{-3}$ measured by *Morita* et al. [9.15]. The temperature is (1) 0.05 K, (2) 0.15 K, (3) 0.25 K, (4) 0.5 K, (5) 1 K, (6) 2 K, (7) 4.2 K, (8) 6 K, (9) 8 K. The graph shows the expected change from the H^2 to $H^{1/2}$ variations.

In the presence of spin–magnetic field coupling, the series in (9.2.16) or (9.2.19) indicates that electron spin adds $\pm g\mu_B H$ to the orbital energy. For two dimensions, such a spin effect is expected to show up if the magnetic field is parallel to the surface. It actually decreases the magnetoconductivity. If spin–orbit coupling is effective in a perpendicular magnetic field, the conductivity is decreased also. It is given by replacing τ_i in (9.2.18) by an appropriate relaxation time τ_H. On the other hand, for three dimensions the spin effect is expected to depend on $g\mu_B H/k_B T$. The relative magnitude of the orbital and spin effects is then determined by a dimensionless ratio $(4\varepsilon_F\tau/\hbar)(k_B T\tau_i/\hbar)$. When this ratio is large, that is if $\varepsilon_F\tau/\hbar \gg 1$, the orbital effect is dominant.

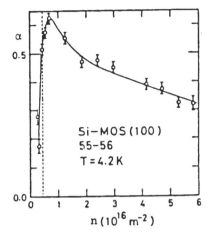

Fig. 9.8. Variation of α with electron density [9.16]

9.3 Interaction Effect

The magnetic field dependence of the magnetoconductivity of 2D and 3D electron systems has been used to test localization theory. As shown in Figs. 9.5 and 9.6, the data have been explained by localization theory very well. Moreover localization theory has been extended to a wider range of magnetic field and also to an angle dependence of the magnetoconductivity of 3D systems. However, close examinations of the fit in Fig. 9.4 have revealed that the parameter a appearing in (9.2.17) is not equal to 1. It is actually not even a constant but depends on electron density. Such a dependence can be due to Coulomb interaction between the electrons. Moreover, the parameter p in (9.1.14) representing the temperature dependence of the inelastic scattering time has also been shown to vary with electron density.

Figure 9.8 illustrates the variation of a with electron density observed by *Kawaguchi* and *Kawaji* [9.16]. Not only is α less than 1 but it also shows a strong density variation with a peak near $\varepsilon_F \tau / \hbar = 0.5$, which is represented by a dotted line. Such a density variation indicates that there is a significant Coulomb effect on localization. In the high density region $n > 3 \times 10^{16}\,\mathrm{cm}^{-2}$ at 4.2 K, α is found to decrease linearly with increasing temperature.

The density dependence of p is illustrated in Fig. 9.9. Note that p is generally less than the theoretically expected value 2, and dips to a minimum value near the point at which $\varepsilon_F \tau / \hbar = 0.5$. The small p values are apparently due to impurities. In the study of the electronic properties of metals the success of Fermi liquid theory and the major role played by the Drude conductivity have long given the impression that interaction effects are not important. The above data show that this is not the case in the regime of weak localization. How, in general, can we understand that Coulomb effects are strong in the regime of weak localization? This question may be answered by recalling that weak localization is related to diffusive motion of electrons.

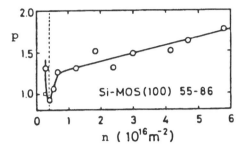

Fig. 9.9. Density dependence of the parameter p for inelastic scattering [9.16]

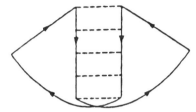

Fig. 9.10. Cooperon propagation

Thus their average time of being at an impurity site can be longer than in the plane-wave case. This causes enhancement of their interaction.

The importance of Coulomb interaction in disordered systems was first pointed out by *Al'tshuler* and *Aronov* [9.17a] in 1979 before the development of scaling theory. Subsequently, Al'tshuler et al. [9.17b] and *Fukuyama* [9.17c] investigated the effect of interaction on weak localization. In view of the diagrammatic methods in these theories, we comment that Fig. 9.3 can be redrawn as in Fig. 9.10 such that the two electrons propagate in the same direction. That is, the maximally crossed diagrams are represented by the ladder diagrams between the two electrons moving in the same direction. We call $\Lambda(q)$ resulting from this type of propagation particle–particle diffusion or Cooper propagation, in analogy with the case of superconductivity.

We must note that there is another type of propagation which can be illustrated by two oppositely directed propagations. This type represents particle–hole diffusion, as it is commonly called. It is expressed by an effective propagator which is similar in form to $\Lambda(q)$. However this type of propagation plays a different role from the Cooper propagation in the presence of a magnetic field. Being neutral, an electron–hole pair does not couple with the vector potential of a magnetic field. Nevertheless, this does not mean that there is no magnetic field dependence of $e - h$ diffusion. It can still couple with a magnetic field through spin. Generally, $e - h$ diffusion affects magneto-conductance only when the magnetic field becomes sufficiently strong. On the other hand, Cooper propagation can affect magnetoconductivity even when the magnetic field is weak.

A diagrammatic approach to conductivity may be made based on a dimensionless parameter

$$\lambda = \frac{\hbar}{2\pi\varepsilon_F\tau}.$$

In terms of the $e - e$ and $e - h$ propagators the above-mentioned authors took into consideration the self-energy corrections to order λ. It has been found that due to Coulomb interaction the lifetime of electrons at the Fermi surface is modified such that

$$\tau \rightarrow \tau \left(1 + \lambda g \ln \frac{\hbar}{4\pi\tau k_B T} \right),$$

where g depends on interaction and diffusion patterns. The corresponding correction to the density of states $g(0) = m/(2\pi\hbar^2)$ at the Fermi surface is given by

$$g(0) \rightarrow g(0) \left(1 - 2\lambda g \ln \frac{\hbar}{4\pi\tau k_B T} \right).$$

Hence, the Pauli susceptibility which depends on $g(0)$ gets a logarithmic correction. The quantum corrections to the include contributions from both Cooper and diffusion channels. The diamagnetic susceptibility is also affected.

A logarithmic temperature dependence of the magnetic susceptibility has been observed in hydrogen doped Ar-Ni metallic glass [9.18]. This anomalous dependence has been attributed to an electron interaction effect based on a fit of the data with the following form for the susceptibility increment:

$$\delta\chi = -AT^{1/2} - BT^{1/2} \ln^{-1}(T_c/T) + C.$$

The temperature dependent terms on the right side can be expected from theory, and A, B, C and T_c are adjustable parameters. The first term represents a correction due to spin splitting in the diffusion channel and the second term is a similar correction in the Cooper channel. The parameters are found to be positive, as expected. Hydrogenation increases A strongly and B slightly.

The conductivity is also changed due to eiectron–electron interaction. It has been shown that for two dimensions this change together with that given by (9.1.14) is expressed for low temperatures by

$$\Delta\sigma(T) = \frac{e^2}{2\pi^2\hbar} \alpha_T \ln T. \tag{9.3.1}$$

Here α depends on g, its simplified form being

$$\alpha_T = p + 4 - 3 \left(\frac{2+F}{F} \right) \ln \left(1 + \frac{F}{2} \right);$$

$$F = \frac{1}{2\pi} \int d\theta \left(1 + \frac{2k_F}{\kappa} \sin \frac{\theta}{2} \right)^{-1}, \tag{9.3.2}$$

where κ is the inverse screening length and k_F is the . For Si [100] inversion layers, typical values of κ and k_F are

$$\kappa = 1.87 \times 10^7 \, \text{cm}, \quad k_F = 1.77 n^{1/2}.$$

Hence, for electron density of order $10^{12} \, \text{cm}^{-2}$ we find $\kappa > k_F$ and F less than 1. To first order in F, α_T is given by

$$\alpha_T = p + 1 - \frac{3}{4} F. \tag{9.3.3}$$

For $F \leq 1$ α_T is positive. Hence, near absolute zero, $\Delta\sigma(T) < 0$. This corresponds to a positive resistivity.

If a 2D system is in a perpendicular magnetic field, the conductivity change $\Delta\sigma(H)$ due to the magnetic field can be expressed as

$$\Delta\sigma(H) = \frac{e^2}{2\pi^2\hbar} \alpha_H \ln H, \tag{9.3.4}$$

where

$$\alpha_H = 1 - F/4. \tag{9.3.5}$$

$\alpha_H < 0$ for $F \leq 1$, and we expect a positive magnetoconductivity or a negative resistivity. One can use for Si inversion layers $\alpha_H = g_v \alpha$, where g_v is the valley degeneracy factor, which is 2 in the [100] direction.

On the other hand, if spin–orbit coupling is strong,

$$\alpha_H = -(1 + F)/2. \tag{9.3.6}$$

If p is small in this case, one finds

$$\alpha_T = 1 - 3F/4. \tag{9.3.7}$$

Thus, $\alpha_T > 0$ even if the corresponding magnetoresistance is positive. Such a behavior has been observed in films of Au and Pd.

Measurements of magnetoresistance of the 2D electron gas in GaAs/ GaAlAs heterostructures have been made by several authors, and the temperature-dependent parabolic magnetoresistance, observed above $0.1T$, has been associated with electron–electron interaction. Figure 9.11 illustrates the data of *Choi* et al. [9–10] for four samples of different widths. The squares, triangles, black circles and white circles correspond respectively to widths of $3.5 \, \mu\text{m}$, $6.2 \, \mu\text{m}$, $15.6 \, \mu\text{m}$ and $34.5 \, \mu\text{m}$. The ordinate is $|\Delta\rho(H)/H^2|n^2 2\pi^2\hbar$, which is proportional to changes in resistance rather than conductivity, and the abscissa represents the difference $\psi(1/2 + \hbar/kT\tau) - \psi(1/2)$ of digamma functions. *Choi* et al. used a Hall geometry in which the measured quantity in the presence of a magnetic field was ρ_{xx} rather than σ_{xx}. For magnetic fields between a few gauss and $1T$, they measured the change in the magnetoresistance, $\Delta\rho(H) = \rho(H) - \rho(0)$. This change is associated with $\Delta\sigma(H)$ by

$$\Delta\rho(H) = \frac{(\omega_c\tau)^2}{\sigma_0^2} \Delta\sigma(T) + [1 + (\omega_c\tau)^2]\frac{\sigma_s}{\sigma_0^2}, \tag{9.3.8}$$

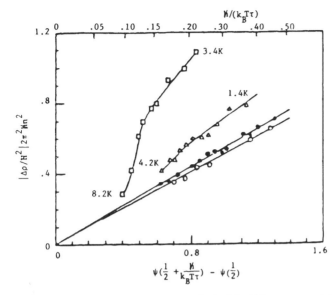

Fig. 9.11. Magnetoresistance in GaAs/GaAlAs measured by *Choi* et al. [9.19]

where σ_0 is the Drude conductivity, and

$$\sigma_s = -\frac{\sigma_0}{1 + (\omega_x \tau)^2} \cdot \frac{2(\omega_c \tau)^2}{1 + (\omega_c \tau)^2} \cdot \frac{2\pi^2 k_B T}{\hbar \omega_c} \cosh\left(\frac{2\pi^2 k_B T}{\hbar \omega_c}\right)$$

$$\times \cos\left(\frac{2\pi \varepsilon_F}{\hbar \omega_c}\right) \exp(-\pi/\omega_c \tau). \tag{9.3.9}$$

σ_s represents SdH oscillations. By omitting the oscillating part, the non-oscillating part is given by

$$\Delta\rho(H) = \left(\frac{e\tau}{m\sigma_0}\right)^2 H^2 \Delta\sigma(H), \quad (2D). \tag{9.3.10}$$

Hence, measurements of $\Delta\rho(H)$ yield $\Delta\sigma(H)$. For high mobility 2D samples in which $k_B T\tau/\hbar \gg 1$, $\Delta\sigma(H)$ can be given by

$$\Delta\sigma(H) = -\frac{e^2}{2\pi^2\hbar}\alpha\left[\Psi\left(\frac{1}{2} + \frac{\hbar}{kT\tau}\right) - \psi\left(\frac{1}{2}\right)\right], \quad (2D). \tag{9.3.11}$$

Since $\Delta\sigma(H)$ is actually independent of H, $\Delta\rho(H)$ varies parabolically in the range 0–0.7T. The two relatively wide samples follow straight lines, passing through the origin, as expected from the theory. From the slopes, the values of α were found to be 0.54 ± 0.02 and 0.51 ± 0.02. These values are close to the theoretical value of 0.7 for $F = 0.45$ and $n = 5 \times 10^{15}$cm^{-2}. However, it has been found more recently that the experimentally determined F is temperature dependent [9.20]. This dependence may be due to inelastic scattering, but it is a matter to be clarified.

Within their experimental uncertainties, *Choi* et al. observed no width dependence. This indicates that the data correspond to two dimensions. For narrower cases, the data showed increasing deviations from the 2D behavior, particularly at higher temperatures. They found also that a narrow sample of width $1.1 \, \mu$m follows the expected 1D behavior in which $|\Delta\rho(H)/H^2|$ is proportional to $T^{-1/2}$,

$$\Delta\rho(H) = -\frac{\alpha'}{n^2 2^{1/2} \pi^2 \hbar w} \left(\frac{\hbar D}{k_B T} \right)^{1/2} H^2, \quad (1D), \tag{9.3.12}$$

where w is the width and α' is the interaction constant for this case. Slightly wider samples with width $3.0 \, \mu$m and $3.5 \, \mu$m showed the same behavior below $1.1 \, $K, while for width $6.2 \, \mu$m the ID behavior was found to be not fully developed.

The parabolic variation of the magnetoresistance and its temperature variation confirm that electro–electron interaction plays an important role in weak localization. Their data show an agreement with the interaction theory not only in the functional dependence but also in the magnitude of the corrections. The data also indicate a dimensional crossover when the sample dimension becomes comparable to the thermal diffusion length, i.e., $w \sim \pi(\hbar D/k_B T)^{1/2}$. If the sample dimension is less than 1.8 times the diffusion length, even zero-dimensional behavior is observed as predicted by theory. The magnetoresistivity for the case with length L is given by

$$\Delta\rho(H) = -\frac{0.558g'L}{n^2 2^{1/2} \pi^2 \hbar w} H^2, \quad (0D), \tag{9.3.13}$$

where g' is a parameter which includes the Hartree interaction and the ratio of the relaxation time τ to that in the Drude conductivity and L is the probe length. One cannot neglect boundary scattering in small samples. In fact boundary scattering can be as effective as the impurity scattering, and is expected to cause irregular oscillations. Moreover large quantum fluctuations appear in narrow channels. It is known that the amplitude of the fluctuations in the conductance of narrow channels at absolute zero is of order e^2/h independent of the sample size, shape and the conductance itself.

10. Hopping, Percolation and Conductance Fluctuations

Unusual conductive properties of electron systems have been observed, particularly in low dimensions.

10.1 Hopping and Percolation

Hopping of electrons between localized states at finite temperatures can cause conduction. In 1968, *Mott* [10.1] derived a formula for the temperature variation of hopping conduction of a disordered system in a remarkably simple way. We start this section with his derivation of a conductivity formula for variable-range hopping.

The wave function of a localized electron may be assumed to be proportional to $\exp(-\alpha R)$, where R is the distance from the localized position. The probability of hopping of the electron between two localized states with an energy difference $\Delta\varepsilon$ should be proportional to the overlapping of the two relevant electron waves and the Boltzmannn factor of the energy difference. Hence,

$$\sigma \propto \exp(-2\alpha R - \Delta\varepsilon/k_B T). \tag{10.1.1}$$

If the density of states per unit volume per unit energy is g near the Fermi energy, the total density of states contained in the volume in a d-dimensional space with a characteristic length R is gR^d. Hence, using an appropriate constant a, we write

$$\Delta\varepsilon = \frac{a}{gR^d}.$$

The distance R can be variable. The particular R that yields the maximum contribution to a is determined by

$$\frac{d}{dR}\left(2\alpha R + \frac{a}{gR^d k_B T}\right) = 0.$$

The distance R satisfying this equation is introduced in (10.1.1) with the above form of $\Delta\varepsilon$. We arrive at the Mott formula for the conductivity at low temperatures due to variable range hopping,

$$\sigma \propto \exp[-(T_0/T)^{1/(d+1)}], \tag{10.1.2}$$

where the constant T_0 is proportional to $\alpha^{d/d+1}$. It may be expressed in terms of a relaxation time as

$$T_0 = \hbar/\tau k_B.$$

Note that the temperature exponent is $1/4$ for bulk disordered systems. The exponent is $1/3$ and $1/2$ for 2D and 1D respectively. These exponents have been observed experimentally [10.2]. *Fowler* et al. [10.3] observed variable range hopping in Si accumulation layers for temperatures between $0.2\,\mathrm{K}$ and $3\,\mathrm{K}$. They found that for gate voltages below around $6\,\mathrm{V}$, the temperature exponent $1/2$ gave the best fit while above $6\,\mathrm{V}$ the exponent changed to $1/3$. That is, they observed a transition in variable range hopping from 1D to 2D. This change may be due to a decrease in the most probable hopping distance because the higher the gate voltage the larger the density of states. However, other mechanisms associated with a transition from strongly localized to weakly localized 1D transport cannot be ruled out.

It is clear that in order to sustain a finite conductivity hopping processes must be continued from one end to the other of a given system. We then face a process called *percolation*.

In general, electrons can diffuse through regular lattices, experiencing a succession of random scatterings. They can also percolate through random media in regular motion like the case of coffee coming out of a percolator through coffee powder. Percolation can be an important process for conduction in disordered systems. For instance, in systems such as In-Ge and Pb-Ge metallic and insulating atoms are distributed randomly so that percolation takes place. A superconductor–normal metal mixture provides another interesting example.

Both continuous and discrete systems can show percolation. In the case of discrete lattices there can be *site percolation* or *bond percolation*. In the former case electrons cannot pass through any bond leading immediately to a defect site. In the latter case particular bonds cannot be passed through. In both cases it is clear that percolation depends on the concentration ratio of the conductive to nonconductive parts. If the probability p of finding conductive parts is small, they are scattered like islands in a nonconductive ocean. As this probability increases these islands will start clustering together to form larger and larger clusters and finally macroscopic clusters connecting one end of the system to the other appear and conductive channels are opened at a certain concentration.

Therefore we expect that the probability of percolating through the medium varies with p as in Fig. 10.1. Its variation may be expressed as

$$P(p) = a(p - p_c)^b.$$

This represents a power law in which the threshold probability p_c is proportional to the critical concentration of the conductive parts.

It has been found experimentally and numerically that the conductivity follows a similar power law at a threshold probability p_c. Figure 10.2 illu-

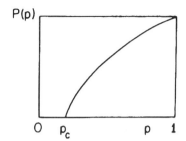

Fig. 10.1. Schematic variation of the probability of percolation with the concentration p of metallic atoms

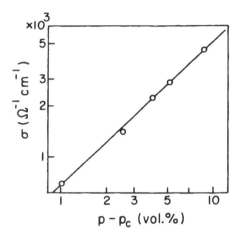

Fig. 10.2. Normal state conductivity of Pb as a function of $p-p_c$ [10.4b]

strates *Deutscher*'s data [10.4] on the normal state conductivity σ of a mixture of metallic atoms Pb with nonmetallic atoms Ge as a function of $(p - p_c)$ in volume percentage p of Pb atoms. The data show clearly the existence of a critical concentration p_c about which σ varies with a certain power of $(p-p_c)$.

Near p_c, the conductivity is expected to behave as follows:

(i) $p > p_c$

$$\sigma(p) = \sigma_0(p - p_c)^\beta,$$

(ii) $p = p_c$

$$\sigma(p_c) = \sigma_0(\sigma_1/\sigma_0)^\gamma, \qquad (10.1.3)$$

(iii) $p < p_c$

$$\sigma(p) = \sigma_1(p_c - p)^{-\alpha}.$$

Here a set of constants (σ_0 and β) or (σ_1, α) characterizes the concentration variation on each side of the threshold p_c. It is expected that

$$\gamma = \frac{\alpha}{\alpha + \beta}. \qquad (10.1.4)$$

Table 10.1. Critical constants for percolation processes

	p_c	α	β	$\alpha/(\alpha+\beta)$	γ
Bond perc. (3D)	0.25	1	1.6 ± 0.1	0.615 ± 0.15	0.67 ± 0.08
Bond perc. (2D)	0.48 ± 0.02	1	1.1 ± 0.1	0.52 ± 0.02	0.51 ± 0.01
Bethe lattice	$1/(z-1)$	1	2	0.66	≥ 0.6
(z branches)					
EMT (3D)	0.33	1	1	0.5	0.5

These relations represent a scaling character near the percolation threshold when the onset of conductivity is viewed as a second-order phase transition. The constants appearing in these relations have been determined by Monte Carlo and other methods. Some of the results for the critical constants are listed in Table 10.1 [10.5]. They are obtained mostly numerically, but a Bethe lattice has been treated exactly. We shall discuss EMT, i.e. effective medium theory and the Bethe lattice shortly. An effective medium theory has also been developed for the dielectric constant of inhomogeneous systems. We shall outline percolation theory in the remainder of this section. The scaling character of percolation can be expressed in a more abstract way by considering the conductivity σ as a function of two variables,

$$\varepsilon = p - p_c,$$

$$x = \sigma_1/\sigma_0,$$

where x is the volume fraction of the conductive component, p_c is the percolation threshold, and σ_0 and σ_1 are the conductivities on either side of the threshold. Of these two variables, ε measures the distance from the threshold, and x determines the magnitude of conductivity. We postulate [10.5]

$$\sigma(\varepsilon, x) = \sigma_0 \lambda \Psi(|\varepsilon|\lambda^{-s}, x\lambda^{-t}), \quad \varepsilon > 0,$$
$$\sigma(\varepsilon, x) = \sigma_1 \lambda \Psi(|\varepsilon|\lambda^{u}, x\lambda^{-v}), \quad \varepsilon < 0, \tag{10.1.5}$$

where s, t, u and v are scaling parameters. These equations imply that if σ_0 and σ_1 are scaled by the same numerical factor, the conductivity σ is multiplied by that same factor since the function Ψ is invariant under such a scaling transformation.

Under the scaling assumption, let us investigate important special cases. First, in the limit $x \to 0$, λ can be chosen so as to obtain

$$\sigma(\varepsilon, 0) = \sigma_0 \varepsilon^{1/s}\Psi(1, 0), \quad \varepsilon > 0,$$
$$\sigma(\varepsilon, 0) = \sigma_1(-\varepsilon)^{-1/u}\Psi(1, 0), \quad \varepsilon < 0. \tag{10.1.6}$$

Second, if $\varepsilon \to 0$ we find

$$\sigma(0+, x) = \sigma_0 x^{1/t}\Psi(0, 1),$$
$$\sigma(0-, x) = \sigma_1 x^{1/v}\Psi(0, 1). \tag{10.1.7}$$

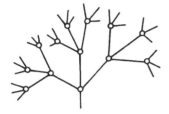

Fig. 10.3. A Bethe lattice

Since these two must coincide at $\varepsilon = 0$, we obtain

$$1 + v^{-1} = t^{-1}. \tag{10.1.8}$$

Moreover, if the derivative $\partial\sigma/\partial\varepsilon$ of the conductivity is also continuous in the limit $\varepsilon \to 0$, we obtain

$$(1 - s)/t = (1 + u)/v + 1. \tag{10.1.9}$$

From (10.1.8) and (10.1.9), we find

$$1 + v^{-1} = s^{-1}/(s^{-1} + t^{-1}).$$

Hence, changing the parameters such that

$$\gamma = 1 + v^{-1}, \quad p = s^{-1}, \quad q = u^{-1}$$

we arrive at

$$\gamma = \frac{p}{p + q}. \tag{10.1.10}$$

At $p = p_{\mathrm{c}}$, one obtains a power law

$$\sigma \propto \sigma_0 x^{\gamma} \tag{10.1.11}$$

in agreement with (10.1.3).

A *Bethe lattice* of degree z is an infinite tree in which each vertex has the same number z of branches. Figure 10.3 illustrates the case of $z = 4$. Such a lattice has been studied extensively in connection with lattice statistics and the percolation probability has been obtained exactly by *Fisher* and *Essam* [10.6]. Let $P(p)$ be the percolation probability for the fraction p of conductive sites. Instead of working on $P(p)$ it is convenient to use $\overline{P}(p) = 1 - P(p)$, which is the probability that all open walks from a chosen vertex are of finite length. We subdivide the walks according to the direction of the first step and introduce the probability $Q(p)$ that all open walks in the chosen direction are finite. We find

$$\overline{P}(p) = [Q(p)]^z. \tag{10.1.12}$$

On the other hand $Q(p)$ satisfies the recurrence relation

$$Q(p) = q + pQ(p)^{z-1}, \tag{10.1.13}$$

where $q = 1 - p$.

Obviously (10.1.13) has a particular solution $Q(p) = 1$. Since $P(p) = 0$, no percolation takes place in this case. For a linear chain in which $z = 2$ only this solution exists. If $z = 3$, there is another solution

$$Q(p) = \frac{q}{1-q} = \frac{1-p}{p}. \tag{10.1.14}$$

We find that $p = 1/2$ leads to $Q(p) = 1$. If $p < 1/2$, the solution in (10.1.14) is meaningless because Q is larger than 1. Hence, we take $Q = 1$ for $p \leqq 1/2$ so that $p_c = 1/2$ is the percolation threshold. When a current flows to a certain junction there are only two directions to continue flowing. For percolation at least one of the two must be open, resulting in $p_c = 1/2$.

In general there are $z - 2$ other solutions. The solution which vanishes at $p = 1$ can be used for $p > p_c$. When $z = 4$, the solution is given by

$$Q(p) = \frac{[p(4-3p)]^{1/2} - p}{2p}. \tag{10.1.15}$$

In this case $Q(1/3) = 1$ so that $p_c = 1/3$. Extending these particular cases we find in general

$$p_c = \frac{1}{z-1} \tag{10.1.16}$$

as in Table 10.1. Since Q is close to 1 near $p = p_c$, a Taylor expansion of (10.1.13) results in

$$Q \sim 1 - \frac{2(p-p_c)}{(z-2)p}. \tag{10.1.17}$$

That is, the percolation probability vanishes linearly following

$$P(p) \sim \frac{2z}{(z-2)} \left(1 - \frac{p_c}{p}\right), \quad (p \to p_c^+). \tag{10.1.18}$$

The percolation and Ising lattice statistics have many common features, and as might be expected the percolation problems of certain lattices have been solved analogously to the corresponding lattice statistics. In particular a duality transformation is a powerful method to determine the threshold probability as in the case of the Ising lattice. For example, the square lattice which is self-dual has

$$p_c = 1/2. \tag{10.1.19}$$

There is also a *renormalization group* approach to percolation [10.7]. For percolation, the average conductive cluster size is expected to diverge, so we expect that a characteristic correlation length ξ diverges like

$$\xi \propto a|p - p_c|^{-\nu}, \tag{10.1.20}$$

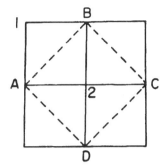

Fig. 10.4. Scaling of a square lattice

where a is the lattice constant. The exponent ν is related to the conductivity exponent β. For d dimensions one might expect

$$\beta = (d-1)\nu + \gamma,$$

where $\gamma > 0$.

Let us discuss such a relation by scaling a square lattice with a unit cell 1A2B to another square lattice with ABCD as in Fig. 10.4 [10.7]. The new lattice constant is larger by a factor $r = 2^{1/2}$. Let p be the probability for bond percolation in the original lattice. Let it be transformed into p' in the new lattice. If only the unit lattice is considered in both cases, neglecting all other neighbors, p' is given by the probabilities for A1B and A2B minus the probability of closing the unit cell. That is,

$$p' = 2p^2 - p^4. \tag{10.1.21}$$

$p' = p'(p)$ is the renormalization equation. The percolation threshold is given by the *fixed point* at which no change in the probability takes place. Hence, it is given by replacing p' by p, resulting in $p^* = 0.618$. Near the fixed point a linearized transformation between $\Delta p' = p' - p^*$ and $\Delta p = p - p^*$ leads to

$$\delta p' = \lambda_1 \delta p, \tag{10.1.22}$$

where the eigenvalue $\lambda_1 = 4p^*(1 - p^{*2}) = 1.528$. Since the ratio of the lattice constants is $2^{1/2}$ the exponent ν for the correlation length is given by

$$(2^{1/2})^\nu = \lambda_1.$$

This leads to $\nu = 0.818$. The correct exponent is 1.34 ± 0.02, indicating that the above approximation based on a single cell is rather poor. Taking into consideration distant neighbors one can of course improve the result.

Let us now discuss how the above approach can generate the conductivity exponent. For a mixture of conductive and nonconductive bonds we introduce a binary distribution

$$P(\sigma) = (1-p)\delta(\sigma) + p\delta(\sigma - \sigma_0). \tag{10.1.23}$$

σ_0 is a constant conductance. This represents a binary distribution: either $\sigma = \sigma_0$ or $\sigma = 0$. However, when scaling is introduced such a binary character may not be preserved. Therefore, in order to derive the conductivity exponent, we introduce a more general requirement that the distribution function $P(\sigma)$ is transformed into $\tilde{P}(\sigma)$ in accordance with

$$\tilde{P}(\sigma) = (1-p)\delta(\sigma) + pP(\sigma). \tag{10.1.24}$$

$P(\sigma)$ is normalized such that

$$\int P(\sigma)\mathrm{d}\sigma = 1.$$

Equation (10.1.23) is a special case of (10.1.24). Within the framework expressed by (10.1.24) and together with the scaling transformation which causes the change from p to p' we assume that the distribution $P'(\sigma)$ is transformed into $\tilde{P}'(\sigma)$ in accordance with

$$\tilde{P}'(\sigma) = (1-p')\delta(\sigma) + p'P'(\sigma). \tag{10.1.25}$$

For convenience, let the four bonds in the original lattice have conductances σ_i $(i = 1, 2, 3, 4)$. The series and parallel connections of the four conductances can be represented by

$$\tilde{P}'(\sigma) = \int \prod_{i=1}^{4} \mathrm{d}\sigma_i \tilde{P}(\sigma_i)\delta\left[\sigma - \left(\frac{1}{1/\sigma_1 + 1/\sigma_2} + \frac{1}{1/\sigma_3 + 1/\sigma_4}\right)\right]. \tag{10.1.26}$$

Introducing $\tilde{P}(\sigma_j)$ of the form of (10.1.25) into this expression and integrating the result over σ_i, we find

$$\begin{aligned}
p'P'(\sigma) = &\ 2p^2(1-p^2)\int \mathrm{d}\sigma_1 \mathrm{d}\sigma_2 P(\sigma_1)P(\sigma_2)\delta\left(\sigma - \frac{\sigma_1\sigma_2}{\sigma_1+\sigma_2}\right) \\
&+ p^4 \int \prod_{i=1}^{4} \mathrm{d}\sigma_i P(\sigma_i)\delta\left[\sigma - \left(\frac{\sigma_1\sigma_2}{\sigma_1+\sigma_2} + \frac{\sigma_3\sigma_4}{\sigma_3+\sigma_4}\right)\right]
\end{aligned} \tag{10.1.27}$$

with the identification

$$p' = 2p^2 - p^4.$$

We now have a set of transformations given by (10.1.21) and (10.1.27) which are of the forms

$$p' = R(p),$$
$$P'(\sigma) = W[P(\sigma)]. \tag{10.1.28}$$

We look for a fixed point from $p^* = R(p^*)$ and a solution in the form

$$\lambda_t P^*(\lambda_t \sigma) = W[P^*(\sigma)]. \tag{10.1.29}$$

This represents an integral equation to determine λ_t. Under scale-changes in length or in δp or in σ_0, if P behaves like

$$P(\sigma, \delta p, \sigma_0, a) = \mu f(\mu \sigma, b^{1/\nu} \lambda \delta p, \mu b^{-\beta/\nu_\lambda - \beta} \sigma_0, a/b) \tag{10.1.30}$$

we find that the average conductivity $\langle \sigma \rangle$ and the coherence length ξ behave as in (10.1.3i) and (10.1.20) respectively:

$$\langle \sigma \rangle = 1/\mu - \sigma_0 (\delta p)^\beta,$$

$$\xi \sim \frac{b}{a} \sim (\delta p)^{-\nu}.$$

At the fixed point, the two lattices with a and with b have the same correlation length so that

$$\frac{(\delta p')^{-\nu}}{(\delta p)^{-\nu}} = \lambda_1^{-\nu} = \frac{1}{r}, \tag{10.1.31}$$

where $\lambda_1 = \delta p'/\delta p$ and r is the ratio of the lattice constants. In the above example, $r = \sqrt{2}$. Equation (10.1.31) yields

$$\nu = \frac{\ln r}{\ln \lambda_1}. \tag{10.1.32}$$

For the average conductivity of the form $(\delta p)^\beta$, we expect

$$\frac{1}{\lambda_t} = \frac{(\delta p')^\beta}{(\delta p)^\beta} = \lambda_1^\beta. \tag{10.1.33}$$

This enables us to evaluate the exponent β from λ_t and λ_1. We shall not give further details based on (10.1.27) since we already know that a single cell approximation is rather poor. Therefore, it is desirable to go beyond this approximation but it becomes increasingly complicated to take into consideration distant neighbors. Nevertheless, improved results have been achieved.

While the Bethe lattice and renormalization group theories are applicable near the percolation threshold, the mean field theory or effective medium theory can be used only away from the threshold. The effective medium theory is similar to the coherent potential approximation discussed in Chap. 8. Some of the relevant results of effective medium theory are compared in Table 10.1 with some other theoretical results.

Hopping conduction in a random conductance network has been studied by several investigators. We may assign a form

$$g_{ij} = g_0 \exp[-s(\varepsilon_i, \varepsilon_j, R)] \tag{10.1.34}$$

for the conductance between sites i and j, where $\varepsilon_i(\varepsilon_j)$ is the energy at site $i(j)$ and R is a hopping distance. The quantity s is a stochastic variable and g_0 is a constant. It has been found that an appropriate average of s is determined by the percolation threshold [10.8]. As a result the dc conductivity is determined essentially by

$$s(\varepsilon_1, \varepsilon_2, R) = s_p, \tag{10.1.35}$$

where s_p is the critical percolation exponent.

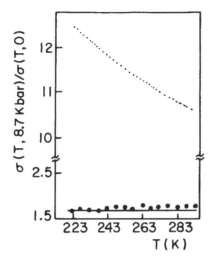

Fig. 10.5. Conductivity ratio of poly-acetylene as a function of temperature. *Black dots*: data. *Line*: Kivelson's theory. *Dotted line*: variable range hopping [10.12]

For hopping in narrow nondegenerate energy bands *Butcher* et al. [10.9] adopted the form

$$s(\varepsilon_i, \varepsilon_j, R) = p(R) + q(\varepsilon_i, \varepsilon_j) \tag{10.1.36}$$

with $p(R)$ given by

$$p(R) = 2\alpha R - u\ln(\alpha R), \tag{10.1.37}$$

where α depends on the wave function and u is a parameter which depends on the band anisotropy. u can be 2 for an isotropic band and 3/2 for an anisotropic band [10.10]. For a sharp distribution of s about s_p the average conductivity assumes the form

$$\langle \sigma \rangle = \phi \exp[-2\alpha s_p(R)], \tag{10.1.38}$$

where ϕ is a slowly varying function of the parameters characterizing the system. It also depends on temperature particularly because one can write $\exp[-q(\varepsilon_1, \varepsilon_2)] \sim \exp(-\varepsilon/k_B T)$ where ε has the meaning of an activation energy. An average hopping distance may also depend on temperature.

Equation (10.1.38) may be considered as a generalization of Mott's expression given by (10.1.1). Its application to n- and p-type Ge resulted in good agreement with experiments. It has been modified for intersoliton hopping in polyacetylene by *Kivelson* [10.11] with a temperature-dependent prefactor and an exponential factor which depends on pressure. For variable range hopping σ does not separate out in temperature and pressure dependences, while it does in the Kivelson formula. This separation has been confirmed experimentally by *Heeger* and *Macdiarmid* [10.12] on polyacetylene. Figure 10.5 illustrates their result in which the ordinate represents $\sigma(T, 8.7\,\text{kbar})/\sigma(T, 0)$, that is, the ratio of σ at 8.7 kbar to that at zero pressure. The dotted curve, representing variable range hopping, is far from the experimental points given by dots. The bottom line represents Kivelson's formula.

10.2 Universal Conductance Fluctuations in 1D Systems

In 1982, *Fowler* et al. [10.3b] observed random but reproducible conductance fluctuations in narrow 2D systems in Si MOSFETs when the gate voltage was changed at low temperatures of order 0.1 K. Their experimental device is shown in Fig. 10.6. The upper graph is an idealized top view of a sample. The two N^+ regions are the source and drain, and P^+ regions are the control electrodes. The width between the controls is 1–2 μm, and the length of the controls is 14 μm. The substrate is 10 Ωcm n-type [100] silicon. The lower graph shows a cross section along the dotted line. The electrons at the surface are pinched into the narrow region between the controls due to the electric field applied from these control-electrodes and arising from the built-in potential between the P^+ regions and the n-type substrate. Figure 10.7 illustrates their conductance data at three temperatures. The structure in this graph is not noise but real. Similar observations have been made of the conductance fluctuations in metallic wires of Au, AuPd and GaAs [10.13]. In 1985, *Webb* et al. [10.13b] performed conductance experiments on very small single gold rings illustrated in Fig. 10.8. These rings had average diameters of 825 and 245 nm, and a lone wire of length 300 nm was attached to them. The resistance was measured with a four-probe bridge operated at 205 Hz and 200 nA (rms). Their experiments clearly demonstrated magneto-conductance oscillations with the principal period of ch/e plus a second set of oscillations with a flux period of $ch/2e$. Figure 10.9 depicts their data on the magnetoresistance of a gold ring at 0.01 K.

On the other hand, the interference experiment conducted by Sharvin and Sharvin showed only one period of $ch/2e$ as discussed in Sect. 9.2. The absence of the periodicity ch/e was long a puzzle, but it is now attributed to the device size relative to the phase coherence length. They used a cylinder which was longer than the phase coherence length. Such a cylinder can be considered to be equivalent to a superposition of very many single rings. Since each ring has its own starting phase the Aharonov-Bohm period ch/e was wiped out while the coherent backscattering period $ch/2e$ was still preserved.

Numerical simulations of variable-range hopping in a $1D$ random resistor network by *Lee* [10.14] based on the Mott formula (9.1.11) and of magnetoresistance by *Stone* [10.15] based on a two-probe generalization of the *Landauer* formula [10.16] by *Büttiker* et al. [10.17] resulted in similar fluctuations. However, the physics behind their formulas did not really correspond to the experimental situation. It has been found that experimentally observed fluctuations are not related to energy differences but rather to quantum interference and that their proper descriptions are given by a four-probe formula. Voltage fluctuations were also calculated by *Maekawa* et al. [10.18] based on linear response theory.

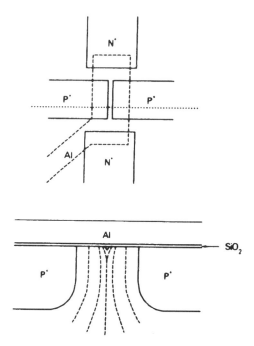

Fig. 10.6. *Fowler* et al.'s experimental device [10.3]

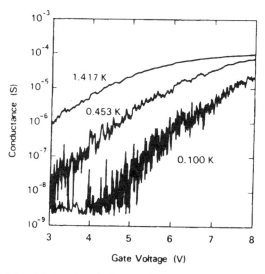

Fig. 10.7. Aperiodic conductance oscillations in narrow 2D systems [10.3]

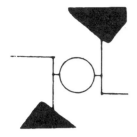

Fig. 10.8.   *Webb* et al.'s experimental device [10.13]

Before discussing conductance fluctuations, it is desirable to mention multichannel conductance, because even though it is relevant to experimental observations, some theoretical controversy has existed. For this purpose, it is convenient to follow the historical development starting with the Landauer formula [10.16] for a 1D conductor. This formula determines current in terms of the transmission and reflection coefficients, T and R, of a conductor acting as a single scattering center in recognition of the wave character of the electrons. His conductance formula is simple,

$$G = \frac{e^2}{h} \frac{T}{R}. \tag{10.2.1}$$

This formula was derived originally for a single barrier which reflects and transmits electron waves. Let us try to evaluate the resistance due to such a barrier. When incident waves hit the barrier a fraction R of the incident electrons will be reflected, decreasing the current. In order to maintain the current an additional incident fraction $2R$ must be brought in. However, this additional fraction is also reflected in the amount $R(2R)$. An infinite succession of such reflections results in a series

$$2(R + R^2 + R^3 + \ldots) = \frac{2R}{1 - R}.$$

This yields (10.2.1) with the proper factor $e^2/\pi\hbar$ for conductivity and $T = 1 - R$.

Büttiker et al. [10.17b] considered a two-probe case in which current flows between two reservoirs acting as source and sink. They assumed that the local chemical potential difference between the two ends of the sample was not equal to that between the reservoirs. The conductance was given as a function of reflection and transmission coefficients T_{ij} and R_{ij} from channels i to j in the following form:

$$G = \frac{2e^2}{h} \left(\sum_{ij} T_{ij} \right) \left(\sum_i v_i^{-1} \right) \bigg/ \left[\sum_i \left(1 + \sum_j R_{ij} - \sum_j T_{ij} \right) v_i^{-1} \right],$$

$$\tag{10.2.2}$$

where v_i is the longitudinal velocity in channel i. The same result had been given earlier by *Azbel* [10.17a]. However *Langreth* and *Abrahams* [10.17c]

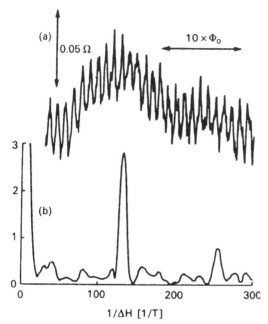

1/ΔH [1/T]

Fig. 10.9a,b. Magnetoresistance at $0.01\,\mathrm{K}$ of a gold ring [10.13]. (a) Superimposed on a slowly varying background. (b) Fourier power spectrum in arbitrary units. Two peaks are visible at $1/\Delta H = 131$ and $260\,T^{-1}$, which corresponds respectively to ch/e and $ch/2e$

adopted a different assumption, namely, that there was a single chemical potential for all the channels in each lead, and arrived at a different formula.

Büttiker [10.19] investigated more recently a four-probe case with four reservoirs having different chemical potentials. The current between two reservoirs 1 and 2 was simply associated with their chemical potential difference such that $(e^2/h)T_{12}\Delta\mu$, where T_{12} is the transmission coefficient as in the one-channel case of Landauer. His simple consideration resulted in a multiprobe generalization of the Landauer formula:

$$G_{ij} = \frac{e^2}{h}\mathrm{Tr}\{T_{ij}T_{ij}^*\}, \quad (i \neq j),$$

$$G_{ii} = \frac{e^2}{h}\mathrm{Tr}\{R_{ii}R_{ii}^* - 1\}, \tag{10.2.3}$$

where $T = (T_{ij})$ and $R = (R_{ij})$ are the transmission and reflection matrices between leads i and j. It has been shown that (10.2.3) is valid even in the presence of a magnetic field. Moreover it is consistent with Onsager's reciprocal relation.

Haug et al. [10.20a] investigated experimentally multichannel magnetotransport through a barrier in two dimensions. They demonstrated that the quantum Hall effect can be used experimentally to distinguish different multichannel formulas which arise from various assumptions about the nonequilibrium distributions of carriers in the leads and models for the voltage measurement process. Their analysis supported (10.2.3) at least for strong magnetic fields. *Washburn* et al. [10.20b] also used a barrier in the quantum Hall region of a narrow 2D system to find that such an obstruction induces tunneling between the states on opposite sides of the channel. This results in backscattering and quantized plateaus in any four-probe measurement of resistance, in general agreement with theoretical considerations [10.21].

Concerning the numerical simulations of conductance fluctuations mentioned earlier, we remark that no ensemble average over impurity configurations was taken so that the results were sample specific. *Stone*'s analysis [10.15] over a small field range revealed no structure in the magnetoresistance unless the change in the magnetic flux through the wire was roughly of order 2 or 3 times the magnetic flux ch/e. This indicated that the phenomenon was due to the Aharonov-Bohm effect. His use of (10.2.2) resulted in an asymmetry in the magnetoconductance but the magnitude of the symmetry was much less than that observed.

There are more theoretical works on conductance fluctuations. Based on a weak-scattering diagrammatic approach, *Al'tshuler* [10.22] showed that sample-dependent fluctuations arise from sample-specific interference conditions. Such an approach is possible because the experiments are carried out in the highly conducting diffusive regime of weak localization in which the mean free path is much larger than the reciprocal Fermi wave number. *Lee* and *Stone* [10.15b] also adopted weak-scattering diagrammatic techniques to conclude that the conductance of any small metallic samples fluctuates universally as a function of chemical potential or magnetic field in the sense that the fluctuations are of order $e^2/h \sim 4 \times 10^{-5}\,\Omega^{-1}$, independent of sample size and degree of disorder, as long as the temperature is low enough.

Licini et al. [10.23] worked experimentally on a weak localization regime of p-Si MOSFETS in which the metal gate electrode was narrow (70×50 nm in size). Below 1 kOe, they observed in consistence with localization theory. In fact, at 2.2 K, the magnetoresistance agreed with the 2D localization formula with an inelastic diffusion length of 120 nm. The agreement suggests that their devices are highly disordered but homogeneous. There were clear deviations from the 2D behavior below 1.5 K. They reported that the resistance was superposed at 0.14 K by aperiodic oscillations which were symmetric under reversal of field direction, persisted up to fields of order 25 kOe and were reproducible. For small changes in the gate voltage, the oscillations retained their phase from one gate voltage to another. Since no gradual shifts with gate voltage took place in the oscillations they concluded that the fluctuations were not the result of a shift with magnetic field of the energies of electronic states but rather arose from changes in the phase of the wave functions. Indeed the typical period of oscillations was around 1 kOe, corresponding to a flux quantum penetrating an area $ch/2eH \sim (W/2)^2$, where W is the

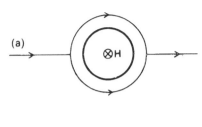

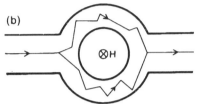

Fig. 10.10. Electron interference in vacuum (a) and in a disordered system (b) The width of the ring in the latter is enlarged for clarity

width of the channel. The temperature dependence of the oscillations was strong when the inelastic diffusion length was of order W as expected for the Aharonov-Bohm effect for a path of order W.

Kaplan and *Hartstein* [10.24] employed a tilted magnetic field on Si MOS-FETs because if the Aharonov-Bohm effect plays a key role the fluctuations should depend on the amount of magnetic flux threading the sample. They varied the field between 0 and $1.5T$ at gate voltages between 10 and 12 V. The peaks and dips in the magnetoconductance were observed at larger magnetic fields as the tilt angle was increased. When the conductance was plotted against the perpendicular component of the applied magnetic field the fluctuations were found to stay nearly constant in magnitude. This showed that the major cause of the fluctuations was orbital in nature, as expected from the Aharonov-Bohm effect. When the source drain voltage was increased the conductance became nonlinear and asymmetric. The shape of the conductance–voltage characteristic changed aperiodically with the applied magnetic field, indicating quantum interference.

The interference mechanism is illustrated in Fig. 10.10, in which graph (a) [(b)] corresponds to the case in vacuum [disordered media]. In [b] no scattering is shown in the leads, and the width of the 2D system is relatively large for clarity.

It is now established experimentally that the fluctuations are ubiquitous in small systems with a typical dimension $L \leq 1\,\mu\text{m}$ and are due to the quantum-mechanical interference which takes place at low temperatures when the carrier wave functions go around impurities, i.e., a random succession of Aharonov- Bohm oscillations at impurity sites. Hence, the fluctuations are reproducible but are sample dependent. Since electronic interference takes place between states that are phase-coherent, its length scale L_ϕ, should be compared with the size L. The phase-coherence length L_ϕ, is defined by

$$L_\phi = (D\tau_d)^{1/2}, \tag{10.2.4}$$

where D is the diffusion constant and τ_d is the lifetime. Samples much larger or much smaller than L_ϕ, may exhibit fluctuations but the magnitude of fluctuations can differ from the universal value of e^2/h. Samples much larger than L_ϕ, are expected to involve a large number of phase-breaking processes and thus to show no strong interference effects. Samples shorter than L_ϕ will have amplitudes which increase with decreasing sample length. Only the samples of size $L \sim L_\phi$ can show fluctuations of order e^2/h.

According to *Al'tshuler* et al. and *Larkin* and *Khmel'nitskii* [10.25b] the current-voltage $(I - V)$ relation in small samples in the diffusive regime in weak localization is nonlinear so that the dimensionless conductance

$$g(V) = G(V)/(e^2/h) \tag{10.2.5}$$

is not constant but is a random function of V. It is described by the correlation function $\langle g(V)g(V + \Delta V)\rangle$. The fluctuations in the $I - V$ curves are determined by a characteristic voltage given by

$$V_c = \hbar D/eL^2. \tag{10.2.6}$$

If $V \ll V_c$, the fluctuations are small. When $V \sim V_c$, the fluctuations are of order 1. If $V \gg V_c$, the fluctuations decay as $(V_c/V)^{1/2}$. Moreover, the $I - V$ curves are asymmetric about $V = 0$ and

$$\langle [g(V) - g(-V)]^2\rangle \sim 1. \tag{10.2.7}$$

The experiment of *Webb* et al. [10.26] on very pure and phase-coherent gold wires is in qualitative agrcement with these predictions. The fluctuations in such wires are found not to be symmetric about zero current:

$$\Delta g(I) \neq \Delta g(-I).$$

The autocorrelation $\langle (\Delta g)^2\rangle$ is of order 0.23.

10.3 Conductance of TTF/TCNQ

One-dimensional organic conductors such as TTF–TCNQ show a variety of interesting properties, including high conductivity, anomalous spin suscepti-bility which is temperature dependent, phase transitions, etc. As discussed in Chap. 1, they consist generally of stacks of flat molecules and are conductive due to charge transfer between the stacks. They are quasi one-dimensional in the sense that the conductivity is highly anisotropic and that electrons in different stacks can interact with each other. Electron–phonon or electron–hole coupling is singularly strong in these systems at low temperatures and causes new electron states. Hence, these systems show a variety of interesting and complex properties, contrary to our expectation from the simple 1D electron model which was introduced around 1955. That is, such a 1D electron model was attractive theoretically for its simplicity but now it is the complexity of actual 1D systems which is stimulating strong interest.

The Fermi "surface" in one dimension consists of two points in k space, one at k_F and the other at $-k_F$. The narrowness of the electron band results in strong electron correlations. Neglecting electron interaction and using the tightbinding approximation with nearest neighbor electron transfer, we can express the energy of an electron measured from the center of the unperturbed band as

$$\varepsilon_k = -2t \cos(ka), \tag{10.3.1}$$

where a is the lattice constant, and t is the transfer integral.

The electron phonon interaction may be expressed by

$$\mathcal{H}_{ep} = \sum \frac{g(k,q)}{\sqrt{N}} a^*_{k+q,s} a_{k,s} (b_q + b^*_{-q}), \tag{10.3.2}$$

where $g(k,q)$ is a coupling parameter, a^*_{ks} (a_{ks}) is the creation (annihilation) operator of an electron with spin s and b_q (b^*_q) is that for a phonon. The electron–phonon coupling is strong for electron scattering from $\pm k_F$ to $\mp k_F$. That is, there is a strong interaction of the electrons with phonons with $2k_F$. This produces a dip (softening) in the phonon spectrum, called the *Kohn anomaly*, at the $2k_F$ phonons. Thus coupling of electron states with k and with $k - 2k_F$ becomes important. Due to mixing of these two states, the energy of an electron with k in the CDW or SDW ground state is given near the Fermi energy by

$$E_k = \tfrac{1}{2}(\varepsilon_k + \varepsilon_{k-2k_F}) \pm [\tfrac{1}{4}(\varepsilon_k - \varepsilon_{k-2k_F})^2 + \Delta^2]^{1/2}, \tag{10.3.3}$$

where Δ is determined by $g(-k_F, 2k_F)$ and by an average occupation number $\langle b_q \rangle$ of phonons at $q = 2k_F$.

The above result shows that an energy gap 2Δ is created at the Fermi level. For absolute zero it indicates that the metallic system becomes unstable at $2k_F$. Thermal excitations of electrons above the Fermi level take place at finite temperature. The gap Δ corresponding to this case becomes temperature dependent. In fact a BCS-type expression for $\Delta(T)$ has been derived. Accordingly the 1D metal is expected to become unstable at a certain temperature, below which it enters into a new phase with a gap. It has been found that the 1D conductor TTF/TCNQ becomes semiconducting below 54 K. This transition has been associated with a periodic variation of the density of conduction electrons on TCNQ chains. Such periodic variations represent *charge density waves* (CDWs) with wave number $2k_F$. Another transition, associated with holes in TTF chains, takes place at a lower temperature of 38 K. Such CDWs will in turn move the ions, resulting in distortion of the lattice (*Peierls distortion*).

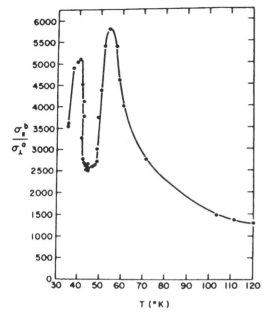

Fig. 10.11. Conductivity ratio of TTF/TCNQ [10.27a]

On the other hand, $(TMTSF)_2PF_6$ at ambient pressure becomes super-conducting below around 1 K. In this case, the gap opening is due to the formation of what are called *spin-density waves* (SDWs) with wave number $2k_F$. In this case, two independent waves are formed by two spin states at 180° phase difference. No charge moves and no lattice distortion takes place because the two waves are out of phase.

In correspondence with the above two transitions in TTF/TCNQ, the conductivity shows anomalies. Figure 10.11 illustrates the data of *Cohen* et al. [10.27a] in which the ratio of conductivities in the directions parallel and perpendicular to the principal axis is plotted against temperature. On the other hand, the activation character of the conductivity becomes clear if $\ln \sigma$ is plotted against $1/T$.

Figure 10.12 illustrates such a plot taken from [10.27b]. The data show a linear variation, i.e., a constant activation energy (~ 200 K), for a fairly wide region. In this range TTF/TCNQ is an intrinsic semiconductor. Below around 10 K, the slope decreases almost continuously, although in a narrow region another linear variation is approximately followed. Such a decrease indicates that smaller gaps are opening up at low temperatures. The number of carriers in TTF/TCNQ has been estimated [10.28] to be around $10^9/cm^3$ at 10 K based on the known value, 215 ± 25 K, of Δ in TTF/TCNQ. With an experimental $\sigma \sim 10^{-5}$ ohm^{-1} cm the mobility is estimated to be around 10^4 cm^2/Vs for an electron or hole in TTF/TCNQ. This is a rather high

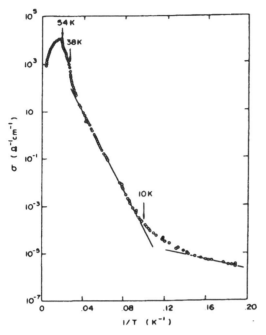

Fig. 10.12. Logarithmic plot of the conductivity of TTF/TCNQ [10.27b]

mobility, due to a small energy gap. Note that using (10.3.1) at k_F in the energy expression (10.3.3) yields an effective mass for $\Delta \gg kT$,

$$m^* = (\hbar^2/2ta^2)[(2t/\Delta)\sin^2 k_F a \pm \cos k_F a]^{-1},$$

where the first factor $(\hbar^2/2ta^2)$ is the effective mass at the bottom of the unsplit band. The $+(-)$ sign corresponds to an electron (hole) at the upper (lower) subband. If terms of order 1 or less are ignored, m^* in the presence of splitting depends on the ratio Δ/t, which is rather small. In fact, the ratio is of order 0.1 in TTF/TCNQ and 0.001 in $(TMTSF)_2PF_6$. Hence, small effective masses can be expected for small energy gaps. Such small masses cause high mobilities and a nonlinear increase of conductivity at high electric fields (hot electron effect). As a function of electric field, this mobility increase was found to be quadratic.

The Peierls description of 1D conductors neglects Coulomb interaction. More recently *Su* et al. [10.29] showed that many of the properties of polyacetylene, which is the simplest organic chain polymer, can be understood in terms of a model of coupled electrons and phonons, also neglecting Coulomb interaction. Since in actual systems the long-range Coulomb interaction can be strong, efforts have been made to refine the theories with electron interaction [10.30]. In brief summary, it has been revealed that the lattice distortion (dimerization) in the noninteracting or Hartree description is suppressed by zero-point motion, intersite correlations, and correlation corrections to the

Hartree-Fock approximation while it is enhanced by on-site correlations of the Hubbard type. A theory based on a local density approximation to the total energy of *trans*-polyacetylene has shown no appreciable energy minimum for nonzero dimerization amplitude [10.31]. Within such an approximation, it has been suggested that the Peierls picture is inadequate for describing actual systems.

11. High-T_c Superconductivity

High-T_c superconductivity may be due to electron–phonon interaction or some other more exotic mechanisms.

11.1 Electron Pairing in BCS Theory

Since its discovery by Kamerlingh Onnes in 1911, superconductivity has attracted considerable attention. This is a fascinating phenomenon which includes many challenging problems. Among them is the problem of searching for a new material with a high critical temperature. Despite much effort, the progress in this search was slow. Starting from Onnes' 4 K in mercury, it took nearly 60 years to reach the 20 K range in Nb compounds with the A15-type structure. Hence, it appeared as if the limit had arrived with T_c of 23.3 K in Nb$_3$Ge. This situation suddenly changed upon the discovery of LaBaCuO with T_c of 35 K by *Bednorz* and *Müller* [11.1] in 1986, which was the 75th anniversary of the discovery of superconductivity. This discovery has triggered strong interest in the search because not only the critical temperature but also the material is different from the previously known superconductors. Indeed a series of new materials with T_c of order 90 K was soon found, causing recognition of the existence of new types of superconductors. It is the purpose of this chapter to study the so-called high-T_c superconductors. First however, classical superconductivity will briefly be discussed.

There are a few important properties which led to understanding ordinary superconductivity. First, the Meissner-Ochsenfeld effect shows that superconductors are perfectly diamagnetic, expelling magnetic induction from their interior. The penetration depth of a magnetic field is of order 10^{-5}–10^{-6} cm. Second, the isotope effect demonstrates that the critical temperature varies as

$$T_c \sim M^{-1/2},$$

where M is the isotopic mass. This effect suggests that lattice vibrations are crucial for the occurrence of superconductivity. Third, tunneling, sound attenuation, thermal conductivity, electromagnetic absorption and other measurements indicate that there is a temperature-dependent energy gap in the elementary excitation spectrum of superconductors. Corresponding to this gap, the electronic specific heat jumps at T_c. Finally, the coherence length is of order 10^4 Å in Type I and 10^2 Å in Type II superconductors. This length governs the variation of the order parameter in space. It is usually defined by

$$\xi_0 = a\hbar v_F / k_B T_c,$$

where v_F is the velocity of electrons at the Fermi surface. The constant a is approximately the same for most superconductors, indicating that superconductivity is independent of the details of the band structure.

Cooper [11.2] suggested in 1956 that a pair of electrons above the Fermi sea forms a bound staste (Cooper pair) via . This concept was immediately taken up in the BCS theory [11.3]. In this theory the negative coupling energy due to virtual exchange of phonons plays an important role. This interaction is proportional to the energy $\hbar\omega_k$ of the zero-point lattice vibrations, which in turn is inversely proportional to the square root of the ion mass.

If two electrons with opposite spins form a pair, new pair-creation and -destruction operators can be introduced such that

$$b_{\mathbf{k}} = a_{-\mathbf{k}\downarrow} a_{\mathbf{k}\uparrow}.$$

This pair is designated by the wave vector $\mathbf{k}$, indicating that if $\mathbf{k}\uparrow$ is occupied so is $-\mathbf{k}\downarrow$. Note that in the ground state in an isotropic system, the total momentum and spin of such a pair are expected to be zero. If the interaction of pairs is negative, the parallel spin pairs can be neglected because their exchange interaction reduces the magnitude of the interaction energy.

Assuming that one pair of states $(-\mathbf{k}'\downarrow, \mathbf{k}'\uparrow)$ is scattered into another pair of states $(-\mathbf{k}\uparrow, \mathbf{k}\downarrow)$ we introduce an effective Hamiltonian,

$$H = \sum_{\mathbf{k}\sigma} \varepsilon a_{\mathbf{k}\sigma}^* a_{\mathbf{k}\sigma} + \sum_{\mathbf{k}\mathbf{k}'} V_{\mathbf{k}\mathbf{k}'} a_{\mathbf{k}\uparrow}^* a_{-\mathbf{k}\downarrow}^* a_{-\mathbf{k}'\downarrow} a_{\mathbf{k}'\uparrow}. \tag{11.1.1}$$

Here the interaction $V_{\mathbf{k}\mathbf{k}'}$ is expected to be negative. If the number of such pairs is large, we may consider that $b_{\mathbf{k}}$ is close to its average $b_{\mathbf{k}}^0$ which is a c-number given by

$$\begin{aligned} b_{\mathbf{k}}^0 &= \langle b_{\mathbf{k}} \rangle \\ &= \mathrm{Tr}[e^{-\beta H} a_{-\mathbf{k}\downarrow} a_{\mathbf{k}\uparrow}]/\mathrm{Tr}[e^{-\beta H}]. \end{aligned} \tag{11.1.2}$$

That is, if we write

$$b_{\mathbf{k}} = b_{\mathbf{k}}^0 + (a_{-\mathbf{k}\downarrow} a_{\mathbf{k}\uparrow} - b_{\mathbf{k}}^0)$$

the difference term in the brackets representing fluctuations may be assumed to be small.

We define an order parameter by

$$\Delta_{\mathbf{k}} = - \sum_{\mathbf{k}'} V_{\mathbf{k}\mathbf{k}'} b_{\mathbf{k}'}^0. \tag{11.1.3}$$

$\Delta_{\mathbf{k}}$ is actually unknown but will be determined in a self-consistent way. The Hamiltonian becomes

$$\mathcal{H} = \sum_{\mathbf{k}\sigma} \varepsilon_{\mathbf{k}} a^*_{\mathbf{k}\sigma} a_{\mathbf{k}\sigma} - \sum_{\mathbf{k}} (\Delta_{\mathbf{k}} a^*_{\mathbf{k}\uparrow} a^*_{-\mathbf{k}\downarrow} + \Delta^*_{\mathbf{k}} a_{-\mathbf{k}\downarrow} a_{\mathbf{k}\uparrow} - \Delta^*_{\mathbf{k}} b^0_{\mathbf{k}}). \tag{11.1.4}$$

For a given $\Delta_{\mathbf{k}}$ the right side is quadratic. Hence, $\mathcal{H}$ can be diagonalized by a linear transformation of $(a_{\mathbf{k}}, a^*_{\mathbf{k}})$. As a result, we arrive at an excitation energy given by

$$E_{\mathbf{k}} = (\varepsilon^2_{\mathbf{k}} + |\Delta_{\mathbf{k}}|^2)^{1/2}. \tag{11.1.5}$$

This corresponds to a pair excitation. At the same time, from (11.1.2) and (11.1.3), we get an equation for the gap parameter,

$$\Delta_{\mathbf{k}} = - \sum_{\mathbf{k}'} V_{\mathbf{k}\mathbf{k}'} \Delta_{\mathbf{k}'} [1 - 2f(E_{\mathbf{k}'})]/2E_{\mathbf{k}'}. \tag{11.1.6}$$

Equations (11.1.5) and (11.1.6) are coupled with each other. Since they are highly nonlinear, certain approximations are necessary to obtain the solutions. We assume simply that $\Delta_{\mathbf{k}} = \Delta$ is constant and that

$$V_{\mathbf{k}\mathbf{k}'} = \begin{cases} -V, & |\varepsilon_{\mathbf{k}}| < \hbar\omega \\ 0, & |\varepsilon_{\mathbf{k}}| > \hbar\omega. \end{cases} \tag{11.1.7}$$

Note that the interaction is given a minus sign, in anticipation of attraction. We then find

$$\Delta = \Delta V \sum_{\mathbf{k}} \frac{[1 - 2f(E_{\mathbf{k}})]}{2E_{\mathbf{k}}}. \tag{11.1.8}$$

Because the sum is always nonnegative, this equation has a meaningful solution only if V is positive. That is, only if $V_{\mathbf{k}\mathbf{k}'}$ is indeed attractive.

For nonzero Δ, (11.1.8) is solved for T_c which enters the Fermi distribution function. The sum over $\mathbf{k}$ in (11.1.8) can be converted into integration with a cutoff at $\hbar\omega$. We assume that $\hbar\omega \gg k_B T_c$, integrate by parts and make use of a known integral,

$$\ln 2\gamma/\pi = 2 \int_0^\infty dx \ln \frac{d}{dx} \left(\frac{1}{e^x + 1} \right),$$

where γ is Euler's constant. Assuming that $f(\hbar\omega)$ is small we arrive at

$$k_B T_c = 1.13 \hbar\omega \exp[-1/g(0)V], \tag{11.1.9}$$

where $g(0)$ is the density of states per spin at the Fermi surface. The magnitude of $g(0)V$ is expected to be around 0.01–0.1 so that T_c determined by this equation is low, as in the ordinary superconductors. Moreover, it is proportional to $M^{-1/2}$ as in the isotope effect. For the coherence length, the BCS theory yields $a = 0.18$.

This compares favorably with the experimental value of 0.15 for Sn and Al. The jump in the electronic specific heat is proportional to $d\Delta^2/dT$ below

T_c and is given by $\Delta c = 1.43\gamma T_c$. Experimental data are generally in good agreement with this jump. The temperature dependence of the penetration depth is found to follow

$$\lambda(T) = \lambda(T = 0)[1 - (T/T_c)^4]^{-1/2}. \tag{11.1.10}$$

This relation is also in agreement with experiments.

Thus the BCS theory gives a satisfactory description of superconductivity. However it applies only to weak coupling superconductivity; it does not provide a good account of strong coupling superconductors such as Pb and Hg. An extension of the BCS theory to strong electron–phonon coupling has been made in several ways.

We shall not go into details of the BCS and related theories because excellent reviews are available [11.4]. Instead, we point out that T_c from these theories are low, as one might expect from the low energy of lattice vibrations at low temperatures. Moreover, the coherence length is of order 10^4 Å, which is much larger than that in the high-T_c superconductors. Since even the atomic structure of these materials is different, as we shall discuss in the next section, they can be a completely new type.

11.2 Properties of High-T_c Superconductors

Since the discovery of superconductivity in LaBaCuO in 1986 some 40 new high-T_c materials have been reported. Although there are varieties in these materials, we focus our attention on three representative families.

11.2.1 2-1-4 Compounds

This group has the general atomic structure $La_{2-x}M_xCuO_{4-y}$, where x is of order 0.15, y is close to 0, and M can be Ba or Sr or Ca. Of these choices in M, the case Sr yields the highest T_c. The name 2-1-4 corresponds to the atomic composition ratio in the basic case of $x = y = 0$.

All the 2-1-4 superconductors have the body-centered tetragonal K_2NiF_4 structure at high temperatures. This structure is shown in Fig. 11.1. The white, black and hatched circles represent respectively oxygen, Cu and La. This structure is known as perovskite. Each Cu atom is at the center of an CuO_6 octahedron but only the one at the center of the unit cell in the diagram is clear in the figure. The octahedra at the top and bottom are displaced relative to the central one. At the top and bottom are the Cu–O1 planes. At the corners of the unit cell are La atoms, which may be replaced by M in $La_{2-x}M_xCuO_{4-y}$, perhaps randomly.

The nonsuperconductive parent compound La_2CuO_4 transforms itself into a single-face-centered orthorhombic structure below about 500 K. At this transition, the CuO_6 octahedra rotate by $5°$ from their neighboring octahedra

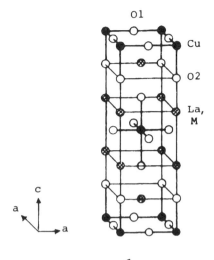

Fig. 11.1. Unit cell of $La_{2-x}M_xCuO_4$

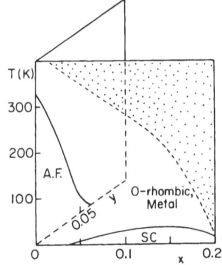

Fig. 11.2. Phase diagram of $La_{2-x}Sr_xCuO_{4-y}$. Between $x = 0$ and the edge of the superconducting state, there can be a spin glass state. The shaded area represents the tetragonal phase

alternating in the direction of rotation. This distortion causes a corrugation of the Cu–O layers in the tetrahedral [110] direction. In the plane $x = 0$ and below about 340 K, the orthorhombic phase is an antiferromagnetic insulator.

A schematic phase diagram of $La_{2-x}Sr_xCuO_{4-y}$ is shown in Fig. 11.2. The superconducting phase appears in a limited region of x which is shown in the plane $y = 0$. The critical temperature T_c depends on x, and the highest T_c occurs near 40 K at around $x = 0.15$. A metal–insulator transition takes place at a low concentration of x and the compounds become insulating. The graph shows an insulating antiferromagnetic phase in the plane $x = 0$. This magnetic phase is extremely sensitive to the parameter y which controls the oxygen content. The dotted region represents the tetragonal phase.

In the ground state of La_2CuO_4 the charges on ions La^{3+} and Cu^{2+} are balanced by 0^{2-}. The magnetic phase may be considered to correspond to the Heisenberg-type anisotropic interaction between Cu^{2+} ions. When doped with M, i.e., in $La_{2-x}M_xCuO_{4-y}$ there are $(x - 2y)$ holes per cell. The existence of oxygen holes has been shown by spectroscopic measurements of the energy loss associated with the excitation of the electrons from 1s to 2p states. This excitation does not take place in O^{2-} since $O(2p)$ states are occupied. In fact in La_2CuO_4 such an excitation is not observed.

Let us denote the energies to add a hole into the $Cu(3d)$ and $O(2p)$ states as ε_d and ε_p, and the Coulomb interactions between the holes by U_d, U_p and V in $Cu(3d)$, $O(2p)$ and neighboring copper and oxygen sites respectively. These energies are estimated to be as follows:

$$\varepsilon = \varepsilon_p - \varepsilon_d = 1 \sim 2\,eV, \quad U_d \sim 8 - 10\,eV, \quad U_p \sim 5 - 7\,eV, \quad V \sim 1 - 2\,eV.$$

These are the parameters entering a model Hamiltonian to be discussed in the next section.

At first, the holes will go into the lowest energy states in $Cu(3d)$, resulting in Cu^{2+}. Additional holes produced by doping are considered to go into $O(2p)$ states because their energy $\varepsilon + 2\,V$ is lower than U_d. The antiferromagnetic phase is considered to be due to the superexchange interaction between the Cu^{2+} spins which is induced by hopping of holes. The $O(2p)$ holes may move around to form a band, but $Cu(3d)$ holes are probably localized. X ray spectroscopy shows that Cu^{2+} states continue up to $x \sim 0.3$ of Sr. Beyond this value the system becomes an insulator.

11.2.2 1-2-3 Compounds

This family has the general structure $RBa_2Cu_3O_{7-\delta}$, where R represents Y, Eu, Gd, etc., and the members are termed 1-2-3 compounds after the composition ratio of the first three atoms. The critical temperature depends on δ, and 93 K for $\delta = 0$ has been reported. In a modified compound, $Y_1Ba_2Cu_3F_xO_y$, T_c as high as 155 K has been reported.

Figure 11.3 illustrates the unit cell of the orthorhombic structure of $RBa_2Cu_3O_7$. Cu1 and O1 atoms form Cu–O chains along the b axis. The central CuO_4 is flat but the top and bottom ones are buckled slightly in opposite directions. The basic structure corresponds to a cubic perovskite with one of the axes triply elongated. This is an oxygen-deficient structure. Along the b axis are the Cu1–O1 chains in which each Cu1 atom has two O1 neighbors at 1.94 Å and two O4 at 1.85 Å. There are two more Cu atoms, Cu2, which form Cu–O2 planes together with two O3 atoms. Four Ba atoms and one O4 atom form a plane which is similar to the plane of four La and O in the case of the 2-1-4. Above 500 °C the b axis contracts while the a axis expands, and finally both become equal and the structure becomes tetragonal.

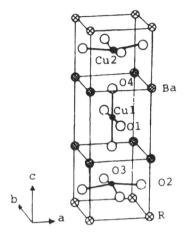

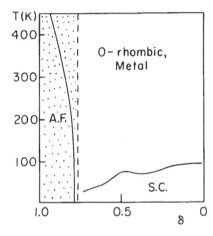

Fig. 11.3. Unit cell of RBa$_2$Cu$_3$O$_7$

Fig. 11.4. Phase diagram of YBa$_2$CuO$_{7-\delta}$. The tetragonal phase is dotted

A decrease in δ causes an increase in hole concentration, and at a certain concentration an insulating antiferromagnetic phase appears as in the 2-1-4 group. Figure 11.4 illustrates a typical phase diagram in which δ and T are used as variables. The tetragonal phase (dotted region) is insulating, which becomes antiferromagnetic at low temperatures. Below around $\delta = 0.7$, the compounds are metallic and the structure is orthorhombic. In this structure, these compounds become superconducting below a relatively high T_c.

Despite the difference in the abscissa, Fig. 11.4 is remarkably similar to Fig. 11.2 in the way the antiferromagnetic and superconducting phases appear. The fraction δ controls the oxygen content in the Cu1–O1 layers. However, it may also affect the distribution of holes between Cu–O$_2$ planes and Cu–O chains. The hole concentration is given by $(1 - 2\delta)$ per cell relative to the purely ionic configurations of Ru^{3+}, Ba^{2+}, Cu^{2+} and O^{2+}. The superconductivity is probably associated with the Cu–O_2 planes because it is sensitive to substitutions of Cu2 atoms more than Cu1 atoms.

In addition to the 1-2-3 compounds there are bismuth-containing superconductors such as $Bi_2Sr_2Cu_1O_{6+x}$, $Bi_2Ca_1Sr_2Cu_2O_{8+x}$ and $Bi_2Ca_2Sr_2Cu_3O_{10+x}$. These oxides have perovskite-like units containing one, two, or three CuO_2 planes sandwiched between Bi-O bilayers. Such bilayers are characteristic of these compounds instead of the linear chains in the 1-2-3 family.

It has also been discovered that Tl–Ba–Cu–O compounds such as $Tl_1Ba_2Cu_1O_5$, etc. are superconductors with T_c as high as 125 K. They consist of Cu perovskite-like units containing one, two or three CuO_2 planes separated by Tl-O bilayers.

11.2.3 Copperless Oxides

It has been discovered recently that several oxides not containing copper can be superconducting above 10 K. Although their critical temperatures are relatively low they can be important for the study of High-T_c superconductivity. Among these oxides are $BaPb_{1-x}Bi_xO_3$, $Ba_{1-x}K_xO_3$ and $LiTi_2O_4$. A lead substitution on the bismuth site in the parent compound $BaBiO_3$ provides the first oxide and a potassium substitution on the barium site produces the second.

$Ba_{1-x}K_xBiO_{3-\delta}$ becomes a superconductor at 30 K. This type of compound has a cubic perovskite structure without the 2D Cu–O_2 planes characteristic of the typical high-T_c superconductors. It consists of a 3D BiO network and does not have magnetic order. In $BaBiO_3$, Ba^{2+} and Bi^{4+} balance with three O^{2-}. Since disproportionation $2Bi^{4+} \rightarrow Bi^{3+}+Bi^{5+}$ takes place in the monoclinic crystal structure, a half-filled band can be expected. $BaBiO_3$ has a monoclinic distorted perovskite structure at room temperature in which octahedra of two sizes form a staggered configuration with two Bi–O distances, i.e. 2.25–2.27 Å and 2.13–2.15 Å. With a small addition of potassium, the Bi–O backbone is straightened, but the two different Bi–O distances still remain, as shown in Fig. 11.5. This diagram, taken from [11.5], represents subtle structural changes which takes place when x is increased from 0 to 0.04 and to 0.4. The octahedra form the unit cell, and the lower figures give key bond distances and angles. Note that CuO_6 octahedra have long and short bonds. $BaBiO_3$ is nonsuperconducting.

The electronic structure of doped $BaBiO_3$ is different from that of the cuprates. With the unequal Bi–O bond lengths and the valence charge disproportion, charge density waves can be expected. Indeed a CDW gap of order 1 eV has been revealed by spectroscopy.

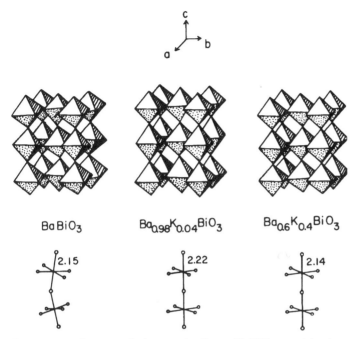

Fig. 11.5. Structural changes in $Ba_{1-x}K_xBiO_{3-\delta}$ with changes in K content. [11.5]

11.2.4 Characteristic Properties

In addition to the structural features, characteristic physical properties of high-T_c superconductivity must be addressed. However, since there are many oxides only some general properties can be discussed.

First, the coherence length ξ_0 is very small $\xi_0 \sim 0.2\hbar v_F/k_B T_c \sim 10$ Å. Hence, it is of the order of the dimension of the unit cell, and is about 1/10 to 1/100 of the values of ordinary superconductors. The smallness of the coherence length means that high-T_c superconductivity is sensitive to small scale compositional and structural changes. Correspondingly, the upper critical field becomes very high. From measurements of H_{c2}, ξ_0 values 12, 7 and 34 Å have been inferred respectively in the directions of c, a and b axes of single crystals of $Y_1Ba_2CuO_7$. That is, the coherence length is anisotropic. In comparison, the lattice parameters are $a = 3.83$, $b = 3.89$, and $c = 11.71$ Å. However the coherence length determined from H_{c2} may not be accurate when the width of the transition region is wide.

Second, the magnetic penetration depth follows the BCS relation given by (11.1.10). Figure 11.6 illustrates the relative incremental penetration depth $[\lambda(T) - \lambda(T = 0)]/\lambda(T = 0)$ determined by *Harshman* et al. [11.6] by transverse muon-spin-relaxation measurements on $YBa_2Cu_3O_{9-\delta}$ ($\delta = 2.1 \pm 0.05$). The dashed line represents the BCS relation for the ordinary superconduc-

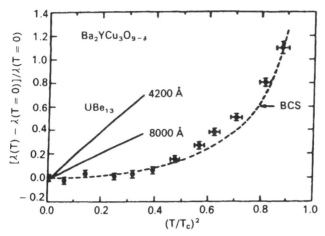

Fig. 11.6. Penetration depth of $YBa_2Cu_3O_{9-\delta}$ determined through muon-spin-relaxation [11.6]

tors. Their data indicate that $\lambda \sim 1400\,\text{Å}$ at $6\,\text{K}$. In the limit of extreme anisotropy, this yields $\lambda \sim 1065\,\text{Å}$. The solid lines correspond to the heavy-electron superconductor UBe_{13} with assumed $\lambda(T = 0)$ of $4200\,\text{Å}$ and $8000\,\text{Å}$. Similar results have been obtained by susceptibility measurements [11.7].

Let us assume that the superconducting carriers in $YBa_2Cu_3O_7$ are in CuO_2 planes. The classical London expression in this case reads

$$\lambda_L = (mc^2d/4\pi ne^2)^{1/2}$$
$$= c/\omega_p, \tag{11.2.1}$$

where n is the two-dimensional density of carriers, m is the effective mass, c is the light velocity, and d is the average spacing between CuO_2 planes. Measurements of the penetration depth have shown that it is inversely proportional to the square root of the carrier concentration, i.e., ω_p^2 is proportional to the carrier concentration. This is an important information because it implies that the holes on the Cu sites are quite well localized. On the other hand if $d = 6.5\,\text{Å}$ and if we associate the critical temperature with the Fermi energy of 2D electrons, we get T_c of order $4500\,\text{K}$. This is of course very high, although there can be factors which reduce this high temperature.

The specific heat jumps at T_c in general agreement with the BCS theory, which predicts $\Delta c = 1.43\gamma T_c$, where γ is the proportionality constant of the linear specific heat. Indeed, in the 1-2-3 materials, $\Delta C/\gamma T_c = 1.5$ and 1.23 have been reported, indicating the applicability of the BCS weak coupling theory. However it is difficult to separate out the electronic contribution from the lattice specific heat since T_c is high, and $\Delta C/\gamma T_c$ of around 3.5–4 is more common. Moreover, the linear specific heat has been observed even in the superconducting state. In the ordinary superconductors the specific heat drops exponentially following $\exp(-\Delta/kT)$. According to *Kumagai* et al. [11.8] the heat capacity of $La_{2-x}Ba_xCuO_4$ lacks the linear term γT in the antiferromagnetic region where $x \leq 0.02$. However, γ increases abruptly from $x = 0.02$ to $x = 0.04$ where the system is still insulating and then approaches a saturation value of $5\,mJ/mol\,K^2$ in the superconducting region of $x \geq 0.05$. This γ is close to $4.66\,mJ/mol\,K^2$ for polycrystalline samples [11.9]. However, more recently $\gamma \sim 16\ mJ/mol\,K^2$ has been obtained for $YBa_2Cu_3O_7$ by *Phillips* et al. [11.9b], who have found that a mjaor role is played by Cu^{2+} in impurity phases such as $BaCuO_2$ and on the YBCO lattice.

The determination of an energy gap $\Delta(T)$ for superconductive transitions is extremely important. According to BCS theory $\Delta(T)$ follows

$$2\Delta(T) = 3.53k_BT_c(1 - T/T_c)^{1/2}$$

so that $\Delta(T_c) = 0$. In the case of high-T_c materials, Δ can be anisotropic.

Tunneling experiments [11.10] have yielded values of around 5 for $2\Delta(0)/kT_c$. For instance, in directions parallel and perpendicular to the Cu-O plane in $YBa_2Cu_3O_{7-\delta}$, $2\Delta(0)/k_BT_c = 5.9 \pm 0.2$ and 3.6 ± 0.2 respectively. These values are consistent with photoemission and nuclear magnetic resonance data. More recently *Friedl* et al. [11.11] have reported a precise Raman scattering value,

$$2\Delta(0)/k_BT_c = 4.95 \pm 0.10.$$

Their observation of the anomalous softening and linewidth increase of the Raman active mode in $YBa_2Cu_3O_{7-\delta}$ at $340\,cm^{-1}$ below the transition point shows that the material is a strong-coupling (type II) superconductor. However, deviations from the mean field behavior may occur if the heat capacity is divergent at T_c.

The upper critical field is consistent with the behavior of type II superconductors. The critical current density is very low and approaches the de-pairing limit at least in high-purity thin films at low temperatures. In a magnetic field well below H_{c2}, the current is extremely depressed. The temperature dependence of H_{c2} is

$$H_{c2} \propto (T_c - T)^n, \tag{11.2.2}$$

where n is around 1.5.

In ordinary superconductivity, the isotope effect represented by

$$T_c \propto M^{-\alpha} \tag{11.2.3}$$

is well obeyed with $\alpha = 1/2$. Deviations from this value of α which are observed in d-band elements can be atributed to Coulomb interaction. In high-T_c superconductors, α is small. In fact, early observations reported $\alpha = 0$. However, more recent experiments have given small but finite values of α. Due to its smallness and because T_c is extremely sensitive to the oxygen concentration, these values are spread. One recent value obtained by *Morris* et al. [11.12] for the 1-2-3 oxides with ^{18}O substitution is

$$\alpha = 0.019 \pm 0.005. \tag{11.2.4}$$

The quantization of magnetic flux has been observed by *Gough* et al. [11.13]. According to their observation on $Y_{1.2} BaO_{0.8}CUO_4$, the quantization unit is $h/2e$. This implies that the charge carriers are electron pairs as in the case of BCS superconductivity. The Knight shift experiments by *Market* et al. and others [11.14] and the nuclear spin relaxation experiments by *Kitaoka* et al. and *Mehring* [11.15] also favor BCS superconductivity.

A series of neutron scattering experiments conducted by *Birgeneau* et al. [11.16] shows antiferromagnetic spin correlations with a scattering amplitude corresponding to a fully occupied Cu^{2+} square lattice in $La_{2-x}Sr_xCuO_4$ for x varying between 0.02 and 0.18. The principal role of the Sr^{2+} doping appears to be to destroy the long-range $Cu^{2+}-Cu^{2+}$ antiferromagnetic spin correlations rather than the Cu^{2+} moment itself. The spin–spin correlation length ξ is quite short, varying from -35 to ~ 8 Å as x increases from 0.02 to 0.18. Approximately, it is given by

$$\xi = 3.8/\sqrt{x} \text{ [Å]}. \tag{11.2.5}$$

This is just the average separation between the holes introduced by the Sr^{2+} doping. Although ξ varies, the local order remains the same as in La_2CuO_4.

The magnetic excitations in $La_{1.85}Sr_{0.15}CuO_4$ at several energies have been studied by neutron scattering experiments [11.16]. The sample used was a single crystal which exhibits a sharp transition at 33 K with almost complete flux exclusion. For magnetic excitation energies of 9 meV or lower, the intensity drops dramatically with decreasing temperature below around 150 K. However, the intensity is nearly independent of temperature in the range between 5 and 300 K at 12 meV. *Shirane* et al. compared their single-crystal results (for $T_c = 33$ K) with that obtained from a sample grown in Japan which has $T_c = 10$ K. Although the weak coupling BCS mechanism may not be applicable, the two samples may be associated with the values 3 and 10 meV for the BCS parameter $2\Delta = 3.5k_BT_c$. Thus one is above and the other is below the BCS value. Their data indicate that a gap may be opened in the Cu^{2+}-hole spin excitation spectrum with an energy comparable to the BCS gap. The apparent gap however develops well above T_c. Similar effects have been observed in NMR and optical studies.

The magnetic susceptibility of the $La_{2-x}Sr_x CuO_4$ compounds increases with x in the superconducting phase. It decreases for $x > 0.3$ where superconductivity disappears. The magnitude of the susceptibility is of order

$$\xi \sim 3 \times 10^7 \text{ emu/g}.$$

The temperature variation of the magnetic susceptibility depends also on x. In the superconducting phase where $0.05 \lesssim x \lesssim 0.3$, the susceptibility increases with T near 10.0 K. It appears that the susceptibility reaches a maximum at a certain temperature, perhaps determined by $k_BT \sim J$, J being the exchange energy.

Such a behavior resembles the case of finite Heisenberg chains. Numerical calculations of 1D chains show that the susceptibility has a broad maximum at such a temperature. In a number of crystals, such as copper tetramine sulfate monohydrate, the magnetic ions are arranged in chains with strong intrachain but weak interchain interactions. Therefore they can be treated at a relatively high temperature by a 1D model. Moreover there are experimental indications that various cupric quinone complex salts have independent finite chains of 10–20 magnetic ions. An interesting aspect of 1D Heisenberg chains is that spin pairs can move around. The case of the 2-1-4 compounds is two-dimensional, but certain similarities appear to exist.

The dynamic susceptibility $\chi(q, \omega \to 0)$, which depends on wavelength q, is enhanced near the Brillouin zone boundary, indicating the existence of strong antiferromagnetic correlations.

The superfluid density ρ_s determined from $H_{c1}(T)/H_c(0)$ or $\lambda^2(T = 0)/\lambda^2(T)$ increases more strongly than the BCS prediction toward 0 K, where it becomes nearly temperature independent. It appears as if a nonzero gap exists over the entire Fermi surface.

The dynamic conductivity $\rho(\omega, T > T_c)$ shows that the scattering time depends on T as well as frequency ω. Hence, at optical frequencies, there must be excitations which are above phonon energies [11.17a–d]. Moreover, the imaginary part of the Raman scattering function, when it is plotted against energy shift, reveals that electronic excitations are continuous, broad and nearly constant except for phonon peaks [11.17e]. However, below $k_B T$ the function decreases linearly to zero.

11.3 Theoretical Consideration

The nature of high-T_c superconductivity is still not fully understood, although a strong-coupling mechanism as in type II superconductivity appears to occur. Therefore, instead of attempting a comprehensive discussion, we will present qualitative descriptions of some important models. We first describe the electronic structure of the 2-1-4 material as a typical material.

The parent material La_2CuO_4 is an antiferromagnetic insulator below 240 K before doping. With doping the compound becomes a "spin liquid" and then transforms into a metal. The material $YBa_2Cu_3O_6$ is also an insulating antiferromagnet with a high Néel temperature. As the oxygen value increases towards $YBa_2Cu_3O_7$ the compound becomes a spin liquid and then a metal. The electrons are tightly bound to their atoms in the insulating phase. Their energy bands are narrow and kninetic energy is too small to allow hopping.

Oxygen has four electrons in the $2p$ shell, and copper ten electrons in the $3d$ shell. The latter consists of the $d(x^2-y^2)$ and $d(z^2)$ orbitals as in Fig. 11.7. The former (white lobes) has four lobes in an xy plane which are directed toward the four two oxygen atoms in the same plane and the latter (dotted)

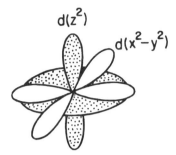

Fig. 11.7. Cu $d(x^2 - y^2)$ orbital and $d(z^2)$ orbital (*shaded*)

has two lobes in the perpendicular (z) direction, which point toward the two oxygen atoms above and below the plane, and one circular orbital in the xy plane.

The single $4s$ electron and one of the ten $3d$ electrons of copper are transferred to oxygen to make La_2CuO_4. Hence, the Cu $d(x^2 - y^2)$ orbital is paritally empty, while the $d(z^2)$ shell is completely filled. The remaining electrons in the $d(x^2 - y^2)$ orbital are strongly hybridized with the oxygen p electrons. The nine electrons in the $d(x^2 - y^2)$ orbital will push away the oxygen atoms above and below the plane and pull in the oxygens within the plane, thus distorting the cubic symmetry. This distortion creates a 2D planar structure.

The antiferromagnetic phase of La_2CuO_4 is due to the unpaired spin of Cu electrons because eight out of the nine d-electrons are paired. The unpaired spins on Cu sites alternate directions and form an antiferromagnetic lattice. These spins are localized and carry no current. However doping creates holes which can move. Doping can be considered to destroy the antiferromagnetic structure to form a "spin liquid".

The Hubbard model or its extension has commonly been used as a basis for theoretical considerations. In this model, holes created by doping hop on the CuO_2 planes. These holes can be considered essential for superconductivity in view of the phase diagrams. They are presumably on the O^- sites in the CuO_2 planes, as shown by X-ray photoemission spectroscopy [11.18].

The resonating-valence-bond (RVB) model developed by *Anderson* [11.19] is based on a single-band Hubbard model with strong on-site correlations. Due to strong correlations, the band is split into two subbands separated by a Coulomb gap. Hence, no current flows. Indeed undoped high-T_c materials are antiferromagnetic insulators. However, doping creates holes, which are considered to decay into two types of quasiparticles, a *spinon* which carries only spin $1/2$ and a *holon* which has a positive charge e but no spin. These quasiparticles move on the same plane in the background of the quantum spin liquid created by doping.

The spinons are paired. That is, an up-spin and a down-spin form a pair. They do not carry any charge but can scatter holons, which carry current, causing resistivity. This resistivity may be approximately proportional to temperature.

The holones move in a highly correlated background so that their motion causes rearrangements of the entire correlated spins. Hence, the system can no longer be antiferromagnetic. The holons can form pairs to cause super-conductivity as in the case of Cooper pairs. They can then hop from plane to plane even though in-plane motion is still favorable. Holon pairs can tunnel through different planes like Josephson tunneling of electron pairs in ordinary superconductors.

In *Emery's* model [11.20] the holes added to oxygen sites are considered to form polaron-like quasiparticles with the neighboring Cu holes, organizing their spins and enhancing their zero-point motion. These quasiparticles are Fermions with spin and charge. Their attractive interaction is provided by a combination of the enhanced superexchange interaction between the copper spins of a pair of quasiparticles separated by one vacant oxygen site and the zero-point motion of holes on nearby Cu sites.

The Hamiltonian representing a single Cu–O_2 plane may be expressed by

$$\mathcal{H} = \sum_{ij\sigma} \varepsilon_{ij} a_{i\sigma}^{\dagger} a_{j\sigma} + \tfrac{1}{2} \sum_{ij\sigma\sigma'} U_{ij} a_{i\sigma}^{\dagger} a_{i\sigma} a_{j\sigma'}^{\dagger} a_{j\sigma'}, \tag{11.3.1}$$

where i labels a copper site $\mathbf{m}$ or an oxygen site $\mathbf{m}+\mathbf{u}$. The vacuum consists of all the $Cu^{+}d$ and $O^{2-}p$ states occupied. The operators $a_{i\sigma}^{\dagger}(a_{i\sigma})$ create (destroy) holes with spin σ in the $3d(x^2 - y^2)$ states of Cu or in the $2p_x$ or $2p_y$ orbital of oxygen. These are the ones most strongly hybridized by overlap integrals. The diagonal interaction constants, (E_i, U_{ii}) correspond to (ε_p, U_p) and (ε_d, U_d) for $O(2p)$ and $Cu(3d)$ states respectively. Between nearest neighbor Cu-O there is hopping represented by $\varepsilon_{ij} = t$, and between holes on neighboring Cu there is interaction $U_{ij} = V$. The estimated orders of magnitude of these parameters have been given in the previous section.

It is clear that the above Hamiltonian is reduced to a single-band Hubbard model given by (1.5.10) when $\varepsilon_{ij} = t$, $U_p = U_d = U$, $U_{pd} = 0$. Calculations [11.21] show that for a half-filled band there is a gap between the energies of occupied and unoccupied states of the Hubbard model. Hence, any additional hole will go into the $O(2p)$ states if the site energy lies inside the gap.

If $t = 0$ and if there is one hole per cell such that there is exactly one hole on each of the N Cu sites, there is a 2^N-fold spin degeneracy. This degeneracy is lifted by superexchange between near-neighbor copper spins. The corresponding effective Hamiltonian is of the form

$$H_s = J_c \left(\sum_{ij} \right) (\mathbf{S}_i \cdot \mathbf{S}_j - \tfrac{1}{4}). \tag{11.3.2}$$

To order t^4, the exchange constant is given by

$$J_c = \frac{4t^4}{(\varepsilon + V)^2} \left(\frac{1}{U_d} + \frac{2}{U_p + 2\varepsilon} \right). \tag{11.3.3}$$

The system is an antiferromagnetic insulator with holes primarily on Cu sites.

For $U_d > (\varepsilon + 2V)$ any additional holes due to doping will go onto oxygen sites. An effective Hamiltonian for a single added hole may be derived by perturbation theory. To second order in t this Hamiltonian is given by

$$\mathcal{H}' = (t_1 + t_2)\sum a^{\dagger}_{\mathbf{m}\sigma'}a_{\mathbf{m}\sigma}a^{\dagger}_{\mathbf{m}+\mathbf{u},\sigma}a_{\mathbf{m}+\mathbf{u}',\sigma'} + t_2\sum a^{\dagger}_{\mathbf{m}+\mathbf{u},\sigma}a_{\mathbf{m}+\mathbf{u}',\sigma}$$

$$+J\sum(\mathbf{S_m}\cdot\mathbf{S_{m+u}} - 1/4)n_{\mathbf{m+u}} - 2\Delta E_c. \qquad (11.3.4)$$

The first two sums are over $\mathbf{u} \neq \mathbf{u}'$, $\mathbf{m}, \sigma$ and σ' and the last sum is over $\mathbf{m}$ and $\mathbf{u}$. $\mathbf{S_m}$ is the spin $1/2$ operators at the Cu site $\mathbf{m}$. $n_{\mathbf{m+u}}$ is the occupation number at the oxygen site $\mathbf{m+u}$, $\mathbf{u}$ being a displacement $[(\pm 1/2, 0), (0, \pm 1/2)]$ for an oxygen site. ΔE_c is the change in the self-energy of a single Cu hole due to the presence of the oxygen hole. To second order in t the parameters are given approximately by

$$t_1 = t^2/\varepsilon, \quad t_2 = t^2/(U_d - 2V - \varepsilon), \quad \varepsilon = \varepsilon_p - \varepsilon_d,$$
$$J = 2t^2/(\varepsilon + U_p - V) + 2t^2/(U_d - 2V - \varepsilon) - J_D, \qquad (11.3.5)$$

where t represents a hopping integral between Cu–O neighbors and J_D is a direct ferromagnetic exchange between holes on neighboring Cu and O sites. The energies t and J_D are estimated to be of order 1 and 0.2–0.5 eV respectively.

A single added hole is expected to be at an oxygen site. If $t = 0$ the hole is stationary at either the x or y orbital of the unit cell. In the absence of Cu–Cu exchange, the ground state spin configuration of a single stationary oxygen hole and its surrounding copper spins is represented by a spin $1/2$ state such that

$$|\Psi\uparrow\rangle = [(\uparrow\downarrow - \uparrow\downarrow)\uparrow - \uparrow(\downarrow\uparrow - \downarrow\uparrow)]\sqrt{6}. \qquad (11.3.6)$$

The two singlet components on the right side are not orthogonal to each other, resulting in an average spin of $-1/3$ on the oxygen site. If $\mathbf{s}$ and $\mathbf{S}_i$ $(i = 1, 2)$ represent respectively the spins of the oxygen hole and neighboring copper holes the interaction energy between the three spins is of the following form apart from a constant self-energy of $-J/2$:

$$h_1 = J\mathbf{s}\cdot(\mathbf{S}_1 + \mathbf{S}_2). \qquad (11.3.7)$$

The Hamiltonian $\mathcal{H}'$ allows hopping of the added oxygen hole. The lowest energy state is expected to be represented by a superposition of states of the form (11.3.6) with equal amplitude at each site. As the quasiparticle energy increases, the spin deviations on surrounding copper spins become more delocalized, and the average value of the spin on adjacent coppers is reduced while the average spin at the oxygen site decreases. The Coulomb repulsion will cause enhancement of the zero-point motion of the adjacent copper holes.

When two oxygen holes are on the same site they interact with each other, having the Coulomb repulsion U_p. An additional energy is needed because their surroundings will no longer be polarized. When they are well separated such a polarization takes place and lowers the energy. Hence, the total energy increase is $U_p + 2J$.

Two oxygen holes with spins s_1 and s_2 can be on adjacent sites, bonding to the same copper hole with spin S_2. The interaction energy can be expressed in a form similar to (11.3.7),

$$h_2 = J[(S_1 + S_2)\cdot s_1 + (S_2 + S_3)\cdot s_2]. \tag{11.3.8}$$

The state of the five spins involved in this Hamiltonian has spin $1/2$. As in the case of h_1, this energy is repulsive. However, lowering of the energy takes place when the hole at the center delocalizes.

When there is an empty oxygen site between two copper spins, the two copper spins have a negative exchange energy. This energy depends on t and ε. Calculations for the case with $U_d = 9$, $U_p = 6$ and $t = 1\,\mathrm{eV}$ show that the difference of the zero-point energies of surrounding copper holes is negative when two oxygen holes are not adjacent if V exceeds approximately $1\,\mathrm{eV}$ since the exchange parameter is enhanced. To order t^4 the exchange parameter for this case is given by

$$J^* = \frac{4t^4}{\varepsilon^2}\left(\frac{1}{U_d} + \frac{2}{U_p + 2\varepsilon - 2V}\right). \tag{11.3.9}$$

In comparison with J_c of (11.3.3), there is an overall enhancement by a factor $[(\varepsilon + V)/\varepsilon]^2$. The reduction in the denominator of the second term also introduces enhancement. For $U_d = 9$, $U_p = 6$, $\varepsilon = 1.5$, $V = 2$ and $t = 1\,\mathrm{eV}$ J_c is approximately $0.1\,\mathrm{eV}$ and $J^*/J_c = 8.3$.

The enhanced superexchange may overcome Coulomb repulsion to cause s-taste superconductivity. The critical temperature estimated from the BCS relation is

$$T_c = 4.52\bar{t}k_F^2 a^2 \exp[4\pi\bar{t}/\lambda_l(k_F)], \tag{11.3.10}$$

where a is the in-plane lattice constant, $\lambda_l(k_F)$ is an effective scattering amplitude for angular momentum l and relative momentum k_F. The Fermi energy $k_B T_F = \bar{t}k_F^2 a^2$ is estimated to be around $0.15\,\mathrm{eV}$ for $La_{1.85}Sr_{0.15}CuO_4$ and $0.4\,\mathrm{eV}$ for $YBa_2Cu_3O_7$. Here,

$$\bar{t} \sim (t_1 + 2t_2)/3. \tag{11.3.11}$$

For $\varepsilon = 1.5$ and $t = 1$, $t_1 = 0.43\,\mathrm{eV}$, and with the same U_d, U_p as before, $t_2 = 0.22\,\mathrm{eV}$. These values lead to $\bar{t} \sim 0.29\,\mathrm{eV}$. This value is not very far from an experimentally inferred value $0.15\,\mathrm{eV}$.

In the above model the key attractive forces are directly related to the low density of oxygen holes. The low density $O(2p)$ holes may move about the entire system to form a band, while the strong Coulomb repulsion localizes the $Cu(3d)$ holes. Since the Fermi momentum associated with oxygen holes is low, the strong short-range Coulomb repulsion is not probed. At the same time the low density is favorable to s-state pairing. However, since the exciation energy is high, the pairing force may not be retarded. In contrast, in the ordinary phonon-mediated case the relevant energy is low and the lattice is slow to relax so that electrons may feel its influence at different times and avoid Coulomb repulsion.

The effect of doping has been considered by *Birgeneau* et al. [11.22] by a different model. In this model, holes cause the O^- and two neighboring Cu^{2+} spins to form an $S = 3/2$ quartet. A local ferromagnetic exchange coupling between Cu spins frustrates the nearest-neighbor Cu^{2+}–Cu^{2+} antiferromagnetic order. It eventually destroys the antiferromagnetic state and provides a pairing mechanism.

Schrieffer et al. [11.23] proposed a model in the itinerant limit. In this case, the holes depress the magnitude of the moment in their immediate environment. Two holes attract each other by sharing such a depression, called a spin bag. *Varma* et al. [11.24] used a two-band model with charge-transfer excitons in which $Cu^{2+}O^{2-} \rightarrow Cu^+O^-$. Doping is expected to reduce the charge-transfer energy, and charge-transfer resonances provide a mechanism for superconductivity. In this model, the existence of local moments on Cu is not assumed.

The linear specific heat of high-T_c superconductors has been considered to be consistent with the RVB theory. However, for the 1-2-3 compounds, *Phillips* et al. [11.9b] have observed that Cu^{2+} magnetic moments in impurity phases such as $BaCuO_2$ and in the $YBaCO_2$ lattice contribute significantly to the specific heat constant and these moments act as pair-breaking centers. A similar pair-breaking effect is well established in conventional superconductors, but YBaCO differs in that T_c is nearly independent of Cu^{2+} on YBaCO. Superconductivity is suppressed by the pair breaking interaction, but due to small ξ, normal regions in the vicinity of the magnetic moments separate more perfectly superconducting regions. This weak-link model is consistent with the small values of $\Delta C(T_c)$, the low critical current, a zero-bias conductance measured in tunneling experiments and the low temperature upturn in C/T and Curie–Weiss behavior of the susceptibility.

The antiferromagnetic configuration which appears in the insulating phase of high-T_c materials may still remain in small local regions even after the system has entered the metallic state, and spin fluctuations may be expected to increase toward T_c. *Pines*, *Scalapino*, and others [11.25] have proposed that these spin fluctuations cause $d_{x^2-y^2}$ pairing and explained the T^3 dependence of the NMR relaxation time [11.26], the reduction in neutron scattering [11.27], the temperature proportionality of the microwave penetration depth [11.28], etc.

However, *Anderson* and his colleagues [11.29] have opposed the formation of $d_{x^2-y^2}$ pairs on the grounds that antiferromagnetic spin fluctuations cannot be treated by perturbation from the Fermi liquid and that the experiments do not necessarily support such pairs but rather only anisotropic pairing. The temperature and frequency dependencies of the conductivity and other transport phenomena suggest that the high-T_c superconductors are not Fermi liquids, while photon emission and magnetism indicate that they are Fermi liquids. Therefore, *Anderson* has proposed that they are *Luttinger* liquids. *Luttinger* [11.30] suggested some time ago that in a 1D conductor in which the electrons are bound to each other like a string, the quasiparticles may not be of a single particle type above the Fermi surface but rather of a collective type which may correspond to the vibration of such a string. The possibility exists that this type of quasiparticles occurs even in two dimensions. Since the infrared conductivity in the c direction remains low even in the super-conductive phase, the quasiparticles may not move towards that direction. Thus, Anderson and his collegues have proposed a BCS-type S-wave theory in which the Josephson tunneling of the quasiparticles from one CuO_2 plane to the adjacent CuO_2 planes is considered.

Despite these theoretical efforts, the pairing mechanism of high-T_c super-conductivity has not yet been unveiled. Nevertheless, it is known that in the copper oxide materials the CuO_2 planes play a primary role for superconductivity. These planes are separated in the unit cell by intercalated layers which include yttrium, calcium, etc. and serve as charge reservoirs. Added oxygen atoms cause a charge redistribution between the conductive planes and reservoir layers. The oxygens suppress or enhance superconductivity depending on whether they are in the conductive planes or reservoir layers.

As mentioned in Sect. 11.2.3, there are copperless high-T_c materials. More-over, C_{60} crystals doped with rubidium–thallium become superconductors with T_c as high as $42.5\,K$. It is unknown whether there is any relationship between various types of high-T_c superconductors. When unveiled, the pairing mechanism in any of these materials would add a new dimension to condensed matter physics.

12. Integral Quantum Hall Effect

The integral quantum Hall effect provides us with a new resistance standard and the fine structure constant with extreme accuracy.

12.1 Quantization of Hall Conductivity of 2D Electrons

In 1980 *von Klitzing* et al. [12.1] discovered that the Hall resistance R_H, of the electrons in Si MOSFETs takes on plateau values while the magnetoresistance drops to zero for certain intervals of gate voltage. These plateaus correspond to quantization of the conductivity component σ_{xy} in units of e^2/h such that

$$\sigma_{xy} = -\frac{e^2}{h}i, \qquad (12.1.1)$$

where i is an integer. At the Hall plateaus, the longitudinal magnetoconductivity σ_{xx} stays nearly zero: it is less than 10^{-7} times its maximum value. In such intervals of gate voltage the variations in the Hall plateau values are very small:

$$\Delta\sigma_{xy}/\sigma_{xy} < 10^{-8}.$$

In two dimensions the transverse magnetoresistivity ρ_{xx} and Hall resistivity ρ_{xy} are related to σ_{xx} and σ_{xy} by

$$\sigma_{xy} = -\frac{\rho_{xy}}{\rho_{xx}^2 + \rho_{xy}^2}, \quad \sigma_{xx} = \frac{\rho_{xx}}{\rho_{xx}^2 + \rho_{xy}^2}$$

Hence, if ρ_{xx} is zero, σ_{xx} also vanishes. Conversely the Hall resistivity is given by the conductivity components such that

$$\rho_{xy} = -\frac{\sigma_{xy}}{\sigma_{xx}^2 + \sigma_{xy}^2}, \qquad (12.1.2)$$

but since ρ_{xx} is nearly zero at the Hall plateaus ρ_{xy} is approximately given by

$$\rho_{xy} \sim -\sigma_{xy}^{-1}. \qquad (12.1.3)$$

Hence, the Hall resistance can be expressed as

$$R_H = \frac{h}{e^2 i},$$ (12.1.4)

where i is an integer as before.

Their discovery is very important because the plateaus enable a very accurate determination of R_H. Its reliable expression is [12.2]

$$R_H = 25812.8063(77)\Omega/i \quad (\pm 0.03\text{ppm}).$$

Note the extreme accuracy of the determined value. Since typically $i = 2-8$, the magnitude of R_H falls in a relatively convenient range for determination. Therefore the new effect has been used as a resistance standard.

The plateaus also make it possible to determine, very accurately, the fine structure constant α and to confirm the theoretical QED results. This fundamental constant is given by

$$\alpha = \frac{e^2}{h} \frac{\mu_0 c}{2}.$$

Note that $\mu_0 = 4\pi \times 10^{-7}$ H/m is the permeability of vacuum and c is the speed of light which is accurately known, i.e., 299792458 m/s.

Figure 12.1 shows the data of *von Klitzing* et al., where U_H and U_{pp} are the Hall voltage and the voltage drop between the potential probes. The inset shows the device used with length $400\,\mu$m and width $50\,\mu$m. The distance between the potential probes is $130\,\mu$m. The field strength is $18\,T$ and the temperature is 1.5 K. They investigated the details of the plateau region for the lowest Landau level. The plateau in R_H has a value $6453.3 \pm 0.1\Omega$. The field strength in this case is $13\,T$ and the temperature is 1.8 K.

The quantization of the Hall conductivity attracted very strong attention immediately after its discovery, and considerable efforts have been made to further investigate this new phenomenon, called the *integral quantum Hall effect* (IQHE). It was found shortly afterwards that the same effect can also be observed in GaAs/GaAlAs quantum wells when the magnetic field is changed. In contrast, in Si MOSFETs, the gate voltage is used to change the density of electrons. In actuality the effect is determined by the filling factor ν defined by

$$\nu = n/(eH/ch)$$
$$= 2\pi l^2 n,$$

where l is the magnetic length and n is the electron density. The quantity eH/ch is the density of states per Landau level.

The Hall plateaus depend on a number of factors. For instance, the horizontal plateaus can be destroyed by increasing the temperature. Indeed, it is possible to measure finite slopes $d\rho_{xy}/dV_g$ in Si MOSFETs and $d\rho_{xy}/dH$ in

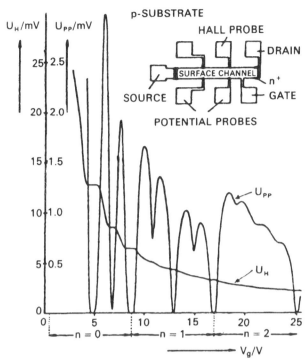

Fig. 12.1. Hall voltage U_H and the voltage drop U_{pp} between the potential probes as functions of the gate voltage V_g of a Si MOSFET at 1.5 K [12.1]. The inset is a top view of the device which has length 400 μm and width 50 μm. The distance between the potential probes is 130 μm

GaAs/GaAlAs when the temperature is approximately 0.2 K or higher. Here, V_g is gate voltage.

  Tausendfreund and *von Klitzing* [12.3a] investigated the temperature dependence of such slopes to observe an activated behavior $\exp(-T_0/T)$ even when ρ_{xx} is only weakly temperature dependent. For high-mobility Si MOSFETs, the activation energy $k_B T_0$ is close to one-half of the cyclotron energy. The minimum of ρ_{xx} is found below 1 K to deviate from a thermally activated behavior due to an additional contribution from variable range hopping. However, such a contribution is not visible in their Hall effect measurements. They found also that the minimal slope of the Hall resistance is proportional to ρ_{xx},

$$\frac{1}{\rho_{xy}}\frac{\mathrm{d}\rho_{xy}}{\mathrm{d}n} \propto \rho_{xx}. \tag{12.1.5}$$

Figure 12.2 represents their data showing this proportionality. The data correspond to the 4th plateau in a Si inversion layer in the temperature range between 1.5 and 4.2 K.

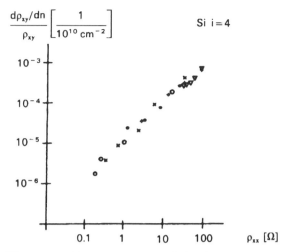

Fig. 12.2. Proportionality of $(1/\rho_{xy})\mathrm{d}\rho_{xx}/\mathrm{d}n$ to ρ_{xx} [12.3a]

Branch et al. [12.3b] have carried out activation studies of the IQHE for a wide range of integer filling factor ν values in a number of GaAs/GaAlAs heterostructures. Their data show that the conductivity component σ_{xx} near the centers of the Landau levels follows the Arrhenius law at low temperatures, with a prefactor which is close to a universal value e^2/h independent of sample and Landau level index.

The motion of a classical electron in an electric field E and a magnetic field H may be described in a single relaxation time approximation by

$$\dot{\mathbf{v}} = -\frac{e}{m}\left(\mathbf{E} + \frac{\mathbf{v}}{c}\times\mathbf{H}\right) - \frac{\mathbf{v}}{\tau}.$$

Since

$$\mathbf{J} = \sigma\mathbf{E}; \quad \mathbf{E} = \rho\mathbf{J},$$

the equation of motion yields, in steady state, the conductivity components given by

$$\sigma_{xx} = \sigma_{yy} = \frac{\sigma_0}{1 + (\omega_c\tau)^2}, \quad \sigma_{xy} = -\sigma_{yx} = -\frac{\sigma_0\omega_c\tau}{1 + (\omega_c\tau)^2}, \tag{12.1.6}$$

where

$$\sigma_0 = \frac{ne^2\tau}{m}. \tag{12.1.7}$$

Therefore one obtains

$$-\sigma_{xy} = \omega_c\tau\sigma_{xx}. \tag{12.1.8}$$

One also finds such relations as

$$-\sigma_{xy} = \frac{nec}{H} - \frac{\sigma_{xx}}{\omega_c \tau},$$

$$\rho_{xy} = \omega_c \tau \rho_{xx}. \tag{12.1.9}$$

The latter equation states that ρ_{xy} and ρ_{xx} are proportional to each other.

A relation similar to (12.1.5) has been observed. The temperature change of ρ_{xy} expressed by the difference

$$\Delta\rho_{xy} = \rho_{xy}(T) - \rho_{xy}(0) \tag{12.1.10}$$

has been found to be proportional to $-\rho_{xx}$,

$$\Delta\rho_{xy} \propto -\rho_{xx}. \tag{12.1.11}$$

In fact, *Yoshihiro* et al. [12.4a] used Si MOSFETs to find that

$$\Delta\rho_{xy} = -0.1\rho_{xx}^{\min}(T).$$

Cage et al. [12.4b] used GaAs heterostructures to arrive at a similar proportionality. The proportionality constant in such a relation may vary from sample to sample. $\Delta\rho_{xy}$ can be much larger than the variations in the value of ρ_{xy} across each step if $\rho_{xx}^{\min}$ is not sufficiently small. Several possible mechanisms have been examined for the above proportionality, including finite size, misalignment of the Hall probe, bulk leakage, variable range hopping, thermal activation, etc. However, none appears to give a completely satisfactory explanation. In any case, note that Hall steps can be very flat and yet the values of ρ_{xy} may deviate significantly from their zero-temperature values.

To understand the IQHE it is very important to recognize that the quantization of Hall resistivity occurs for an *interval* of gate voltage or magnetic field. In this respect it deviatives from what we might expect classically. Let us discuss this point in more detail. If i Landau levels are completely filled at absolute zero, there will be no current. Under the condition $\sigma_{xx} = 0$, the first relation in (12.1.9) suggests

$$\sigma_{xy} = \sigma_H = -\frac{nec}{H}. \tag{12.1.12}$$

Since the density of electrons is given by

$$n = i(eH/ch),$$

we obtain (12.1.1). However this occurs only at a discrete point instead of an extended density range. Only at the middle of adjacent Landau levels do the classical and quantized values of σ_{xy} agree with each other.

Attempts have been made to determine the plateau width. This determination is by no means simple because observational errors are involved and because the width depends on several factors, including the current, mobi-

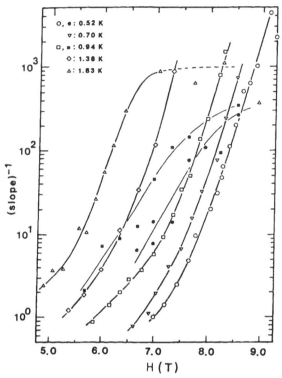

Fig. 12.3. Inverse Hall slopes at the $i = 4$ plateau as functions of magnetic field in Si MOSFETs at various temperatures [12.6]

lity, temperature, magnetic field, etc. Nevertheless a plateau width may be defined, for instance, by a condition such as

$$\Delta\rho_{xy}/\rho_{xy} \lessgtr \pm 0.1\%.$$

In general, the width decreases linearly with increasing current, temperature, and mobility. There is a critical value of these quantities above which the width disappears. In detail however, more complex behaviors may take place. For instance the width disappears at very high mobilities and yet small mobilities are not necessarily favorable. That is, there are optimum mobilities for which the effect is well manifested. Such mobilities are around $10^4 \, \text{cm}^2/\text{Vs}$ in Si inversion layers and $10^5 \, \text{cm}^2/\text{Vs}$ in GaAs heterostructures.

The ρ_{xx} minima are only moderately current dependent in GaAs heterostructures, and a critical current of order $340 \, \mu\text{A}$ has been reported. The critical current density has been estimated to be around $0.7 \, \text{A/m}$ but a higher value of order $1.16 \, \text{A/m}$ has also been observed in a different GaAs device. On the other hand a smaller value of $0.2 \, \text{A/m}$ has been observed for Si MOSFETs.

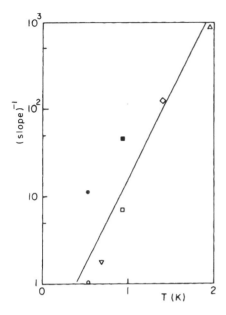

Fig. 12.4. Inverse Hall slopes at $7T$ as a function of temperature. The data points are taken from Fig. 12.3

Concerning the breakdown of the IQHE, *Kinoshita* et al. [12.6] observed on Si MOSFETs that a reduction of magnetic field from 8.9 to $8.3T$ caused abrupt disappearance of the $i = 4$ plateau at 0.5 K. This observation suggests that a critical magnetic field exists. The field reduction corresponds to a decrease of electron density from 8.5×10^{11} to 8.0×10^{11} cm^{-2}. The critical magnetic field increased when the temperature was reduced. They measured Hall resistivity as a function of gate voltage at the $h/4e^2$ plateau. As this plateau started to disappear it became inclined. Hence, the inverse of this slope could be used as a measure to assess the quantization.

Figure 12.3 illustrates *Kinoshita* et al's data on such slopes at different temperatures in units of the corresponding slope of the H/ne vs V_g curve of classical electrons. Note that for a fixed magnetic field, the lower the temperature the smaller the inverse-slope, even though quantization is generally expected to be better toward absolute zero. In the limit of absolute zero, the inverse-slope might become very small even if it still exists, as indicated more clearly in Fig. 12.4 in which their data points at $7T$ are plotted against temperature. The straight line is only a guide to the eye. Although the data points are spread at low temperatures, it is worth investigating the existence of the IQHE near absolute zero and at a finite magnetic field.

Figure 12.5 specifies the region in which quantized plateaus are observable. Such a region depends on the accuracy of width determination, as shown by solid and dashed lines. Below the thermal limit line, σ_{xx} becomes finite and the quantization of σ_{xy} is less complete due to thermal excitation across the energy gap between Landau levels. Accurate measurements of plateaus are

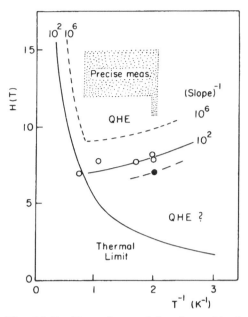

Fig. 12.5. Dependence of the observable domain of the IQHE on Ihe magnitude of the inverse slope [12.6]

possible in the hatched area. When the plateaus are about to disappear, random fluctuations in both ρ_{xy} and ρ_{xx} are observed in resemblance to the conductance fluctuations in narrow 2D systems. Perhaps some of the localized electrons at ρ_{xx} minima hop between localized sites as they are delocalized. Inhomogeneity in samples and electron heating may contribute to such fluctuations. Inter-Landau-level scattering by impurities may cause the breakdown of the quantum Hall effect [12.6].

By fabricating multiple 2D electron layers in close proximity, additional degrees of freedom associated with their coupling can be obtained. The double quantum well is the simplest of these devices in which tunneling may play an important role. It has been found that in a double quantum well, the odd-integer quantized Hall state is absent in a high magnetic field [12.7]. For a pair of identical quantum wells with inter-well tunneling, the lowest subband states of each well mix to form symmetric and antisymmetric states. It has been shown [12.8] that the symmetric-antisymmetric energy gap is destroyed by Coulomb effects. This explains the observed absence of the IQHE.

The IQHE has been observed at microwave frequencies as high as 30 GHz. *Kuchar* et al. [12.9] used GaAs/GaAlAs heterostructures with mobility $20\,000\,\mathrm{cm^2Vs}$ at 2.2 K. They examined their data in terms of the classical formula

$$\sigma_{xy} = \frac{ne^2}{m^*} \frac{\omega_c \tau^2}{(1 - \mathrm{i}\omega\tau)^2 + \omega_c^2 \tau^2}. \tag{12.1.13}$$

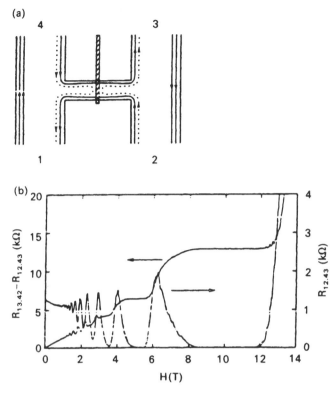

Fig. 12.6. Integral quantum Hall effect in narrow 2D systems [12.10]

Using τ as an adjustable parameter, the classical curve was found to give the overall shape of the data curve, in particular at two crucial points. As in the static case, the data were found to deviate from the classical curve at the plateaus except for the centers of the plateaus.

The IQHE has also been observed in narrow 2D systems. Using four-probe measurements, *Washburn* et al. [12.10] observed that some states are reflected by a barrier into the oppositely directed states at the outer edge of the sample as in Fig. 12.6. In (a), a schematic view of the four-probe GaAs/GaAlAs wire with a narrow cross gate (hatched) is shown. The thin solid lines that run around the edges of the device represent the Landau level paths and their directions. For a large enough gate voltage, the barrier due to the gate will induce breakdown of the IQHE in which some of the paths are reflected back as shown by dotted lines. In effect, the barrier causes quantized plateaus in any four-probe measurements. In (b), resistance $R_{12,43}$ (right ordinate) and resistance difference $R_{12,42} - R_{12,43}$ (left ordinate) are shown. These quantities resemble σ_{xx} and σ_{xy} in large samples. The resistances are defined as the ratio of voltage, measured across a pair of the leads, to the current through the sample. That is,

$$R_{ij,kl} = V_{kl}/I_{ij}, \tag{12.1.14}$$

where V_{kl} is the voltage drop measured from lead k to lead l.

It is worth mentioning that by molecular-beam epitaxy 3D anisotropic structures can be fabricated. Such structures have a weak periodic potential in the direction of the growth and can be regarded as multi-quantum-well systems with extremely thin and/or low barriers. *Störmer* et al. [12.11] observed quantization of the Hall effect and concomitantly vanishing magnetoresistance in such structures.

12.2 Localization and Scaling

The IQHE is described in this section in terms of localization and scaling. Let us note that in general the Landau levels of 2D electron systems have the natural width eH/ch per unit area and their density of states depends on impurities. The observed behavior of the longitudinal conductivity σ_{xx} in the IQHE suggests that the electron states are extended near the center of each impurity-broadened Landau level and localized near its edges. It is worth mentioning that a measurement of the diamagnetic susceptibility of a 2D electron system in the quantized Hall regime of a GaAs/GaAlAs superlattice showed only ordinary de Haas–van Alphen oscillations without any sign of localization, even though these oscillations are also due to the electron's orbital motion. The existence of optimum mobilities in the manifestation of Hall plateaus and the observed small increase in the plateau width with impurity concentration suggest that impurities affect the IQHE in a subtle way.

The concept of scaling has played a very important role in understanding electron transport in disordered systems in which localization can take place. In fact the scaling theory of weak localization predicts that electrons are localized in the macroscopic limit in two dimensions. However, the IQHE occurs in the presence of a strong magnetic field, and the finiteness of σ_{xx} suggests that extended states should exist near the center of a Landau level.

The scaling theory of weak localization is a one-parameter theory. In relation to weak localization theory, *Khmel'nitskii* [12.12a] proposed a scaling theory of the IQHE with σ_{xx} and σ_{xy}, two independent parameters, which vary with the linear dimension L of a given sample. With the conductivity components measured in units of e^2/h, these parameters are considered to vary with length L following

$$\frac{\partial \sigma_{xx}}{\partial \ln L} = \beta_{xx}(\sigma_{xx}, \sigma_{xy}), \tag{12.2.1}$$

$$\frac{\partial \sigma_{xy}}{\partial \ln L} = \beta_{xy}(\sigma_{xx}, \sigma_{xy}), \tag{12.2.2}$$

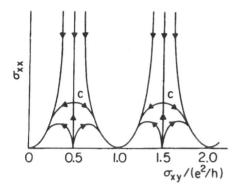

Fig. 12.7. A schematic flow diagram

The solutions of these equations may be represented by flow curves with σ_{xx} plotted against σ_{xy}. Figure 12.7 shows such flow curves schematically. In the metallic region where $\sigma_{xx} \gg 1$, σ_{xx} is independent of σ_{xy}. Hence, the curves are vertical. Starting from an initial point $(\sigma^0_{xy}, \sigma^0_{xx})$ the curves converge into localization fixed points $(j, 0)$ where j is an integer or zero. This convergence represents the IQHE. The arrows indicate the direction of increasing localization length. In addition to the localization fixed points, there are delocalization fixed points such as C. These points are unstable because they are saddle points. They are associated with the extended states near the center of each band. According to (8.2.28), the maximum value of σ_{xx} at the center of the jth Landau level is given by

$$\sigma_{xx}(0) = j + \tfrac{1}{2}. \tag{12.2.3}$$

In practice, flow curves can be obtained by using temperature because σ_{xx}, σ_{xy} and the inelastic scattering length vary with temperature. Hence, σ_{xx} and σ_{xy} will show temperature driven flows even at high temperatures. Such flows are strong when $k_B T$ is comparable to the Landau level broadening. On the other hand, flow lines move at low temperatures downward towards the fixed points. Wei et al. [12.12d] observed flows in InGaAs/InP in between 770 and 50 mK which are consistent with the scaling theory. In contrast, the temperature driven "classical" behavior was observed above 4 K.

Power-law singularities are observed at finite temperatures in the quantum Hall regime of the lowest Landau level. Here both ρ_{xy} and ρ_{xx} show a characteristic power law dependence on T. In a GaAs/InP heterostructure between 0.1 K and 4.2 K, the maximum slope of the ρ_{xy} curve is found to diverge as

$$(\partial \rho_{xy}/\partial H)_{\max} \propto T^{-\kappa}. \tag{12.2.4}$$

ρ_{xx} vanishes in an interval of a magnetic field with the half-width Δ which follows

$$\Delta \propto T^{\kappa}, \tag{12.2.5}$$

where

$$\kappa = 0.42 \pm 0.04. \tag{12.2.6}$$

These results, consistent with the scaling theory, show that the characteristic power law behavior in the transport coefficients is a universal feature of delocalization in the IQHE. In this respect, it is important to remark that essentially the same behavior with nearly the same exponent κ has been observed for the fractional quantum Hall effect in GaAs/GaAlAs between $1/3 < \nu < 2/5$, where ν is the filling factor of the lowest Landau level.

The above observations indicate that the singular parts of the resistivity components $\rho_{\alpha\beta}(T, H)$ in the delocalized or metallic regime are functions of a single variable x as follows [12.12]:

$$\rho_{\alpha\beta}(T, H) \propto f(x); \quad x \propto (H - H^*)T^{-\kappa}, \tag{12.2.7}$$

where H^* represents a singular point in the free-electron spectrum of the Landau level. The exponent κ is related to the exponent p of the inelastic scattering (Thouless) length such that

$$\kappa = p/(2\nu), \quad L_{in}(T) \propto T^{-p/2}, \tag{12.2.8}$$

where ν represents the exponent for the localization length ξ of the levels near the Fermi energy which diverges near H^* such that

$$\xi \sim |H - H^*|^{-\nu}. \tag{12.2.9}$$

A combination of (12.2.8) and (12.2.9) leads to

$$|x| = [L_{in}(T)/\xi(H)]^{1/\nu}. \tag{12.2.10}$$

The Hall plateaus are well defined if $|x| \gg 1$. On the other hand, for a metallic region in which the mean free path between inelastic collisions does not exceed the localization length one can impose the condition

$$|x| \lesssim 1. \tag{12.2.11}$$

Near the plateau $\sigma_{xy} = j + 1/2$, one can use the following scaling variables:

$$\theta = \sigma_{xy} - j - 1/2, \quad \sigma = \sigma_{xx} - \sigma_{xx}^*. \tag{12.2.12}$$

Under a renormalization-group transformation $L \to bL$, these variables are transformed such that $\theta \to b^s\theta$, $\sigma \to b^{-t}\sigma$, and the localization length follows

$$\xi(\theta, \sigma) = b\xi(b^s\theta, b^{-t}\sigma). \tag{12.2.13}$$

This implies a form such as

$$\xi = |\theta|^{-1/s} A(|\theta|^{t/s}\sigma). \tag{12.2.14}$$

The exponent $\nu = 1/s$ has been predicted to be universal.

Chalker and *Daniell* [12.13] have investigated the behavior of the two-particle spectral function defined by

$$S(r, \varepsilon, \omega) = \left\langle \sum_{i,j} \delta(\varepsilon - \omega/2 - \varepsilon_i)\delta(\varepsilon + \omega/2 - \varepsilon_j)\psi_i(0)\psi_i^*(r)\psi_j^*(r)\psi_j(0) \right\rangle.$$

$$(12.2.15)$$

Here, $\varepsilon_i(\varepsilon_j)$ is the energy of the state $\psi_i(\psi_j)$. For all arguments its Fourier transform takes the form

$$S(q, \varepsilon, \omega) = \frac{\rho(\varepsilon)h(q, \omega)}{\pi[\omega^2 + h^2(q, \omega)]}, \qquad (12.2.16)$$

where $g(\varepsilon)$ is the density of states per volume in d dimensions and

$$h(q = 0, \omega) = 0.$$

When a wave packet spreads diffusively for long times and large distances with diffusion constant D, the function $h(q, \omega)$ is given for small q and ω by

$$h(q, \omega) = \hbar D q^2.$$

At the mobility edge $\varepsilon = \varepsilon_c$, $h(q, \omega)$ may be assumed to be homogeneous. The length scale can be set by the linear size of a system if the size is larger than the inelastic scattering length. The energy scale ω must be less than the bandwidth. The mean level spacing, which is proportional to L^{-d}, can be used as an energy scale, d being the dimension. Invariance under the scale change $L \rightarrow bL$ implies

$$h(q, \omega) = b^d h(b^{-1}q, b^{-d}\omega) = \hbar q^d f(q^d/\omega). \qquad (12.2.17)$$

For 2D electrons in the lowest Landau level and for small q and ω, the function f is given by

$$f(q^2/\omega) = \begin{cases} D, & q^2/\omega < a, \\ (\omega a/q^2)^{\eta/2} D, & q^2/\omega > a, \end{cases} \qquad (12.2.18)$$

where a is a constant which is proportional to the density of states $g(\varepsilon_c)$ and

$$\eta = 0.38 \pm 0.04.$$

Figure 12.8 illustrates results obtained by extrapolation from numerical results for finite numbers of electrons. A Gaussian white-noise impurity potential was assumed such that

$$\langle V(x, y)V(x', y') \rangle = v^2 l^2 \delta(x - x')\delta(y - y'), \qquad (12.2.19)$$

where v is the strength of the potential and l is the magnetic length. The line segments in Fig. 12.8 represent estimated errors. The unit for f is vl^2h. The large errors near 0.2 of the abscissa reflect the intrinsic difficulty in obtaining f near the maximum of the function S. The collapse of their data onto a single curve reflects the scaling behavior expressed by (12.2.17).

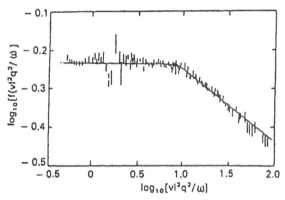

Fig. 12.8. Scaling behavior of $h(q, \omega)$ [see (12.2.17)] in the two-particle spectral function [12.13]

12.3 Theoretical Interpretation

The general behavior of ρ_{xx} or σ_{xx} in the quantum Hall regime suggests that, in order to interpret the IQHE, there must be a localized region between a pair of adjacent Landau levels. The quantization of the Hall conductivity in such a region has been explained in several ways. Let us first discuss the gauge invariance theory of *Laughlin* [12.14] which deals with a single electron in the absence of Coulomb interaction.

Each Landau level which is broadened by impurities is assumed to have an extended region near its center and localized regions at its edges on both sides. When the Fermi energy lies in the ith localized region, below which all the levels are filled, there is no current even if an external electric field E is applied. We adopt a Corbino geometry with an electric field E in the x direction, in which circulation takes place, and y in the perpendicular direction. For the Hall conductivity we use the condition that $E_y = 0$ and apply a magnetic field perpendicular to the disk on which all the electrons are located.

The circumference of the disc is assumed to be so long that x and y constitute a rectangular coordinate system. It is given by $2\pi R$ independent of y. If an electron is in an extended state and moves around the disk once, the magnetic flux passing through its closed orbit at R must be quantized in units of the flux quantum given by

$$\phi_0 = ch/e. \tag{12.3.1}$$

That is, $\pi R^2 H$ is an integral multiple of ϕ_0.

Let us now send in a magnetic flux Φ through a very thin solenoid at the center of the disk. It is confined within the solenoid and does not produce a magnetic field acting on the electron. We can then require

$$\mathrm{curl}\mathbf{A} = 0.$$

where $\mathbf{A}$ is the vector potential. Hence, with a single-valued function χ one can write

$$\mathbf{A} = \text{grad } \chi.$$

One can eliminate the vector potential from the Schrödinger equation by a gauge transformation in accordancc with

$$\psi \to \psi \exp\left(\mathrm{i}\frac{e}{ch}\chi(r)\right). \tag{12.3.2}$$

When the electron moves around the disk once with all other electrons fixed in space, ψ will change such that

$$\psi \to \psi_1 \exp\left(\mathrm{i}\frac{e}{c\hbar}\Phi\right). \tag{12.3.3}$$

Hence,

$$\Phi = \text{integer} \times \phi_0. \tag{12.3.4}$$

The total flux is then also quantized,

$$\pi R^2 H + \Phi = j\phi_0, \tag{12.3.5}$$

where j is an integer.

For a fixed value of the right side of the above equation, i.e., for a given j, an increase of Φ must cause a decrease in R. If Φ increases by ϕ_0, R in the above equation will correspond to $(j-1)$ instead of j. Other than this change, there is no remaining effect of the change in Φ. However, if there is an electric field $E_x = E$ the energy will change by

$$\Delta\varepsilon = eEL, \tag{12.3.6}$$

where L is the distance between the outer and inner edges of the disk. The current density j, due to the electron is given by

$$j_y = -\frac{c}{L}\frac{\Delta\varepsilon}{\Delta t}\frac{\Delta t}{\Delta\Phi}. \tag{12.3.7}$$

Replacing $\Delta\varepsilon/\Delta\Phi$ by $\Delta\varepsilon/\phi_0$ we obtain

$$j_y = -\frac{e^2}{h}E. \tag{12.3.8}$$

Repeating the same process in all i Landau levels, we find the total current to be i times the above value. Hence, for

$$\sigma_{yx} = j_y/E = -\sigma_H$$

we arrive at

$$\sigma_H = \frac{e^2}{h}i. \tag{12.3.9}$$

There are some other interpretations of the IQHE. In the topological theory forwarded by *Thouless* and his colleagues [12.15], the transverse conductivity is evaluated from

$$-\sigma_{xy} = \frac{ie^2\hbar}{L^2} \sum_n \frac{(v_x)_{on}(v_y)_{no} - (v_y)_{on}(v_x)_{no}}{(\varepsilon_0 - \varepsilon_n)^2}, \tag{12.3.10}$$

where i is imaginary, and v_x and v_y are the velocity components. For simplicity the sum over the electrons is omitted. They adopt the boundary conditions for the wave function:

$$\psi(y + L, y) = \exp(ik_x L)\exp(iyL/l^2)\psi(x, y),$$

$$\tag{12.3.11}$$

$$\psi(x, y + L) = \exp(ik_y L)\psi(x, y).$$

For a new wave function defined by

$$\overline{\psi} = \exp(-ik_x x - ik_y y)\psi$$

the Hamiltonian

$$\mathcal{H} = \frac{1}{2m}\left[\left(\frac{\hbar}{i}\frac{\partial}{\partial x}\right)^2 + \left(\frac{\hbar}{i}\frac{\partial}{\partial y} + \frac{eHx}{c}\right)^2\right] + V(x, y)$$

is transformed into

$$\overline{\mathcal{H}} = \frac{1}{2m}\left(\frac{\hbar}{i}\frac{\partial}{\partial x} + \hbar k_x\right)^2 + \frac{1}{2m}\left(\frac{\hbar}{i}\frac{\partial}{\partial y} + \frac{eH}{c}x\right)^2 + V(x, y).$$

The velocities can be expressed in terms of the derivatives of the Hamiltonian. First-order perturbation theory yields (omitting a j-sum)

$$\delta\overline{\psi}_0 = \sum_n{}' \frac{1}{\varepsilon_0 - \varepsilon_n}\left(\frac{\partial\tilde{\mathcal{H}}}{\partial k_j}\right)_{no} \delta k_j \overline{\psi}_n.$$

The transverse conductivity is given by

$$-\sigma_{xy} = \frac{ie^2}{\hbar(2\pi)^2}\int d\mathbf{k}_j \int d\mathbf{r}\left(\frac{\partial\overline{\psi}_0^*}{\partial k_x}\frac{\partial\overline{\psi}}{\partial k_y} - \frac{\partial\overline{\psi}_0^*}{\partial k_y}\frac{\partial\overline{\psi}_0}{\partial k_x}\right),$$

where the integrations are carried out over the unit cells in r and k space. The integral over the unit cell in the k space can be converted into a loop integral by Stokes' theorem. As a result we obtain

$$-\sigma_{xy} = \frac{ie^2}{4\pi h}\oint dk_j \int d\mathbf{r}\left(\overline{\psi}_0^*\frac{\partial\overline{\psi}_0}{\partial k_j} - \frac{\partial\overline{\psi}_0^*}{\partial k_j}\overline{\psi}_0\right). \tag{12.3.12}$$

ψ can be assumed to be a single-valued analytic function in the unit cell in k space and changes only by an r-independent phase factor θ. Hence, the integrand will be $\partial\theta/\partial k_j$ so that the loop integral produces 4π times an integer. That is, each filled Landau level contributes an integer to the Hall conductivity in units of e^2/h. The above type of integral is known to be topologically invariant.

Considerations have been made of cases in which the periodic potential is incommensurate with the flux, or an annulus geometry is used rather than an infinitely extended periodic lattice, or the system involves short-range impurity scattering.

A simple and direct evaluation of the Hall conductivity has been reported by *Heinonen* and *Taylor* [12.16]. They note that the current density carried by the one electron state k is given by

$$j_k = -\frac{ie\hbar}{2m}[(\nabla\psi_k^*)\psi_k - \psi_k^*\nabla\psi_k] - \frac{e^2}{2mc}\mathbf{A}|\psi_k|^2, \tag{12.3.13}$$

where $\mathbf{A} = (0, Hx, 0)$, $k = 2\pi l/L_y$, $l = 0, \pm 1, \pm 2\ldots$ The wave function may be written as $\psi_k = e^{iky}\phi_k$. If ϕ_k is real,

$$j_{k,x} = 0, \quad j_{ky} = -e\omega_c(x + x_k)\phi_k^2, \quad x_k = \hbar k/m\omega_c. \tag{12.3.14}$$

Note that $j_{ky} = j_k$ is independent of y. The energy of state k is given by

$$\varepsilon_k = \int \phi_k \left(\frac{p_x^2}{2m} + \frac{1}{2}m\omega_c^2(x + x_k)^2 + V(x)\right)\phi_k d\mathbf{r}. \tag{12.3.15}$$

This energy is related to the current by

$$\frac{\partial\varepsilon_k}{\partial x_k} = \int \phi_k^2 m\omega_c^2(x + x_k)^2 d\mathbf{r} = \frac{m\omega_c}{e}\int j_k d\mathbf{r}. \tag{12.3.16}$$

Note that due to the variational principle no differentiation of ϕ_k with respect to x_k is necessary. The right hand side integral is equal to $L_y i_k$, where i_k is the total current carried by state k. The total current I carried by the jth Landau level is then

$$I = -\sum_k \frac{e}{m\omega_c L_y}\frac{\partial\varepsilon_k}{\partial x_k}. \tag{12.3.17}$$

Since $x_k = \hbar k/m\omega_c$, the sum over k can be carried out over x_k. For large L_y we have

$$I = -\frac{m\omega_c L_y}{2\pi\hbar}\int \frac{e}{m\omega_c L_y}\frac{\partial\varepsilon}{\partial x_k}dx_k = -\frac{e}{h}\Delta\varepsilon\Big|_{\text{left}}^{\text{right}}, \tag{12.3.18}$$

where $\Delta\varepsilon$ denotes the Fermi energy difference between the right and left sides of the jth Landau level.

Let us now consider for simplicity the lowest two Landau levels. On the right side of each Landau level $\partial\varepsilon/\partial x > 0$, so that the states at the edge carry positive current. On the left side $\partial\varepsilon/\partial x < 0$, and the corresponding current is negative. We add electrons to the system such that the total current I_y is maintained. Initially there is a Fermi energy difference $\varepsilon_R - \varepsilon_L = hI_y/(-e)$. The subscripts R and L represent right and left respectively. Since the Hall voltage measured across the sample is $\Delta\varepsilon/(-e)$ the Hall conductivity is e^2/h.

We add more electrons to the system. They will distribute themselves in the empty states at the edges such that the Fermi energy difference is maintained. Hence, the Hall conductivity will remain the same. Eventually the Fermi energy at the right end of the lower Landau level will reach $3\hbar\omega_c/2$, and electrons will fall into the next Landau level starting from the lowest energy states. These states are located at the center of the sample rather than the edges. Hence, $\partial \varepsilon / \partial x$ will be zero, and they will not contribute to the current. The Fermi energy difference between the edges is still the same and σ_H remains e^2/h.

As the Fermi energy approaches the edges where $\partial \varepsilon / \partial x \neq 0$, both levels will be filled until the Fermi energies of the two levels have been equalized and

$$\varepsilon_R^{(1)} - \varepsilon_L^{(1)} = \varepsilon_R^{(2)} - \varepsilon_L^{(2)}.$$

Here the superscripts (1) and (2) refer to the lower and higher Landau levels, and the subscripts R and L correspond to right and left. Since the total current is kept constant each Landau level must carry a current $I_y/2$ and the Fermi energy difference between the edges will be $hI_y/2(-e)$. The Hall conductivity will be $\sigma_H = 2e^2/h$. The same process will be repeated for further density increases.

The above theories treat an ideal case. In the presence of impurities and Coulomb interaction, the conductivity may be expressed in terms of the memory function. In particular, in the localized regions between adjacent Landau levels, the imaginary part M'' of the memory function may be assumed to vanish. We may then solve the integral equation given by (8.3.17) for the real part M'. Between the Landau levels $(2i - 1)\mu_B H$ and $(2i + 1)\mu_B H$, the filling factor may be expressed as $\nu = i(1 + x)$, where x varies between 0 and 1. One can show [12.17] that in the localized region between the two levels there exists a solution given by

$$M' = \omega_c(1 - \nu/i).$$

This results in $\sigma_{xy} = -(e^2/h)i$ under a certain condition which depends on the impurity potential.

For a complete description of the IQHE it is necessary to show why the extended and localized regions appear in the Landau levels and how they depend on impurities, magnetic field and temperature, although there are some theories [12.18] and experiments [12.19] which show that short-range impurity forces are important. It is desirable to evaluate the plateau widths as functions of the Landau level index and other parameters. Further investigations of the activation energy for ρ_{xx} are also desirable.

In closing this chapter, let us remark on a new type of quasiparticles which has been observed for $\nu = 1$ in the IQHE resume. If the lowest Landau state is fully occupied at $\nu = 1$ in a perpendicular magnetic field, the electrons are in a spin polarized state. If an electron with a reverse spin is added to this state, it may be thought of as forming an electron pair with opposite spins. However, in the presence of spin interaction this is not energetically favorable. Instead, it has been suggested that a so-called *skyrmion* quasiparticle is formed.

Such a quasiparticle, originally introduced by A.H.R. Skyrme in 1958 as a model for an atomic nucleus, corresponds to a spatially spread spin perturbation about the added spin in a way similar to the case of a soliton.

In comparison with a single spin-flip excitation, such an excitation requires more Zeeman energy but the mear parallel configurations of neighboring spins save on exchange energy.

Experimental evidence for the existence of such a particle has been given [12.20] in terms of the Knight shift. The frequency of the nuclear spin-flip transition rate change can be used for a probe. As an alternative probe, the relaxation rate of the polarized nuclei may also be used because the nuclei can relax by putting an electron into a spin-flipped state. If the quasiparticles for the $\nu = 1$ state are spin-1/2 quasielectrons, the Knight shift is constant below $\nu < 1$ and decreases gradually for $\nu > 1$. However, it has been observed that the shift decreased sharply for both quasiparticles ($\nu > 1$) and quasiholes ($\nu < 1$) which were produced by tilting the magnetic field. Thus, the observation supports the existence of skyrmion quasiparticles for $\nu = 1$.

However, it appears as if the states $\nu = 3, 5$, and 7 may not have this type of quasiparticles [12.21]; this has yet to be tested. In the next chapter the quantized Hall states for $\nu \leqq 1$ are discussed. These states, particularly the case of $\nu = 1/3$, are very interesting because of some other types of quasiparticles.

13. Fractional Quantum Hall Effect

The fractional quantum Hall effect occurs when, in a strong magnetic field, electrons fall into highly correlated liquid states.

13.1 Experimental Results

Two years after the discovery of the IQHE, *Tsui* et al. [13.1] observed that for a certain interval of the magnetic field the Hall resistivity of high-mobility GaAs/GaAlAs heterostructures takes on a plateau value $\rho_{xy} = 3h/e^2$ at $\nu = 1/3$, where ν is the filling factor of the lowest Landau level (Fig. 13.1). At the same time the diagonal part of the resistivity tensor ρ_{xx} was found to drop to zero. Since then the same phenomenon, now called the *fractional quantum Hall effect* (FQHE), has been found at many fractions ν with, in general, odd denominators.

The FQHE resembles the IQHE except that it is observed only in high mobility samples for ν which is primarily less than 1. In some cases ρ_{xx} does not drop to zero but only dips down to low but finite values. As in the case of the IQHE the quantization of the Hall conductivity is very good. For $\nu = 1/3$ the quantization of the Hall plateau is as accurate as 3 parts in 10^5.

The variation of ν is achieved for GaAs/GaAlAs heterostructures by magnetic field. In terms of magnetic field the width is around 8.2, 2.3 and 2.9 kG for the 1/3, 2/3 and 2/5 plateaus, respectively. In general, a small increase of impurity concentration reduces the plateau widths, whereas the opposite is observed in the IQHE.

Since the FQHE occurs when the lowest Landau level is partially occupied, it is clear that Coulomb interaction is behind the phenomenon. Despite similarity, the IQHE occurs in the localized region between Landau levels and the quantization of Hall conductivity has been interpreted in terms of noninteracting electrons.

In general the fractional Hall states are characterized by odd denominator ν. For $\nu = p/q$, we talk of the $1/q$ series for various p. At $q = 3$, the 1/3 and 2/3 minima have the same strength. The state 2/3 is considered as the electron–hole conjugate of the 1/3 state. That is, not only electrons but also holes show the FQHE. Except for such a symmetry, the strength of the ρ_{xx} minima decreases generally with increasing p in a given q series.

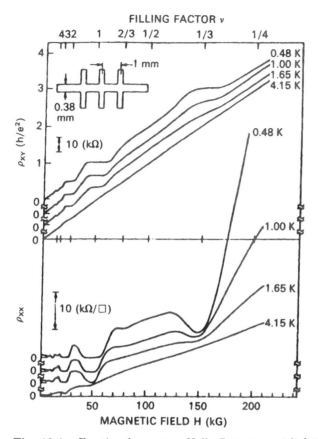

Fig. 13.1. Fractional quantum Hall effect at $\nu = 1/3$ [13.1]

The current–voltage relation is linear at $\nu = 1/3$ and $2/3$. This linearity holds even for very low voltages of order $1\,\mu$V. No pinning of a possible charge density wave has been observed, and therefore a charge density wave state does not seem to represent the ground state of the FQHE.

Both ρ_{xx} minima and ρ_{xy} slopes (with respect to magnetic field) approach zero near absolute zero. As the temperature is increased these quantities tend to approach saturation values. In a limited temperature range, an activated behavior of ρ_{xx} is observed, where

$$\rho_{xx} \propto \exp(-\Delta/2k_{\mathrm{B}}T). \tag{13.1.1}$$

For $\nu = 1/3$ and $2/3$, this activated behavior appears for a region in which ρ_{xx} and temperature vary respectively by one order of magnitude and by a factor of five. However, deviations from this behavior take place at low temperatures. The low temperature deviations are probably due to hopping conduction. The activated regions of $\nu = 2/5$ and $3/5$ are much smaller than that of $\nu = 1/3$.

The activated behavior indicates that there is a gap between the ground state and mobile excited state. The gap Δ depends on mobility μ such that it decreases with increasing disorder. Hence, there is a threshold at which the effect disappears. According to *Kukushkin* and *Timofeev* [13.2], Δ varies like

$$\Delta = 2\overline{\Delta}_\nu(1 - \mu_0/\mu)(e^2/\kappa l), \qquad (13.1.2)$$

where μ_0 is constant and κ, which is the dielectric constant, is approximately 13 in GaAs. The dimensionless activation energy $\overline{\Delta}_\nu$ is found to be 0.018 ($\nu = 1/3$), 0.013($\nu = 4/5$). Equation (13.1.2) shows that Δ varies like $H^{1/2}$. Such a variation has also been observed by *Boebinger* et al. [13.3].

An interesting experiment showing the difference between the $\nu = 2/3$ and 1/3 states was performed by *Haug* et al. [13.4]. They used a rotating gear which enabled a continuous tilting of the specimens relative to the direction of the magnetic field. That is, the magnetic field was tilted from the direction perpendicular to the surface of the sample. They observed that the activation energy for $\nu = 2/3$ increased with the tilt angle, whereas it decreased for $\nu = 1/3$. Concerning other fractional states, the minimum at $\nu = 4/3$ was observed to be less pronounced with increasing tilt angle, and small changes took place for $\nu = 5/3$, 2/5 and 3/5.

Furneaux et al. [13.5] conducted experiments with an added parallel magnetic field. For $\nu = 2/3$, they observed a linear increase of Δ with the total magnetic field, while for $\nu = 1/3$ and 5/3 Δ depends only on the perpendicular magnetic field. For $\nu = 4/3$ their measured gap decreased with increasing parallel field.

The observed difference between $\nu = 1/3$ and 2/3 is important. Even though the 1/3 and 2/3 states are conjugate to each other, their Δ must be interpreted differently. Indications are that spin-flip excitations are important only for certain cases, while changes in the electron motion in the z direction with the parallel magnetic field play a role in certain other cases. At $\nu = 2/3$, the excited quasielectron–quasihole pair may have a different net spin from its ground state. At higher magnetic fields a transition from a spin-flip excitation to a Coulomb gap without involving spin may be expected. For $\nu = 4/3$ the ground state may not be spin polarized, and the observed variation may be due to a slightly enhanced g-factor. At $\nu = 1/3$ and 5/3 a spin-flip excitation probably does not take place.

For $\nu = 8/5$, *Eisenstein* et al. [13.6] observed a sharp change in the dependence of the activation energy on tilt angle. This change is so sharp that it can be considered to represent a phase transition between two fractional states, perhaps one spin-polarized and the other unpolarized. Figure 13.2 shows their data. The activation energy Δ decreases smoothly as tilt angle θ increases from zero. For $\theta > 30\,^\circ$, Δ also increases smoothly. Near $30\,^\circ$ a complication due to splitting of the ρ_{xx} minimum occurs as shown in Fig. 13.2. Here the activation energy in K for $\nu = 8/5$ is plotted against the total field. Solid and open circles refer to low- and high-field cases. For $\theta < 25^\circ$ and $\theta > 40^\circ$, their ρ_{xx} data displayed activated behavior over almost two decades

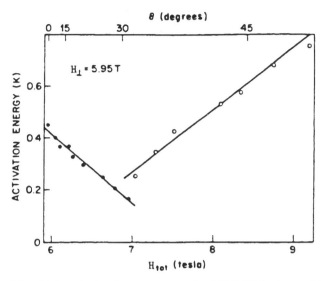

Fig. 13.2. Activation energy at the 8/5 FQHE observed in tilted fields [13.6]

in resistivity. However, around 30°, where the splitting took place, the temperature dependence was more complicated and the determination of Δ was difficult.

The sudden change in the behavior of Δ can be due to a phase transition between two fractional ground states of the same $\nu = 8/5$. However, there can be only one ground state because the activation energy represents the energy gap between the ground and excited states. In any case it is conceivable that for small angles, where the Zeeman energy is small, the ground state is unpolarized. Note that a possible unpolarized ground state for $\nu = 2/5$ has been suggested by *Halperin* [13.7] based on the small g-factor of order 0.5. The Zeeman energy at $\nu = 8/5$ increases with θ so that such an unpolarized state may become unfavorable at large angles.

The FQHE appears generally at odd denominator ν but an even denominator fractional state has been discovered by *Willett* et al. [13.8a] at $\nu = 5/2$. This state has been found to collapse rapidly as the magnetic field is tilted away from the vertical direction even though nearby odd-denominator states remain largely unaffected. Since tilting of the magnetic field enhances the spin-flip energy, the collapse of the 5/2 state indicates a significantly reduced spin polarization. That is, the FQHE may be considered to occur even in spin unpolarized states. They did not observe quantization at $\nu = 1/2$ and 3/2.

Clark et al. [13.8c] reported earlier even-denominator Hall states at $\nu = 5/4$, 5/2, 11/4, 7/2 and 15/4 on n-type GaAs/GaAlAs heterostructures of very high mobility. They attained $\mu = 2.1 \times 10^6 \, \text{cm}^2/\text{Vs}$ by illuminating the sample with light. They also observed several odd-denominator states in addition to the previously observed states.

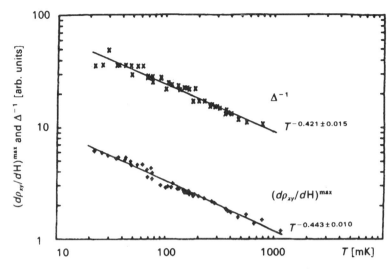

Fig. 13.3. Critical exponent of $(d\rho_{xy}/dH)^{max}$ and Δ^{-1} for $1/3 < \nu < 2/5$ [13.3c]

The FQHE is not restricted to the lowest Landau level. In fact, quantization at $\nu = 4/3$, $5/3$, $7/3$, $8/3$, $7/5$, $8/5$, $9/7$, $13/9$, etc. has been reported. Evidence of the FQHE in a ballistic 1D electron gas has also been reported. [13.9].

An important power law behavior has been found recently. In between $\nu = 2/5$ and $\nu = 1/3$, and for $22\,\mathrm{mK} < T < 1.2\,\mathrm{K}$, *Engel* et al. [13.3c] have found that the maximum value of $d\rho_{xy}/dH$ and the width of ρ_{xx} obey the same power law,

$$(d\rho_{xy}/dH)_{max} \propto T^{-\kappa}, \tag{13.1.3}$$

$$1/\Delta H \propto T^{-\kappa}. \tag{13.1.4}$$

Their data showing the temperature dependences of $(d\rho_{xy}/dH)^{max}$ and Δ^{-1} in a GaAs/GaAlAs heterostructure are illustrated in Fig. 13.3. They studied a total of three different samples. The average exponent is given by

$$\kappa = 0.43 \pm 0.02. \tag{13.1.5}$$

The width is defined by the field-separation between the two extrema of $d\rho_{xy}/dH$. Note that the same power law appears in the IQHE as shown in (12.2.4). The power law indicates that the correlation length for the wave function of the quasiparticles in the FQHE for $1/3 < \nu < 2/5$ is divergent and a localization–delocalization transition takes place. The exponent κ is a universal critical exponent.

13.2 Ground State of the FQHE

Since the FQHE occurs in a strong magnetic field particularly when the lowest Landau level is partially occupied, a certain combination of one electron states

$$\phi_l(z) = (2^{l+1}\pi l!)^{-1/2} z^l \exp(-|z|^2/4), \tag{13.2.1}$$

is expected to represent the ground state. Here $z = x + iy$ is the coordinate of an electron in units of the magnetic length. Such a combination must be antisymmetric with respect to electron exchange and must include ν as a characteristic parameter. The states of the electrons in the lowest Landau level are distinguished by the angular momentum quantum number 1, and the momentum dependence of their Hamiltonian and wave function can be ignored.

For the ground state, *Laughlin* [13.10] proposed a trial wave function

$$\psi_m = \prod_{i<j}(z_i - z_j)^m \exp\left(-\sum_{i=1}^{N}|z_i|^2\right), \tag{13.2.2}$$

where N is the total number of electrons and $m = 1/\nu$ is an odd integer. Note that this wave function is antisymmetric, characterized by a particular $\nu = 1/m$, and consists only of the states of the lowest Landau level. The total angular momentum of this state is determined by the power of z's in the polynomial part and is given by

$$L = \tfrac{1}{2}mN(N-1). \tag{13.2.3}$$

ψ_m is a many electron state and vanishes most strongly for a given m when two electrons overlap each other. Such a strong vanishing property is expected for a δ-function-type potential. Indeed, it has been shown that ψ_m becomes accurate for such a short range repulsive potential.

It is worth noting that the prefactor of the exponential function in ψ_m can be represented by the van der Mond determinant defined by

$$\prod_{i<j}(z_i - z_j) = \begin{vmatrix} z_1^0 & z_1^1 & z_1^2 & \cdots & z_1^{N-1} \\ z_2^0 & z_2^1 & z_2^2 & \cdots & z_2^{N-1} \\ \vdots & & & & \vdots \\ z_N^0 & z_N^1 & z_N^2 & \cdots & z_N^{N-1} \end{vmatrix}$$

$$= D_{0,1,2,\ldots,N-1}(z_1, z_2, \ldots, z_N). \tag{13.2.4}$$

This form shows how the electrons are distributed in the angular momentum space.

Laughlin has found numerically that the ground state energies associated with ψ_m are below the corresponding energies of the charge density wave states for both $m = 3$ and 5. Moreover he has found that ψ_m has an interesting electrostatic analog. This is expressed by

$$|\psi_m|^2 = \exp(-\beta\Phi). \tag{13.2.5}$$

Here, $\beta = 1/m$ and Φ is given by

$$\Phi = -\sum_{i<j} 2m^2 \ln|z_i - z_j| + \frac{m}{2} \sum_{i=1}^{N} |z_i|^2. \tag{13.2.6}$$

With this expression and with the identification $\beta = 1/k_BT$, (13.2.5) is the Boltzmann distribution of the classical OCP consisting of N identical particles with charge m. Φ is the potential consisting of the mutual repulsion in the first term and the attractive interaction in the second term. The latter is due to the presence of a cloud of neutralizing charges. These neutralizing charges are distributed uniformly in the system with a charge density $1/\sqrt{(2\pi)}$.

Monte Carlo simulations [13.11] have indicated that an OCP is a fluid if $\Gamma = (\pi n)^{1/2}e^2\beta$ is less than 128, and is a crystal above this value. In the present case $\Gamma = (2m)^{1/2}$, which is not very large. Hence, Laughlin's state can be considered to represent a liquid-like electron state. Moreover this liquid is incompressible because density changes amount to injecting electrons or holes, resulting in higher energies. With incompressibility, no sound wave excitations will take place in the ground state, assuring a low energy ground state. Indeed the presence of the Hall plateaus means that the quantized states are stable against any perturbation.

The ground state energy per particle ε_g can be obtained from

$$\varepsilon_g = n \int \frac{e^2}{r} [g(r) - 1]\pi r\,dr, \tag{13.2.7}$$

where $g(r)$ is the radial distribution function. Laughlin used a hypernetted chain result [13.11] of $g(r)$ to obtain approximate energies.

An exact formula for ε_g has been derived by *Takano* and *Isihara* [13.12] based on second quantization. For $\nu = 1/3$ they obtained $\varepsilon_g = -0.388855$ ($N = 4$), -0.390255 ($N = 5$) and -0.391517 ($N = 6$) in units of $e^2/\kappa l$. These values may be compared with the results obtained numerically by *Yoshioka* et al. [13.13] under a periodic boundary condition: 0.4152 ($N = 4$), -0.4127 ($N = 5$) and -0.4128 ($N = 6$). On the other hand, the limiting value for $N \to \infty$ is given by [13.14, 15]

$$\varepsilon_g = -0.4100. \tag{13.2.8}$$

For finite N, ε_g is expected to be higher, but its value depends on boundary conditions. For example, *Haldane* and *Rezayi* [13.16] obtained -0.449954 for $N = 4$ based on the spherical representation,

$$\Psi_m^{HR} = \prod_{i<j}(u_iv_j - v_iu_j)^m, \tag{13.2.9}$$

where

$$(u, v) = [\cos(\theta/2)e^{i\phi/2}, \sin(\theta/2)e^{-i\phi/2}].$$

Morf and *Halperin* [13.17] performed extensive Monte Carlo calculations of the ground state energy of 2D electrons in a magnetic field. From the cases $N = 20$, 30, 42, 72 and 144, they derived an empirical formula for the N dependence of ε_g for $\nu = 1/3$. It is expressed as

$$\varepsilon_g = -0.4101 + 0.06006N^{-1/2} - 0.0423N^{-1}. \tag{13.2.10}$$

By a separate Monte Carlo calculation, they obtained -0.410 ± 0.001 for the ground state energy in the thermodynamic limit.

We have been concerned so far with $\nu = 1/m = 1/3$. *Levesque* et al. [13.14] made use of an OCP radial distribution function and reported the following empirical formula for the m dependence of ε_g:

$$\varepsilon_g = -\frac{0.782133}{m^{1/2}}\left(1 - \frac{0.211}{m^{0.74}} + \frac{0.012}{m^{1.7}}\right). \tag{13.2.11}$$

According to this formula, the ground state energy decreases also with m. The first term here is the energy of a triangular Wigner lattice, which is preferred in the low-density limit. For $\nu > 0.1$ or $m < 10$, liquid-like states are more favourable. For $m = 5$ this formula produces $\varepsilon_g = -0.3277$.

Laughlin's wave function has several important features. Moreover numerical analyses of small systems have shown that it is very close to the true ground state. Nevertheless, it is a Jastrow-type trial function represented by a product of pair functions. In an effort to approach the true ground state, *Takano* and *Isillara* [13.15] have constructed a new variational wave function based on states of the lowest Landau level such that many-body correlations are included and yet the total angular momentum is conserved. Their function is given by

$$\Psi = \prod_{i<j}(z_i - z_j)(f_{ij})^p(f_{ji})^p \exp\left(-\sum_k |z_k|^2/4\right). \tag{13.2.12}$$

Here

$$f_{ij} = z_i - z_j + c\sum_{k\neq i,j} z_k, \tag{13.2.13}$$

where c is a variational parameter.

Note that f_{ij} is a linear function of z's. This ensures the conservation of the total angular momentum. In the limit $c \to 0$, Ψ reduces itself to Ψ_m. Note that p can be any integer or zero for an arbitrary odd integer $m = 2p + 1$, in which 1 corresponds to Fermi statistics. The exponent of the first factor in Ψ reflects this common feature while p in Ψ is arbitrary.

Since Laughlin's state is close to the true ground state, c must be small. If we use an OCP analog, the electrostatic potential is given to lowest order in c by

$$\Phi = -m^2\sum_{i<j}\ln|z_i - z_j| + \frac{m}{4}\sum_i |z_i|^2 + \frac{m-1}{2m}\sum_{i>j}\frac{|c|^2|\sum_k' z_k|^2}{|z_i - z_k|^2}. \tag{13.2.14}$$

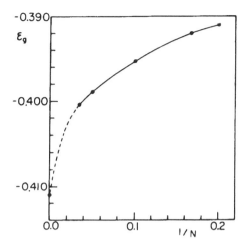

Fig. 13.4. Ground state energy of the variational wave function [13.15] as a function of N

Hence, the c term introduces an effective repulsive potential to the original Laughlin state. This potential depends on the center of gravity of charges other than chosen pairs, and decreases rather quickly. It also depends on the density $(1 - \nu)$ of the background charges. It is appropriate to consider that c is small because it is known that Laughlin's state is close to the true ground state.

Takano and *Isihara* have found that for small c the ground state energy of Ψ is indeed lower than Laughlin's and approaches the limiting value of -0.4101 for $N \to \infty$ as in Fig. 13.4. Moreover, their Monte Carlo calculations indicate that Ψ for $\nu = 1/3$ has three-fold degeneracy. In fact for $N = 30$, cusp-type energy minima with values -0.4008, -0.4008 and -0.4009 appear at three different values of c as shown in Fig. 13.5. These values may coincide with each other in the thermodynamic limit, causing degeneracy. Such degeneracy is desirable for the true ground state of the FQHE as has been discussed by *Su* and others [13.18].

The state Ψ_m for $m = q$ is the parent state of the fractional states at $\nu = p/q$. That is, there is a hierarchy of fractional states. In order to investigate this hierarchy, let us consider perturbing the $1/m$ state by expanding the system very slightly at a fixed magnetic field. The stability of the state requires that the electrons stay in the same state ψ_m even if ν deviates slightly from the exact quantized value of $1/m$. Hence, the expansion will cause a charge deficiency at a certain cost of energy. In order to create a quasihole, we pierce the system at z_0 with an infinitesimally thin solenoid and send in a flux quantum ϕ_0 adiabatically [13.19]. The state with one quasihole at z_0 can be represented by

$$\Psi_m^{+1} = \prod_{i=1}^{N}(z_i - z_0)\psi_m(z_1,\ z_2,\ \ldots,\ z_N), \tag{13.2.15}$$

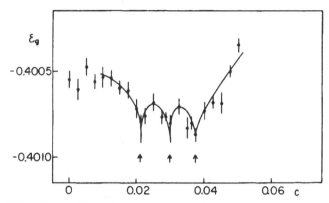

Fig. 13.5. Ground state energy of the variational wave function Ψ given by (13.2.12) as a function of the variational parameter c. The curve is only a guide to the eye. [13.15]

where the normalization constant has been omitted. The reason why this function is appropriate can be seen from an OCP analog. If we write

$$|\Psi_m^{+1}|^2 = \exp(-\Phi^{+1}/m) \tag{13.2.16}$$

we find the "electrostatic potential", Φ^{+1}, to be given by

$$\Phi^{+1} = -2m^2 \sum_{i<j} \ln|z_i - z_j| + \frac{m}{2} \sum_i |z_i|^2 - 2m \sum_i \ln|z_i - z_0|. \tag{13.2.17}$$

This is the same as Φ of (13.2.6) except for the last term, which represents an additional phantom charge 1 at z_0. To screen this charge the plasma will accumulate charge -1, which is equivalent to a real charge $-1/m$ because each charge has been transformed into a charge m in the OCP. Remember that Φ represents the electrostatic potential for particles with charge m.

A similar consideration shows that a quasiparticle at z_0 introduced to the $1/m$ fractional state can be represented by

$$\Psi_m^{-1} = \prod_{i=1}^N \left(2\frac{\partial}{\partial z_i} - z_0^* \right) \psi_m(z_1, z_2, \ldots, z_N). \tag{13.2.18}$$

Note the appearance of differential operators, [see (13.3.12)]. These operators act only on the polynomial part of ψ_m.

When many quasiholes or quasiparticles are created in the $1/m$ state, they can interact with each other in a way similar to the electrons in the original stage. If M quasiholes have been created, the wave function may be expressed analogously to ψ_m by

$$\Psi_m^{+M} = \prod_{i<k}^{M}(Z_i - Z_k)^\theta \exp\left(-\sum_k |Z_k|^2/4l_h^2\right) \times$$

$$\times \prod_{j}^{N}(z_j - Z_k)\Psi_m(z_1, z_2, \ldots, z_N), \tag{13.2.19}$$

where the Z_j and Z_k represent respectively the coordinates of the electrons and quasiholes. The magnetic length of the quasiholes is denoted by l_h. θ is a constant which depends on quasihole statistics. It is 0 (or 1) if the quasiholes obey Bose (or Fermi) statistics.

We can associate the zeros of the above wave function with particles. Let us refer to these zeros as vortices. Under the magnetic field represented by the magnetic length l, there are $1/(2\pi l^2)$ electron vortices per unit area. Since the magnetic field is kept constant in our process of creating quasiholes, the number of electron vortices is given by

$$mn + n_h = eH/ch$$

$$= g_e, \tag{13.2.20}$$

where n_h is the quasihole density. Due to the creation of the quasiholes, the electron density n determined from (13.2.20), is now $(g_e - n_h)/m$. That is, there is a slight decrease from the original value g_e/m. On the other hand, the quasiholes also produce zeros, and the number of quasihole's vortices is given by

$$n + \theta n_h = 1/(2\pi l_h^2)$$

$$= g_h. \tag{13.2.21}$$

When the filling factor of these quasiholes reaches a particular value, they can form a new Laughlin state. This state is given analogously to ψ_m by

$$\psi_{p_1+\theta} = \prod_{j<k}(Z_i - Z_k)^{p_1+\theta} \exp\left(-\sum_k \frac{|Z_k|^2}{4l_h^2}\right). \tag{13.2.22}$$

The exponent p_1 is an even integer in order to maintain the same statistics represented by θ. The filling factor of the quasiholes is given by

$$\nu_h = \frac{n_h}{g_h}$$

$$= \frac{1}{p_1 + \theta}. \tag{13.2.23}$$

From (13.2.20) and (13.2.21) we find

$$n_h = g_e - mn,$$

$$g_h = n + \theta(g_e - mn).$$

Introducing these into (13.2.23) we arrive at

$$\nu = \frac{1}{m + \frac{1}{p_1}}. \tag{13.2.24}$$

This is the electron filling factor which corresponds to ν_h when a new liquid-like state is formed by the quasiholes. Note that this ν is independent of θ.

For $m = 1$ and $p = 2$, (13.2.24) yields $\nu = 2/3$. The state Ψ_1 corresponds to the case in which the lowest Landau level is completely filled. Hence, the $\nu = 2/3$ can be considered to be the hole state which is conjugate to the electron's 1/3 fractional state.

In the case of quasiparticles, a similar consideration yields

$$\nu = \frac{1}{m - \frac{1}{p_1}}. \tag{13.2.25}$$

For example, the 2/5 state can be obtained by choosing $p_1 = 2$ and $m = 3$. One can further extend the above considerations to generate daughter states from the parent $1/m$ state. The result will constitute a hierarchy of the fractional states.

13.3 Elementary Excitations and Off-Diagonal Long Range Order

It costs energy to create a quasiparticle or a quasihole in the fractional quantized Hall states. Since an addition (or a removal) of one electron is equivalent to adding m quasiparticles (or quasiholes) the difference in the derivatives of the total ground state energy is expressed as

$$(\partial E_g/\partial N)_{\nu+} - (\partial E_g/\partial N)_{\nu-} = m(\tilde{\varepsilon}_+ + \tilde{\varepsilon}_-), \tag{13.3.1}$$

where $\tilde{\varepsilon}_-$ and $\tilde{\varepsilon}_+$ represent respectively the energy necessary to create a quasiparticle or a quasihole. This difference should be positive, reflecting the stability at $\nu = 1/m$. The sum

$$\Delta = \tilde{\varepsilon}_+ + \tilde{\varepsilon}_- \tag{13.3.2}$$

is the energy gap for a neutral excitation.

Δ is the energy which characterizes the activated form of ρ_{xx} given by (13.1.1). Experimentally the activation energy at the center of the $\nu = 1/3$ plateau is in the range 0.02–0.03 in units of $e^2/\kappa l^2$. In terms of temperature, this energy corresponds roughly to 6 K. This is very small in comparison with the activation energy for the IQHE. Its smallness is understandable because the excitations in the FQHE take place within the lowest Landau level.

Considerable efforts have been made to evaluate Δ. Unfortunately, this evaluation, which is somewhat difficult, has produced larger values than the experimental Δ.

In a single-mode approximation, *Girvin* et al. [13.20] developed a theory of collective excitations in analogy with Feyman's theory of liquid helium. According to Feyman's theory, the excitation spectrum of liquid ^{4}He is given by

$$\Delta(k) = \frac{f(k)}{S(k)}, \tag{13.3.3}$$

where $S(k)$ is the structure factor and $f(k)$ is the oscillator strength given by

$$f(k) = \frac{1}{2N}\langle 0|[\rho_k^*, [\rho_k, \mathcal{H}]]|0\rangle. \tag{13.3.4}$$

Here, $\mathcal{H}$ is the Hamiltonian and ρ_k is the Fourier transform of the local density,

$$\rho_k = \sum_i \exp(-i\mathbf{k}\cdot\mathbf{r}_i).$$

$f(k)$ given by the double commutator is simply k^2.

In order to develop a similar theory for the FQHE, both $f(k)$ and $S(k)$ must be constructed within the lowest Landau level. That is, one must construct a Hilbert space using functions belonging to the lowest Landau level. We adopt the magnetic length as the unit for length and note that the eigenfunctions of the lowest Landau level are given by $\phi_l(z)$ of (13.2.1),

$$\phi_l = (2^{l+1}\pi l!)^{-1/2}z^l \exp(-|z|^2/4).$$

Apart from the exponential factor, which is independent of l, ϕ_l, is characterized by z^l. Hence, such a Hilbert space may be constructed by using entire functions which can be expanded in powers of z such that [13.21]

$$f(z) = \sum_n a_n z^n. \tag{13.3.5}$$

A scalar product of two such functions f and g is defined by

$$(f,g) = \int f^*(z)g(z)d\mu(z) \tag{13.3.6}$$

where

$$d\mu(z) = \frac{1}{2\pi}dxdy\exp(-|z|^2/2). \tag{13.3.7}$$

Not that (f,f) is nonnegative, and

$$(z^m, z^l) = \begin{cases} 0, & (m \neq l), \\ m!, & (m = l). \end{cases} \tag{13.3.8}$$

We define a differential operator d/dz such that it acts only on the polynomial part of the eigenfunction ϕ_l. Let us consider

$$f_l = \frac{z^l}{(2^l l!)^{1/2}} \qquad (13.3.9)$$

with the l-dependent part of the normalization factor of ϕ_l. We find

$$z f_l = [2(l+1)]^{1/2} f_{l+1} \qquad (13.3.10)$$

and

$$\frac{\mathrm{d}}{\mathrm{d}z} f_l = \left(\frac{2}{l}\right)^{1/2} f_{l-1}. \qquad (13.3.11)$$

Thus we find that

$$a^+ = \frac{z}{\sqrt{2}}, \quad a = \sqrt{2}\frac{\mathrm{d}}{\mathrm{d}z} \qquad (13.3.12)$$

are the creation and annihilation operators. Accordingly we define the projection of z^* into the lowest Landau level by

$$\bar{z}^* = z^\dagger = 2\frac{\mathrm{d}}{\mathrm{d}z}. \qquad (13.3.13)$$

Many variables can be considered similarly, and we can project functions of such variables on the lowest Landau level. The projected density function $\bar{\rho}_k$ is given by

$$\bar{\rho}_k = \sum_{j}^{N} \exp\left(-\mathrm{i}k\frac{\partial}{\partial z_j}\right) \exp\left(-\frac{\mathrm{i}k^*}{2}z_j\right). \qquad (13.3.14)$$

The projected oscillator strength is

$$\bar{f}(k) = \frac{1}{2}\sum_q u(q)(e^{q^*k/2} - e^{qk^*/2}) \times [\bar{S}(q)e^{-k^2/2}(e^{-k^*q/2} - e^{-kq^*/2})$$

$$+ \bar{S}(q+k)(e^{k^*q/2} - e^{kq^*/2})], \qquad (13.3.15)$$

where

$$\bar{S}(k) = S(k) - (1 - e^{-|k|^2/2}) \qquad (13.3.16)$$

is the projected structure factor. Using these results one can show that the excitation energy in the single-mode approximation is given by

$$\Delta(k) = \frac{\bar{f}(k)}{\bar{S}(k)}. \qquad (13.3.17)$$

$\Delta(k)$ can be evaluated numerically by using an appropriate form of the radial distribution function which is needed for $S(k)$. It has been found that each of the $\nu = 1/3$, $1/5$ and $1/7$ states has a particular $\Delta(k)$, which starts with a finite value at $k = 0$, gradually decreases to a roton-type minimum, and then approaches an asymptotic value. Note that the existence of a finite gap at $k = 0$ differs from the case of liquid ^{4}He in which $\Delta(0) = 0$. The excitation energy obtained from the roton minimum is 0.106 for $\nu = 1/3$ and

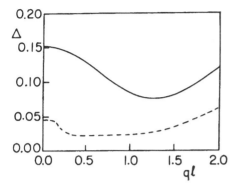

Fig. 13.6. Dispersion for the collective excitation at $\nu = 1/3$ FQHE. *Solid line*: $n = 0$ Landau level. *Dashed line*: $n = 1$. The excitation energy Δ is in units of $e^2/\kappa l$ [13.22]

0.025 for $\nu = 1/5$ in the long wavelength limit. These results are considerably higher than the results based on the HNC approximation but are not very far from Monte Carlo results.

MacDonald and *Cirvin* [13.22] extended the above single-mode theory to include higher Landau levels. Figure 13.6 illustrates their dispersion relation for the 1/3 fractional state. The solid line illustrates Δ for the $n = 0$ Landau level, which agrees with what they obtained earlier. The dashed line corresponds to $n = 1$ Landau level. Note that these two curves are qualitatively different. Due to the spin degeneracy and particle–hole symmetry, the latter curve describes the strong-field limit of the collective excitation spectrum for total filling factor 7/3, 8/3, 10/3 and 11/3. For $n = 1$, the collective excitation energies are smaller than for $n = 0$. As usual, Δ is expressed in units of $e^2/\kappa l$, where l is the magnetic length. For $\nu = 1/5$, Δ is much smaller than the case of $\nu = 1/3$. For $\nu = 0$ and $\nu = 1/5$, a roton minimum of around 0.02 occurs near $ql = 1.0$.

Experimental observation of rotons is rather difficult. However, for $\nu \geqq 1$ an observation of roton structure in the density of states for inter-Landau-level transitions has been reported by *Pinczuk* et al. [13.23]. They performed inelastic light scattering measurements of inter-Landau-level modes of magnetoplasmons and spin-density excitations with the dispersion relation,

$$\omega(q) = \omega_c + \Delta(q, H).$$

For $\nu \geqq 1$, their data on GaAs heterostructures indicated the excitation binding and roton behavior predicted in the Hartree-Fock approximation.

It is interesting to note that the existence of a new type of off-diagonal long-range order in the FQHE [13.24] has been pointed out. To show this, let us use the singlet density matrix given by

$$\rho(z, z') = \frac{N}{Z} \int dz_2 dz_3 \ldots dz_N \Psi^*(z, z_2, \cdots, z_N)\Psi(z', z_2, \cdots, z_N), \quad (13.3.18)$$

where Z is the norm of Ψ. In the second quantization, this one-body density matrix is expressed as

$$\rho(z, z') = \sum_{m,n} \phi_m^*(z)\phi_n(z')\langle 0|c_n^+ c_m|0\rangle, \tag{13.3.19}$$

where ϕ's are the lowest Landau level wave functions given by (13.2.1). If the ground state is isotropic and homogeneous one may assume

$$\langle 0|c_n^* c_m|0\rangle = v\delta_{nm}. \tag{13.3.20}$$

Then, the density matrix becomes

$$\rho(z, z') = \frac{v}{2\pi} \exp\left(-\frac{1}{4}|z - z'|^2\right) \exp\left[-\frac{1}{4}(z^*z' - zz'^*)\right]. \tag{13.3.21}$$

This is short-ranged with a characteristic scale given by the magnetic length.

Let us now introduce the singular gauge field given by

$$A_j(z_j) = \frac{m\phi_0}{2\pi} \sum_{i\neq j} \text{grad}_j[\Im\{\ln(z_j - z_i)\}], \tag{13.3.22}$$

where ϕ_0 is the flux quantum, to change the phase of the wave function such that

$$\psi_{\text{new}} = \exp[-im \sum_{i<j} \Im\{\ln(z_i - z_j)\}]\Psi_{\text{old}}. \tag{13.3.23}$$

Under this transformation, Laughlin's wave function becomes

$$\Psi_m = \prod_{i<j} |z_i - z_j|^m \exp\left(-\frac{1}{4}\sum_k |z_k|^2\right). \tag{13.3.24}$$

Note that the polynomial part is now replaced by its absolute value. Hence, this new wave function is completely real and symmetric regardless of m.

By using the new wave function in (13.3.19), the singlet density matrix in the singular gauge can be expressed as

$$\bar{\rho}(z, z') = \frac{v}{2\pi} \exp[-\beta\Delta f(z, z')]|z - z'|^{-m/2}. \tag{13.3.25}$$

where $\beta = 2/m$ plays the role of $1/k_B T$, and $\Delta f(z, z')$ is the difference in free energy between two impurities of charge $m/2$ at z and z' and a single impurity of charge m at arbitrary position. Due to complete screening of the impurities by the plasma, the free-energy difference rapidly approaches a constant at large distances. This asymptotic behavior is important because then the density matrix decreases algebraically; it proves the existence of ODLRO (off-diagonal long range order) at the plasma temperature corresponding to β^{-1}. For $m = 1$ the exact asymptotic value is given by $\beta\Delta f \rightarrow -0.03942$. For $m = 3$ or 5, numerical analyses show that $\beta\Delta f$ also approaches a constant.

Equation (13.3.25) is based on Laughlin's state. This state is correct for short-range repulsive interactions. For such interactions the zeros of Ψ can be directly attached to the particles and there is a one-to-one correspondence between the existence of ODLRO and the occurrence of the FQHE. As the range of interactions increases, $(m-1)$ of the zeros will move away from the particles. Laughlin's state will then, become inaccurate, and ODLRO may eventually disappear. Then the onset of short-range behavior of $\rho(z, z')$ may take place.

Concerning the general discrepancy between the theoretical and experimental Δ, several factors have been investigated. For instance, according to *Yoshioka* [13.25a], finite thickness reduces the theoretical Δ. Further reductions of Δ take place due to disorder as discussed by *Gold* [13.25b]. Such reductions indicate the existence of a threshold magnetic field. Since the reduction factor depends on magnetic field, certain deviations of Δ from the quadratic field dependence may be expected.

We have discussed that, in order to screen the additional phantom charge represented by (13.2.17), the plasma will have a real charge $-1/m$. Hence, the FQHE at $\nu = 1/m$ is associated with quasiparticles with a fractional charge $e^* = e/m$. An experimental evidence which supports the existence of such fractional charges has been given. *Simmons* et al. [13.26a] used a high-mobility GaAs/GaAlAs heterostructure to observe resistance fluctuations with a quasiperiod $\Delta H = 0.05T$ near the minima of the diagonal resistance at $\nu = 1/3$. This period is three times the corresponding period $0.016T$ at $\nu = 1, 2, 3$ and 4, which is independent of ν. It is also independent of current leads and the temperatures of their experiment. Thus these results indicate transport by quasiparticles of fractional charge $e/3$ at $\nu = 1/3$. *Clark* et al. [13.26b] have also arrived at a conclusive evidence for fractional quasiparticle charges. Consistent with the case of the IQHE, the prefactor of σ_{xx} in an Arrhenius plot at low temperatures has been found to be $(e/q)^2/h$ for fractions $\nu = p/q$, leading to the fractional charges $\pm e/q$.

The Laughlin state explains the appearance of the conjugate states $\nu = 1/p$ and $\nu = 1-1/p$, but does not explain why the states $\nu = p/(2p\pm 1)$ appear strongly and why the $\nu = 1/2$ state, which is the limit of $p \to \infty$, does not occur. Experimentally, anomalies at $\nu = 1/2$ have been observed by *Willett* et al. [13.8c] and *Du* et al. [13.8d]. For instance, for the odd denominator states the sound velocity increases while the sound decay decreases. The opposite has been obeserved for $\nu = 1/2$.

At $\nu = 1/2$, one half of the Landau level is occupied and each electron has two magnetic flux quanta. The electrons, carrying the magnetic quanta, move in the plane freely as if there is no external magnetic field. An electron carrying an even number of flux quanta is called a *composite fermion*. *Jain* [13.27] has associated the $\nu = p/(2p \pm 1)$ state with the IQHE of compisite particles with $2p$ magnitic flux quanta. Halperin et al. [13.28] have shown that the $\nu = 1/2$ state can be transformed to a mathematically equivalent system of fermions interacting with a Chern–Simons gauge field, such that the average effective magnetic field seen by the fermions is zero.

When each composite fermion carries $2p$ flux quanta, it sees a magnetic field given by

$$B^* = B_+^- 2p\phi_0 n,$$

where B is the external magnetic field, $\phi_0 = 2\pi c/e$ is the flux quantum, and n is the density. Accordingly, the filling factor

$$\nu^* = \phi_0 n / B^*$$

of the composite fermions corresponds to the electron filling factor given by

$$\nu = \frac{\nu^*}{2p\nu^* \pm 1}.$$

If $\nu = p'/[2pp' \pm 1]$, the composite fermions fill an integer number p' of quasi-Landau levels. This explains the incompressiblility in a partially filled Landau level as in the case of the IOHE.

14. Phase Transition and Stability of Matter

This chapter deals with such fascination subjects as the crystallization of electrons and the stability of matter. Do electrons crystallize? If so, what are their properties? How can matter consisting of positive and negative charges exist without collapsing? These fundamental and intriguing questions make the theory of electrons very attractive and important.

14.1 Crystallization

The theory of electron liquids is more straightforward than that of ordinary liquids because the Coulomb interaction is well defined, while the interaction potentials of ordinary liquids are more complex. In comparing these two cases, the interesting question of whether electrons crystallize or not arises, since they repel each other and are subject to strong quantum effects.

In 1934, *Wigner* [14.1] predicted that electrons may condense into a crystalline state at a low density near absolute zero. In contrast to molecules in ordinary liquids, the mass of electrons is extremely small and quantum effects are very large so that they are in a gaseous state at high densities. As their density is reduced, the Coulomb interaction becomes stronger than the kinetic energy, as (1.1.7) indicates. Accordingly, the electrons try to be as far apart as possible from each other in order to minimize the electrostatic repulsive energy and eventually crystallize.

Experimental efforts were made to test Wigner's prediction but without success. Theoretical attempts to treat this crystallization were also made without any significant development for several decades, and the study of this subject was stagnating. However, the recent realization of two-dimensional electron systems has changed this aspect because it is now possible to observe a Wigner lattice. The reason is that in 2D the electron density can be varied in a much wider range than in 3D, and the Coulomb interaction is relatively stronger.

An observational achievement was reported in 1979; *Grimmes* and *Adams* [14.2] finally succeeded in observing the melting of an electron crystal which was formed above the surface of liequid helium. It was advantageous to use such a 2D electron system above the surface of liquid helium because of much less impurities than in the semiconductor interface. They observed that a particular spectrum line respresenting the coupling between the ripplons, as the quanta of the surface waves, and the plasmons, as the quanta of the collective motion of the electrons, disappeard at an electron density of 4.4×10^8 cm^{-2} and a temperature of 0.475 K, indicating that melting has taken place. The melting point was found to vary in proportion to $n^{1/2}$, with n as the electron density. This indicates that the parameter defined by

$$\Gamma = \pi^{1/2} e^2 n^{1/2} / kT$$

plays a key role. In Fact, their data show that the melting point corresponds to Γ which is given by

$$\Gamma_m = 131 \pm 7.$$

Similar values of Γ have been obtained numerically by Monte Carlo and molecular dynamics methods. *Kosterlitz* and *Thouless* [14.3a] made an interesting theoretical approach to melting and succeeded in deriving a formula to estimate a theoretical melting point. It is based on a consideration of the excess free energy due to dislocations in the crystal.

Let us suppose that there is an isolated dislocation in a 2D electron crystal which has a surface area L^2. the associated energy is given by

$$E_{id} = \pi J \ln(L/a_c).$$

Here, J represents the magnitude of the energy associated with a dislocation with a core size a_c. In order to obtain the associated excess entropy, we note that the probability of finding such a dislocation is given by b^2/L^2 where b is the magnitude of the Burgers vector **b**. Hence, the entropy varies also logarithmically, and the extra free energy due to the dislocation is given by

$$F_{id} = (\pi J - 2k_B T) \ln(L/a_c). \tag{14.1.1}$$

the melting point is considered to correspond to the temperature at which this extra free energy vanishes:

$$T_m = \frac{\pi J}{2k_B}. \tag{14.1.2}$$

J is determined in ordinary crystals by Lame's constants λ, μ, and b, but λ is infinite in the case of a 2D Wigner crystal. Therefore, *Thouless* [14.3b] used μ for absolute zero. If c is the speed of sound, the value of Γ at melting is given by

$$\Gamma_m = \frac{\rho b^2 c^2}{4\pi k_B}. \tag{14.1.3}$$

This formula results in $\Gamma_m = 78.7$. A value $\Gamma_m = 128.2$, which is closer to the experimental value, has been obtained by *Morf* and others [14.4] by taking the temperature variation of μ into consideration.

According to the experiment of *Glattli* et al. [14.5] the change in entropy at melting seems to be in the order of $0.02\, k_B$ or less. This suggests that the transition is of a continuous type because an entropy change in the order of $0.3k_B$ is expected for a first-order transition.

The melting criterion based on Γ is classical. Near absolute zero, another parameter r_s is expected to play an important role. *Imada* and *Takahashi* [14.6] obtained a phase curve for a wide range of temperatures and r_s by extrapolating their Monte Carlo results. Their result shows that a maximum melting temperature occurs at around $3\,\mathrm{K}$ when the electron density is around 2×10^{12} cm^{-2}. These values can be too large, but in any case the melting temperature decreases towards absolute zero when the density is changed from the maximum value to either higher or lower values.

More recently, technical advances have enabled experimental observations on 2D electron systems in GaAs/GaAlAs heterostructures. However, it is known that a hard-sphere gas condenses into a liquid-like state. Hence, it is interesting to compare the case of hard-spheres to the case of electrons with the long-range Coulomb interaction. Even though the two systems are very different, both a hard-core and Fermi statistics seem to play a somewhat similar role.

An electron crystal cannot be observed directly. As an alternative technique for a visual observation, a macroscopic crystal formed on the surface of liquid helium may be used. The surface of liquid helium is not completely flat but has small depressions. Electrons fall into such depressions when a positively charged metal plate is placed under the surface in order to enhance the effect of the depressions. The depressions can be about $0.1\,\mathrm{mm}$ deep when the electric field is of the order of $2000\,\mathrm{V/cm}$. These depressions can have as many as 10^6 electrons each and form a new macroscopic charged system which can crystallize into a so-called *dimple crystal*. This crystal can be observed by utilizing the total reflection of light at the outer surface of the dimples illuminated from the bottom. A picture showing a hexagonal dimple crystal has been published by *Leiderer* et al. [14.7].

Thus, it is now firmly established that electrons can form a crystal at low temperatures and densities. Since they are in a gaseous state in the opposite limit, there must be an intermediate liquid-like state with strong electron–electron correlations as described in Chapt. 1. However, a rigorous theoretical approach to this intermediate state has not been developed as yet. One must say that electron liquid theory is still in its infancy in spite of remarkable recent developments.

14.2 Properties of an Electron Lattice

Since the formation of a Wigner lattice has been confirmed, let us now investigate its statistical properties. At absolute zero, the energy of an electron lattice is obtained by summing up the Coulomb interactions between the electrons on the lattice points. Therefore, the energy can be calculated in a straightforward way for a given crystal structure. Note that in the evaluation it is necessary to neutralize the system by putting a cloud of positive charges in the background in order to avoid a divergence which is characteristic of the long-range Coulomb potential. Note also that the zero-point vibration of the electron lattice gives rise to a vibrational energy. This vibration is determined by the equations of motion which are given by expanding the $1/r$ potential about the equilibrium point. For a 3D body-centered electron lattice, *Carr* [14.8] treated the harmonic vibrations, first order anharmonic vibrations, and exchange interaction. He obtained the following expression for the energy per electron:

$$\varepsilon_g^L = -\frac{1.79186}{r_s} + \frac{2.628}{r_s^{3/2}} - \frac{0.73}{r_s^2} + \cdots . \tag{14.2.1}$$

This result is for low densities, i.e., it is valid approximately for $r_s \gtrapprox 0.54$. Note in the above series that the first term is proportional to r_s^{-1}, the second to $1/r_s^{-3/2}$, etc.

At high densities where r_s is small, the electron system is close to an ideal gas and has kinetic energy, which is proportional to r_s^{-2}, exchange energy which is proportional to r_s^{-1}, and correlation energy given by (4.2.18):

$$\varepsilon_g^G = \frac{2.21}{r_s^2} - \frac{0.916}{r_s} - 0.094 + 0.0622 \ln r_s - 0.013 r_s \ln r_s - \cdots \tag{14.2.2}$$

The exchange energy varies like the harmonic-vibration energy in (14.2.1), but the latter is twice as large in magnitude.

The above energy expressions represent the two limiting cases. It is very difficult to devise a rigorous approach for electron systems in the intermediate density region, and, in fact, no exact theory has been developed as yet. Hence, *Isihara* and *Montroll* [14.9] evaluated the energy for all densities by interpolating the above two expressions based on a Padé approximation. According to their calculation, the ground-state energy reaches a minimum value of −0.163. Ryd. per electron for $r_s = 3.77$. It is intriguing that their ground state energy curve shows a van der Waals type wiggle at $r_s = 14.4$. If this loop is associated with a phase transition as in the case of a van der Waals gas, perhaps a crystallization with density changes in the order of 40 % takes place.

The ground-state energy of a 2D triangle electron lattice is given in the harmonic approximation [14.10]

$$\varepsilon_g^L = -\frac{2.21}{r_s} + \frac{1.63}{r_s^{3/2}} + \cdots . \tag{14.2.3}$$

For a given r_s, the first term is lower than that of (14.2.1). This is due to the stronger electron correlations in 2D than in 3D. In comparison with 3D, the energy minimum occurs at a smaller r_s with a lower minimum energy in 2D [14.11].

The zero-point vibration produces an electronic specific heat. For the same 2D triangle lattice, the specific heat per area is given by

$$C_A = \frac{3k}{(0.1905)^2} \left(\frac{kT}{\hbar\omega_p}\right)^2 \zeta(3), \tag{14.2.4}$$

where $\zeta(x)$ is the Riemann zeta function, and

$$\omega_p = \left(\frac{2\pi ne^2}{ma}\right)^{1/2}. \tag{14.2.5}$$

Here, a is the lattice constant. Note that the specific heat is proportional to T^2 as a characteristic of 2D. The dispersion relations for the longitudinal and transverse vibrations are given as follows:

$$\omega_l^2 = \omega_p^2 aq - 0.1815(\omega_p aq)^2$$
$$\omega_t^2 = (0.063\omega_p aq)^2. \tag{14.2.6}$$

Here, the first term, ω_l, represents the plasmon frequency given by

$$\omega_l = \left[\frac{2\pi ne^2}{m}q\right]^{1/2}. \tag{14.2.7}$$

Not that at $q = 0$, both ω_l and ω_t vanish as another feature of 2D.

14.3 Stability of Matter

Statistical mechanics is based on the requirement that the free energy of a macroscopic system is proportional to the total number of particles in the system. This requirement, known as the existence of the thermodynamic limit, has been studied for various cases [14.12]. For instance, *Fisher* and *Ruelle* [14.12 b] investigated systems with positive potentials which include a part with an integrable Fourier transform. *Lieb* and *Lebowitz* [14.12c] examined the case of charged particle systems. The closely related and equally fundamental subject of the stability of matter has also been studied. This is described here in statistical terms, although a quantum approach is needed.

When the total Hamiltonian of a given system is divided into a reference part $\mathcal{H}_0$ and the rest as follows:

$$\mathcal{H} = \mathcal{H}_0 + \mathcal{H}_1, \tag{14.3.1}$$

the free energy $\mathcal{F}$ can be decomposed into the corresponding two parts:

$$\mathcal{F} = \mathcal{F}_0 + \mathcal{F}_1. \tag{14.3.2}$$

The following bounds have been derived for the perturbed part of the free energy [14.13]:

$$\langle \mathcal{H}_1 \rangle_0 \geqq \mathcal{F}_1 \geqq \langle \mathcal{H}_1 \rangle, \tag{14.3.3}$$

where the lower (upper) bound is given by the statistical average of $\mathcal{H}_1$ in the actual (reference) system. In what follows, these energy bounds are used for absolute zero.

While a macroscopic system may be at a finite temperature, it is defined to be stable if its ground state energy has a lower energy bound which is proportional to the total number of particles:

$$E \geqq -AN. \tag{14.3.4}$$

It is necessary to clarify why matter, consisting of positive and negative charges, exists in nature without collapsing due to their Coulomb attraction.

The stability of the case of an OCP is clear. For a classical OCP, we note

$$\sum_{i<j}^{N} \frac{1}{r_{ij}} = \sum_{i<j}^{N} \left(\frac{e^{-\lambda r_{ij}}}{r_{ij}} + \frac{1 - e^{-\lambda r_{ij}}}{r_{ij}} \right),$$

where λ is a positive constant. Since the first term on the right-hand side is nonnegative the left-hand side sum is larger or equal to the sum of the last terms on the right-hand side. By adding and subtracting the $i = j$ terms in which we take the limiting values for 0 distance, we arrive at

$$\sum_{i,j}^{N} \frac{1}{r_{ij}} \geqq -N\lambda. \tag{14.3.5}$$

Therefore, the system can be stable. We remark that more generally a lower bound of the ground state energy of a system with repulsive potentials can be obtained in a similar way.

In the case of a quantum-mechanically degenerate OCP we have already discussed the ground-state energy per electron in the high- and low-density limits. Also, a lower energy bound can be obtained by using a charged boson system [14.14]. Thus, we can conclude that the OCP is stable.

The case of two-component plasmas is far more complex. Around 1968, *Lenard* and *Dyson* [14.15a] used a rigorous and elaborate approach to study their stability and obtained a lower energy bound which is proportional to the total number of particles. However, the proportionality constant was unexpectedly large. *Lieb* [14.15d] then developed an approach which is related to the Thomas–Fermi theory and obtained an improved result. His and related works have been collected in the volume edited by *Thirring* [14.16].

The actual evaluation of the lower bound requires a quantum-mechanical approach which may involve mathematical complications. In what follows, a statistical mechanical consideration of stability is presented. For simplicity, we take a neutral system consisting of N positive and N negative charges of the same absolute magnitude e and neglect the kinetic energy. However, the kinetic energy does play a key role, and the distribution functions which we use are quantum mechanical.

For simplicity, let us use $r_{\alpha\beta}$ for the distance between a given charge pair $\alpha - \beta$, where α or β can be either $+$ or $-$. Let us introduce the probability $P_{\alpha\beta}(r)\mathrm{d}\mathbf{r}$ of finding $\alpha\beta$ particle at a distance r from α at the origin. We impose the condition:

$$\int P_{\alpha\beta}(r)\mathrm{d}\mathbf{r} = 1. \tag{14.3.6}$$

We introduce the average energy denoted by

$$\varepsilon_{\alpha\beta} = e^2 \left\langle \frac{1}{r_{\alpha\beta}} \right\rangle = e^2 \int \frac{1}{r} P_{\alpha\beta}(r)\mathrm{d}\mathbf{r} \geq 0. \tag{14.3.7}$$

The above integral is not divergent at $r = 0$, because of the differential volume $4\pi r^2 \mathrm{d}r$. There is also no divergence at large distances, because for $r \geq d$,

$$\int_d^\infty \frac{1}{r} P_{\alpha\beta}(r)4\pi r^2 \mathrm{d}r \leq \frac{1}{d}\int_d^\infty P_{\alpha\beta}(r)4\pi r^2 \mathrm{d}r \leq \frac{1}{d}.$$

Due to the symmetry between the positive and negative charges, we expect

$$P_{+-}(r) = P_{-+}(r); \tag{14.3.8}$$
$$\varepsilon_{++} = \varepsilon_{--}.$$

Let us assume that $P_{\alpha\beta}$ consists of two parts: $P(r)$, which is common to all pairs, and $\Delta P_{\alpha\beta}$, which is intrinsic to a particular pair $\alpha - \beta$:

$$P_{\alpha\beta}(r) = P(r) + \Delta P_{\alpha\beta}(r). \tag{14.3.9}$$

We introduce the following conditions:

$$\int P(r)\mathrm{d}\mathbf{r} = 1, \tag{14.3.10}$$

$$\int \frac{1}{r} P(r)\mathrm{d}\mathbf{r} = 0. \tag{14.3.11}$$

Due to (14.3.11), the probability $P(r)$ does not cause any change in the averages $\varepsilon_{\alpha\beta}$ between various pairs. In this sense, it may be called an auxiliary probability function. If V is the volume, the simplest auxiliary function is given by

$$P(r) = \frac{1}{V}. \tag{14.3.12}$$

This function corresponds to smearing out the charges. At the thermodynamic limit, it satisfies (14.3.11). We shall not treat $P(r)$ further because after all it does not affect the comparison of Coulomb averages of $\alpha - \alpha$ and $\alpha - \beta$ pairs, although one can introduce a more realistic auxiliary function. As for $\Delta P_{\alpha\beta}$, it satisfies

$$\int \Delta P_{\alpha\beta}(r)\mathrm{d}\mathbf{r} = 0, \tag{14.3.13}$$

$$\int \frac{1}{r} \Delta P_{\alpha\beta}(r)\mathrm{d}\mathbf{r} \geq 0, \tag{14.3.14}$$

because of the conditions on $\Delta P_{\alpha\beta}(r)$ and $P(r)$.

In comparison with $\Delta P_{+-}(r)$, $\Delta P_{++}(r)$ is expected to increase more slowly because the plus charge at origin attracts negative charges and repels positive charges. However, there can be similarities between the two distributions. Both may be expected to decrease toward the origin, approach the same asymptotic value and perhaps wiggle in between. The function $P(r)$ represents such common features. As for the intrinsic distribution functions, we assume a relationship:

$$\Delta P_{++}(r) = \mu(r)\Delta P_{+-}(\lambda(r)r). \tag{14.3.15}$$

Here, $\mu(r)$ and $\lambda(r)$ are scaling functions which may depend on r. Hence, the assumption is fairly general. The function $\mu(r)$, which is nonnegative, can satisfy the condition (14.3.14) if

$$\mu(r) = c\lambda^2(r)\frac{\mathrm{d}(\lambda(r)r)}{\mathrm{d}r}, \tag{14.3.16}$$

where c is a constant. In order to relate the two distribution functions to each other, the two scaling functions have been introduced because they play different roles. In particular, the constant c in $\mu(r)$ determines the magnitude of $\Delta P_{++}(r)$. If λ is constant, $\mu = c\lambda^3$. In this form, we see that if c is large, λ is small, and vice versa. Since $\lambda(r)$ is a function of r, the extra r in the argument of ΔP_{+-} is not really necessary.

The difference between the averages of the two relevant Coulomb energies is given by

$$\varepsilon_{++} - \varepsilon_{+-} = 4\pi e^2 \int_0^\infty x\Delta P_{+-}(x)[c\lambda(r) - 1]\mathrm{d}x. \tag{14.3.17}$$

Let us introduce the probability $P_0(r)$ of finding any charge at r about a plus charge which is the average probability defined by

$$P_0 r = P(r) + \frac{1}{2}\left(\Delta P_{++}(r) + \Delta P_{+-}(r)\right), \tag{14.3.18}$$

and the average reciprocal distance given by

$$\frac{1}{r_0} = \frac{1}{2}\left(\left\langle\frac{1}{r_{++}}\right\rangle + \left\langle\frac{1}{r_{+-}}\right\rangle\right). \tag{14.3.19}$$

The total Coulomb energy is given by the pair interaction energies. Under (14.3.16) the following two possibilities can occur:
   Case [1]: $c\lambda(r) \geqq 1$.
   In this case, $\varepsilon_{++} \geqq \varepsilon_{+-}$ and therefore

$$\varepsilon_{+-} \leqq \varepsilon_0 \leqq \varepsilon_{++}, \tag{14.3.20}$$

where $\varepsilon_0 = e^2/r_0$. Hence,

$$E_{\mathrm{c}} \geqq N\varepsilon_0. \tag{14.3.21}$$

The system can be stable in perhaps an atomic or gaseous state with not too large N.

Case [2]: $c\lambda(r) \lessgtr 1$.
In this case, it is possible that

$$\varepsilon_{+-} \gtrless \varepsilon_0 \gtrless \varepsilon_{++}. \tag{14.3.22}$$

Hence,

$$-N\varepsilon_0 \gtrless E_c. \tag{14.3.23}$$

The system prefers to be in a condensed state, but it does not collapse because of the kinetic energy. In OCP, in the Rydberg unit, the kinetic energy varies as $1/r_s^2$, while the Coulomb energy varies as $1/r_s$, where $r_s = (3/4\pi n)1/3$. thus, the Coulomb energy becomes effective only when the average particle distance is rather great and there is no divergence at infinite r.

Since the above consideration is general, it is necessary to use a quantum approach to obtain a lower bound of the ground-state energy explicitly. *Lenard* and *Dyson* [14.15a] developed a complex but rigorous theory and obtained a lower bound, although its magnitude was unexpectedly large. An improved result has been obtained by *Lieb* [14.15d].

The above treatment is not directly applicable to atoms. However, neglecting neutrons and other quantum conditions, we introduce a model atomic system consisting of a nucleus with z protons which are surrounded by z electrons. Let us treat this system in the same way as above and consider the case in which $\langle r_{++} \rangle \leqq \langle r_{+-} \rangle$ and $\varepsilon_{++} \neq \varepsilon_{+-}$. Because of the asymmetry between the charges, our analysis becomes more involved than in the previous symmetric case. However, there can be cases in which the average repulsive energy between the charges of the same kind is larger than the absolute magnitude of the attractive energy between oppositely charged pairs:

$$\frac{1}{2}(\varepsilon_{--} + \varepsilon_{++}] \geqq \varepsilon_{+-}. \tag{14.3.24}$$

This is analogous to Case [1] above. We find then that the Coulomb energy of the model system has a negative lower bound which is proportional to z:

$$E_c \geqq -\frac{z}{2}(\varepsilon_{++} + \varepsilon_{--}). \tag{14.3.25}$$

However, in order to keep the Coulomb energy negative it is necessary to require that z has an upper bound given by

$$z \leqq \frac{\varepsilon_{--} + \varepsilon_{++}}{\varepsilon_{--} + \varepsilon_{++} - 2\varepsilon_{+-}}. \tag{14.3.26}$$

If the average Coulomb energy between the pairs of the same charge is 1% larger than that of oppositely charged pairs, the maximum z is estimated to be 100. In spite of the simple model and approach, this is a reasonable estimate.

The stability of a two-component plasma may depend on dimensionality. This is so because the spatial integration of the Coulomb potential carries a factor $4\pi r^2$ in 3D, eliminating the possibility of a divergence at the origin. In lower dimensions, the situation is different and, in fact, the Coulomb effect is generally stronger.

Even if the stable existence of matter has been clearly established, charged particle systems can take on different phases. The stability of a particular phase and phase transitions are important subjects which invite theoretical treatments. We have discussed that a 3D electron system is in a gaseous phase at high densities and crystalline state at low densities, but an exact all density theory which yields a phase diagram has not been developed as yet. In spite of its long history and much progress, electron theory still leaves a vast, unexplored territory.

We conclude this chapter and also this book with the hope that a breakthrough will take place in this unknown territory in the near future.

A. Appendix

A.1 Pair Distribution Function

For the theoretical study of electron gases and liquids, use of the pair distribution function is frequently made because it not only exhibits the spatial correlations of particles but also yields the X-ray structure factor and thermodynamic functions. Therefore, various theoretical methods have been developed for its evaluation. While it is not the purpose of this section to discuss all these theories, we give a brief account of a linked cluster theory [A.1–3] and its connection with an integral equation approach. We shall then present several basic formulas which can be applied to electron fluids and other many-body systems.

In fluid systems, the probability of finding two particles at a distance r is given by the pair distribution function $\rho_2(r)$. When r is very large, this function is expected to approach n^2, n being the number density. Hence, for a uniform system the radial distribution function $g(r)$ is introduced such that

$$\rho_2(r) = n^2 g(r). \tag{A.1.1}$$

The pair distribution function in a grand ensemble can be expressed exactly and generally in the cluster series [A.1]

$$\rho_2(r) = n^2 + \sum_{l=2}^{\infty} b_l(r) z^l, \tag{A.1.2}$$

where z is the absolute activity and $b_l(r)$ is the cluster integral represented by all the l-particle connected graphs which are labeled by $\mathbf{r}_1$ and $\mathbf{r}_2$, $r = |\mathbf{r}_2 - \mathbf{r}_1|$. The two particles at $\mathbf{r}_1$ and $\mathbf{r}_2$ in such graphs are connected by interaction potentials and/or by quantum exchanges if quantum statistics should be used. When integrated over $\mathbf{r}$, $b_l(r)$ satisfies the normalization condition

$$V \int b_l(r) \mathrm{d}\mathbf{r} = l(l-1) b_l, \tag{A.1.3}$$

where b_l is the quantum cluster integral of 1 particles which are "connected" with each other either by interaction or by quantum mechanical exchanges [A3]. Different from those for $b_l(r)$, the graphs for b_l are not labeled. The above equation results because there are $l(l-1)$ ways in which the pair of particles is placed in any of the l-particle graphs. The equation has actually been derived rigorously [A.1]. Through such a derivation, we learn that b_l is the coefficient in the following cluster series expansion of the logarithm of the grand partition function:

$$\ln \Xi = \sum_{l=1} b_l z^l. \tag{A.1.4}$$

Here, z is the absolute activity. The grand ensemble average of the total number of particles is

$$\langle N \rangle = \sum_{l=1}^{\infty} l b_l z^l. \tag{A.1.5}$$

By differentiating this equation with respect to $\ln z$, we find

$$\sum_{l=1}^{\infty} l^2 b_l z^l = \langle N^2 \rangle - \langle N \rangle^2. \tag{A.1.6}$$

In these equations, $\langle \ldots \rangle$ denotes a grand ensemble average. These relations ensure the normaliztion of the pair distribution function

$$V \int \rho_2(r) \mathrm{d}\mathbf{r} = \langle N(N-1) \rangle. \tag{A.1.7}$$

Let us now investigate the pair distribution function of a classical fluid from a different angle. For homogeneous fluids, the radial distribution function $g(r)$ can be used and can be expressed exactly as

$$g(r) = \exp[-\beta \phi(r) + S(r)], \tag{A.1.8}$$

where $\beta = 1/k_\mathrm{B} T$, ϕ is the interaction potential, and $\exp[S(r)]$ is given by a desnity series

$$\exp[S(r)] = \sum_{s=0}^{\infty} g_s(r) n^s. \tag{A.1.9}$$

The function $g_s(r)$ is represented by those irreducible diagrams that include the pair of particles at $\mathbf{r}_1$ and $\mathbf{r}_2$. These diagrams can be constructed systematically in accordance with the following steps [A2]:

1) Formation of an essentially series diagram

A series connection of Mayer f-bond diagrams or more complicated graphs forms an essentially series diagram. The connecting points of component graphs which must be passed always in going from $\mathbf{r}_1$ to $\mathbf{r}_2$ are called nodal points.

2) Formation of a generalized chain

A series connection of essentially series diagrams forms a generaized chain diagram. One generalized chain is represented by an effective line. In Fig. A.1, (a) is a simple series diagram, (b) is a more complicated series diagram, and

(a)

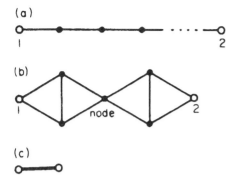

(b)

(c)

Fig. A.1a-c. Essentially series diagrams for the pair distribution function. (a) Simple series diagram. (b) Complex series diagram with a node. (c) Essentially series diagram which combines (a) and (b). The heavy line represents an effective line

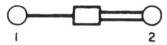

Fig. A.2. Diagram with three effective lines. The box represents an arbitrary complex diagram of unlabeled particles

(c) represents an effective line. In the diagram, white circles are labeled and black circles are unlabeled. For classical fluids, particles can be illustrated by points, but in the next section we shall introduce quantal graphs which are illustrated in a slightly different way.

3) Parallel connections with effective lines

A pair of junctions can be connected by two or more effective lines in order to create topologically different diagrams. A junction can be either a node or one of the two particles. In Fig. A.2, the middle box is a junction.

Since any number of independent effective lines may connect r_1 and r_2, the function $g_s(r)$ can be decomposed into a sum of products of such lines. Accordingly, we have

$$g_s(r)n^s = \sum_{m_l} \prod_l \frac{[\beta_l(r)n^l]^{m_l}}{m_l!}, \qquad (A.1.10)$$

where the sum of products is subject to the condition

$$\sum l m_l = s.$$

Using (A1.10) in (A1.9) we find

$$S(r) = \sum_l \beta_l(r)n^l, \qquad (A.1.11)$$

where the right-hand side is a sum over all the irreducible effective line diagrams of arbitrary numbers of nodes.

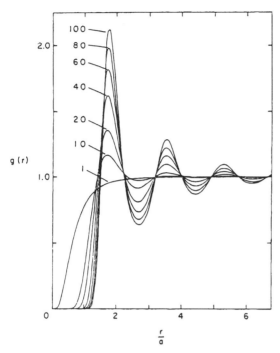

Fig. A.3. Radial distribution function $g(r)$ for two dimensions based on the HNC approximation [A.4]

In general, $S(r)$ consists of the contribution $N(r)$ from the diagrams with nodes and the contribution $E(r)$ from those without. Let us introduce

$$G(r) = g(r) - 1, \quad S(r) = N(r) + E(r), \tag{A.1.12}$$

and

$$X(r) = G(r) - N(r). \tag{A.1.13}$$

Recognizing that $nG(r_{32})dr_3$ gives the probability of finding particle 3 in the volume element dr_3 at a position r_{32} from particle 2, one obtains an integral equation

$$G(r) = X(r) + n \int X(r_{13})G(r_{32})dr_3. \tag{A.1.14}$$

Here, $r_{ij} = r_j - r_j$. This equation shall be called the Ornstein–Zernike equation. Although (A1.14) is exact, $X(r)$ must be given in order to determine $G(r)$.

In the so-called hypernetted chain (HNC) approximation, $E(r)$ is neglected so that $N(r) = S(r)$. Then,

$$X(r) = g(r) - 1 - \ln g(r) - \beta\phi(r), \tag{A.1.15}$$

and (A1.14) becomes

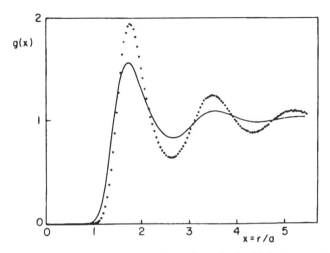

Fig. A.4. Monte Carlo and HNC results [A.5] for $\Gamma = 40$

$$\ln g(r) + \beta\phi(r) = n \int [g(r_{13}) - 1 - \ln g(r_{13}) - \beta\phi(r_{13})][g(r_{32}) - 1]\mathrm{d}\mathbf{r}_3.$$
(A.1.16)

In the Percus–Yevick (PY) approximation, $X(r)$ is taken such that

$$X(r) = g(r)\{1 - \exp[\beta\phi(r)]\}.$$
(A.1.17)

The corresponding integral equation is

$$g(r)\mathrm{e}^{\beta\phi(r)} = 1 + n \int [[1 - \exp\beta\phi(r_{13})]g(r_{13})[g(r_{32}) - 1]\mathrm{d}\mathbf{r}_3.$$
(A.1.18)

Graphically, both HNC and PY approximations are correct only to first-order diagrams in density. To second order, the HNC approximation misses one graph and the PY approximation two. The details concerning these equations can be found in statistical mechanics books [A.1].

The radial distribution function $g(r)$ of a 2D electron gas determined numerically by *Lado* [A.4] based on the HNC approximation is illustrated in Fig. A.3 as a function of r/a, where a is the ion-circle radius determined by

$$\pi a^2 n = 1.$$

The parameter Γ is dimensionless and is defined by

$$\Gamma = (e^2/a)/k_\mathrm{B}T.$$

The role played by this parameter is discussed in Chap. 3. For $\Gamma \lesssim 1$, the radial distribution function increases monotonically as a function of r/a. The short-range order as indicated by the first peak starts at around $\Gamma = 2.8$. The first peak based on the HNC approximation is generally lower than that determined by the Monte Carlo method, as shown in Fig. A.4, which

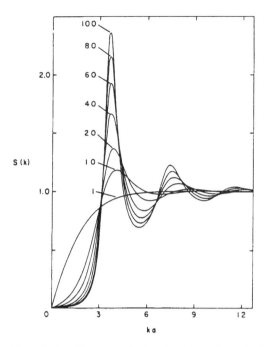

Fig. A.5. Structure factor $S(k)$ based on the HNC approximation [A.4]

represents the results of *Caillol* et al. [A.5] for $\Gamma = 40$. The dots are Monte Carlo results on a sphere, and the curve represents the HNC approximation. They also obtained improved results by modifying the HNC approximation. From these results we learn that for $\Gamma \geq 10$ the radial distribution function of a classical plasma shows oscillations analogous to the case of ordinary liquids. Hence, in such a domain of Γ, the system may be considered as an electron liquid. Figure A.5 illustrates the HNC static structure factor $S(k)$ plotted against ka. This graph corresponds to Fig. A.3. Similar graphs have been obtained for three dimensions [A.6]. Also, some improvements on the HNC approximation have been worked out.

The pair distribution function can be used to evaluate the energy. A convenient formula is obtained by introducing a coupling parameter ξ to the Coulomb potential and noting that

$$\frac{\partial U(\xi)}{\partial \xi} = -\frac{\partial}{\partial \beta}\left[\frac{\partial}{\partial \xi}\ln(\mathrm{Tr}\{e^{-\beta H}\})\right]$$

$$= \frac{V}{2}\frac{\partial}{\partial \beta}\left[\beta \int \phi(r)\rho_2(r,\xi)\mathrm{d}\mathbf{r}\right], \qquad (A.1.19)$$

where V is the total volume. Upon integration over ξ from 0 to 1, the internal energy U can be obtained.

A.2 Classical and Quantum Chains

In evaluating the pair distribution function based on a linked cluster expansion, it is important to try to sum its series to infinite order because the result is exact as far as the series is concerned. However, such a summation is generally not possible and can be achieved for only very limited types of graphs. Among them, the so-called chain diagrams should be discussed because their contribution to the pair distribution function represents long-range correlations which are particularly important to electron fluids because the Coulomb potential is long ranged. The pair distribution function obtained in the chain diagram approximation yields the energy obtained in the so-called ring diagram approximation. That is, it is possible to evaluate the energy directly in consideration of ring diagrams.

Among the connected diagrams for the cluster integral $b_1(r)$ of the pair distribution function, those with a pair of particles at $\mathbf{r}_1$ and $\mathbf{r}_2$, $r = |\mathbf{r}_2 - \mathbf{r}_1|$, connected by $(l-2)$ unlabeled particles in the form of a chain represent the lowest-order connected graphs in interaction for a given number l. In order to derive $b_l(r)$ in the chain diagram approximation, the diagrams can be constructed in the $r - \beta$ space, where $\beta = 1/k_B T$.

The pair distribution function of a fluid is given by the diagonal element of the density matrix which in turn is represented by the propagator of the particles in the $r - \beta$ space defined for $(\mathbf{r}, 0) \to (\mathbf{r}, \beta)$ by

$$K(\mathbf{r}', \beta; \mathbf{r}, 0) = \sum_n e^{-\beta \mathcal{H}} \psi_n(\mathbf{r}') \psi_n^*(\mathbf{r}),$$

where $\mathcal{H}$ is the total Hamiltonian and ψ_n is the eigenfunction. Here, for convenience we used only a single representative variable $\mathbf{r}$ or $\mathbf{r}'$ for the coordinates of all the particles in the system. The propagation in the direction of β starts at $\beta = 0$, where the particles are free, and ends at $\beta = 1/k_B T$. For the ideal case in which the Hamiltonian consists of kinetic energies and $\psi(\mathbf{r})$ is given by a Slater determinant, the propagator can also be expressed by a determinant.

In the presence of Coulomb interaction, the propagator can be determined by iteration starting from the ideal case. Let us consider a propagation in which one electron propagates from $(\mathbf{r}_1, \beta_1)$ to $(\mathbf{r}', \beta_2)$, where it interacts with another electron at $(\mathbf{r}_2, \beta_2)$, and then further propagates to the point $(\mathbf{r}_1, \beta + \beta_1)$. Figure A.6 illustrates such a propagation in which the straight and wavy lines represent respectively the free propagation and the interaction.

For convenience, let us introduce a periodicity β in the β-space such that the destination point $(\mathbf{r}_1, \beta + \beta_1)$ coincides with the starting point $(\mathbf{r}_1, \beta_1)$. This is equivalent to drawing graphs on a torus of circumference length β. Accordingly, one complete cycle $0 \to \beta$ of a particle is given by a loop when the graph is projected on a sheet of paper. That is, one particle is illustrated by one loop. This illustration, which is different from the classical case shown in Fig. A.1 or A.2, is advantageous because the particle can be anywhere in the loop. In the classical case, a point may represent a particle, while

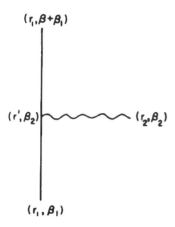

Fig. A.6. Propagation from $(\mathbf{r}_1, \beta_1)$ to $\mathbf{r}_1, \beta + \beta_1)$ for G_1

in the present type of illustration, such a point is "stretched" to make the whereabouts of the particle somewhat vague. This feature becomes clearer when the wave function $\psi(\mathbf{r})$ is represented by a Slater determinant. Since the ideal quantum propagator will also be given by determinant, its expansion yields a product such as $K_0(\mathbf{r}_2, \beta; \mathbf{r}_1, 0) K_0(\mathbf{r}_1 \beta; \mathbf{r}_2, 0)$ of single particle propagators. This product is represented by a two-loop diagram under the convention in which $(\mathbf{r}_i, 0)$ is plotted at the same point as $(\mathbf{r}_j, \beta)$ for $i = 1, 2$. A portion of such a graph may belong to either particle 1 or particle 2. Hence, the illustration by a single two-loop graph is consistent with the uncertainty principle.

In terms of a one-particle free propagator $K_0(\mathbf{r}, \beta; \mathbf{r}', \beta_1)$ and with the Coulomb potential $\phi(r)$, the process in Fig. A.6 is represented by the product

$$-K_0(\mathbf{r}_1, \beta_1 + \beta; \mathbf{r}', \beta_2) \phi(|\mathbf{r}_2 - \mathbf{r}'|) K_0(\mathbf{r}', \beta_2; \mathbf{r}_1, \beta_1).$$

The minus sign originates from the iterative series of the total propagator in which the terms with odd powers of ϕ always carry a minus sign. The integration of this product over all the intermediate coordinates $\mathbf{r}'$ results in

$$F_1(\mathbf{r}_2 - \mathbf{r}_1; \beta_2 - \beta_1) = -\frac{1}{(2\pi)^3} \int G_1(q; \beta_2 - \beta_1) u(q)$$
$$\times \exp[i\mathbf{q}\cdot(\mathbf{r}_2 - \mathbf{r}_1)] d\mathbf{q}, \tag{A.2.1}$$

where $u(q)$ is the Fourier transform of the Coulomb potential and is given by

$$u(q) = \int \phi(r) e^{i\mathbf{q}\cdot\mathbf{r}} d\mathbf{r}.$$

In the natural units in which $\hbar = 1$ and $2m = 1$, where m is the electron mass,

$$G_1(q; \beta_2 - \beta_1) = z\lambda^{-3} \exp[-\alpha(\beta - a)q^2/\beta]. \tag{A.2.2}$$

Here, z is the absolute activity, and the de Broglie thermal wavelength λ in these units is given by

$$\lambda = (4\pi\beta)^{1/2}.$$

Because of the periodicity in β, $G_1(q; \beta_2 - \beta_1)$ can be expanded in a Fourier series as follows:

$$G_1(q : \beta_2 - \beta_1) = \frac{1}{\beta} \sum_j \lambda_j(q) \exp[2\pi ij(\beta_2 - \beta_1)/\beta] \tag{A.2.3}$$

where for Boltzmann statistics $f(p) = z\exp(-\beta p^2)$,

$$\lambda_j(q) = \frac{1}{(2\pi)^3} \int d\mathbf{p} \int_0^\beta f(p) \exp\{\alpha[p^2 - (\mathbf{p} + \mathbf{q})^2]\} d\alpha$$

$$= z\lambda^{-3} \int \exp[-q^2\alpha(\beta - \alpha)/\beta] \exp(2\pi ij\alpha/\beta) d\alpha. \tag{A.2.4}$$

The above process which produced F_1 of (A2.1) can be repeated l times to give

$$F_l(\mathbf{r}_2\beta_2; \mathbf{r}_1\beta_1) = \frac{1}{\beta(2\pi)^3} \sum_j \int d\mathbf{q} [-u(q)]^{l-1} [\lambda_j(q)]^l$$

$$\times \exp[i\mathbf{q} \cdot (\mathbf{r}_2 - \mathbf{r}_1)] \exp[2\pi ij(\beta_2 - \beta_1)]. \tag{A.2.5}$$

The cluster coefficient $b_l(r)$ for $r = |\mathbf{r}_2 - \mathbf{r}_1|$ is then given by

$$b_l(r) = \sum_l z^l F_l(\mathbf{r}_2\beta; \mathbf{r}_1 0). \tag{A.2.6}$$

We arrive at the following chain diagram formula for the pair distribution function,

$$\rho_2(r) = n^2 - \frac{1}{(2\pi)^3\beta} \sum_j^\infty \int \frac{u(q)\lambda_j(q)^2}{1 + \lambda_j(q)u(q)} \exp(i\mathbf{q} \cdot \mathbf{r}) d\mathbf{q}. \tag{A.2.7}$$

This chain diagram formula is useful to study long distance correlations in many-body systems. Although chain graphs appear in classical cases, they are also important for quantum cases. That is, (A2.7) is also useful for quantum systems although the eigenvalues λ_j are different from the classical cases. Note that for electron fluids (A2.7) is represented by all the chain diagrams with $\mathbf{r}_1$ and $\mathbf{r}_2$ at both ends. Those chains in which these coordinates appear in the middle can be forgotten for $\rho_2(r)$ of an electron fluid because their contributions are proportional to $u(0)$ which is usually set to zero,

$$u(0) = 0. \tag{A.2.8}$$

This condition is introduced to avoid a divergence due to the long range Coulomb potential, in view of the charge neutrality of the entire system. That is, for an electron system, it is customary to place neutralizing background charges.

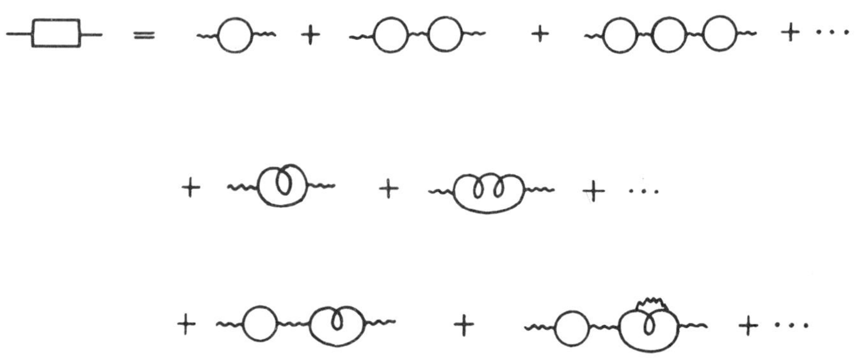

Fig. A.7. Quantum box diagram represents an arbitrary structure of unlabeled diagrams connected with each other by interaction or by quantum exchange

If quantum statistics have to be used, it becomes necessary to use the quantum eigenvalues λ_j which reflect the statistics. The form of quantum λ_j will be given shortly. Also, the graphical connections should be made not only by interaction lines but also by quantum exchanges. Figure A.7 illustrates an effective chain of graphs for quantum cases.

In the so-called exchange-chain diagrams, the two representative particles at r_1 and r_2 have a quantum mechanical exchange. In Fig. A.8, A represents a classical type chain diagram and A^* is an exchange-chain diagram. For diagram analyses, it is convenient to introduce the concept of conjugate diagrams [A.7]. In Fig. A.8 the left side graphs are conjugate to the right side graphs in the same row. Thus, the chain diagrams such as A are conjugate to the exchange-chain diagrams A^* in the same row. Note that A_s^* is conjugate to a separated pair of diagrams A_s in the left column. The three typs A, A^* and A_s^* of diagrams may be classified as one-effective-line diagrams.

When quantum statistics are used, (A2.7) is modified such that

$$\rho_2(r) - n^2 + I_2(r) = A(r) + A^*(r) + A_s^*(r) + \dots , \tag{A.2.9}$$

where $I_2(r)$ and $A(r)$, $A^*(r)$ etc. represent respectively the contributions from the ideal exchange graphs and from A, A^*, etc. It is convenient to combine A and A^* and express $\rho_2(r)$ as

$$A(r) + A^*(r) = \eta(r);$$
$$\eta(r) = -\frac{1}{(2\pi)^3\beta} \int \sum_j \frac{u(q)}{1 + \lambda_j u(q)} [\lambda_j^2 - \lambda_j^2(q, r)] e^{i\mathbf{q}\cdot\mathbf{r}} d\mathbf{q}. \tag{A.2.10}$$

The contribution $I_2(r)$ is given by

$$I_2(r) = \left(\frac{1}{(2\pi)^3} \int \frac{\exp(i\mathbf{p} \cdot \mathbf{r})}{z^{-1}e^{\beta p^2} + 1} d\mathbf{p} \right)^2 = [n^{(0)}(r)]^2$$
$$= [\lambda^{-3} G_{3/2}(z, r)]^2. \tag{A.2.11}$$

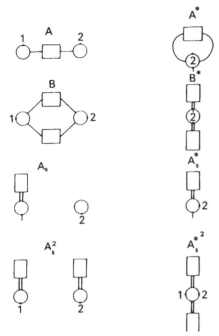

Fig. A.8. One- and two-line diagrams. Those in the same row are conjugate to each other

This contribution is due to the exchange graphs of an ideal electron gas, $n^{(0)}(r)$ being the ideal gas contribution to the number density, see (A3.5), and

$$G_{3/2}(z,r) = -\sum_{s=1} \frac{(-z)^s}{s^{3/2}} \exp(-\pi r^2/s\lambda^2). \tag{A.2.12}$$

The function $\lambda_j(q,r)$ represents the quantum eigenvalue function defined by

$$\lambda_j(q,r) = \frac{1}{(2\pi)^3} \int d\mathbf{p} \int_0^\beta f(p)[1 - f(\mathbf{p}+\mathbf{q})] \exp\{\alpha[p^2 - (\mathbf{p}+\mathbf{q})^2]\}$$
$$\times \exp(i\mathbf{p}\cdot\mathbf{r} + 2\pi i j\alpha/\beta)d\alpha, \tag{A.2.13}$$

where the distribution function is

$$f(p) = \frac{ze^{-\beta p^2}}{1 + ze^{\beta p^2}}.$$

In constructing the integrand of $\lambda_j(q,r)$ in (A2.13) note that $f(p)$ appears when at least one unlabeled particle of momentum p should be present while $[1 - f(p)]$ is used for a propagation without such a requirement.

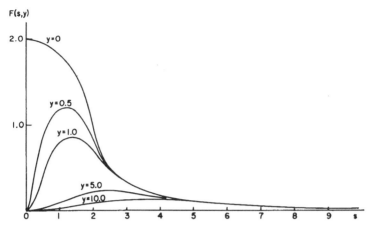

Fig. A.9. Function $F(s, y)$ which characterizes the eigenvalue function of a 3D electron gas near absolute zero. $s = q/k_F$ and $y = 2\pi j/\beta k_F^2$. $F(s, \omega)$ for $2\pi j/\beta = -i\omega + 0$ in the limit $\beta \to \infty$ characterizes a polarization function [A.7b]

Note also that $\lambda_j(q, r)$ is reduced to λ_j of the quantum case in the limit $r \to 0$;

$$\lambda_j(q) = \frac{1}{(2\pi)^3} \int d\mathbf{p} \int_0^\beta f(p)[1 - f(\mathbf{p} + \mathbf{q})] \exp\{\alpha[p^2 - (\mathbf{p} + \mathbf{q})^2]\}$$
$$\times \exp(2\pi i j\alpha/\beta) d\alpha. \tag{A.2.14}$$

At absolute zero, (A2.14) yields

$$\lambda_j(q) = \frac{k_F}{4\pi^2} F(s, y);$$

$$F(s, y) = 1 - \frac{1}{8}\left(s - \frac{4}{s} - \frac{y^2}{s^3}\right) \ln \left|\frac{(s^2 + 2s)^2 + y^2}{(s^2 - 2s)^2 + y^2}\right| - \frac{y}{2s}\left[\tan^{-1}\left(\frac{s^2 + 2s}{y}\right)\right.$$
$$\left. - \tan^{-1}\left(\frac{s^2 - 2s}{y}\right)\right], \tag{A.2.15}$$

where $s = q/k_F$, $y = 2\pi j/\beta k_F^2$ for $\beta \to \infty$, and the natural unit is used. Figure A.9 illustrates $F(s, y)$ as a function of s for several values of y. Similar results can be obtained for 2D, and the corresponding function $F(s, y)$ is shown in Fig. A.10. Note that the case $y = 0$ has a discontinuous derivative at $s = 2$, i.e., at $q = 2k_F$. A graphical comparison of $F(s, 0)$ in the form of the polarization function can be found in Fig. 2.1 in the text.

Note that the eigenvalue expression in (A2.15) reduces to the dielectric function in the random phase approximation if $2\pi j/\beta$ is replaced by $-i\omega + \eta$, where η is a small parameter. The dielectric function is discussed in Chap. 2.

Since $I_2(0) = n^2$ we find from (A2.10) that

$$A(0) + A^*(0) = \eta(0) = 0. \tag{A.2.16}$$

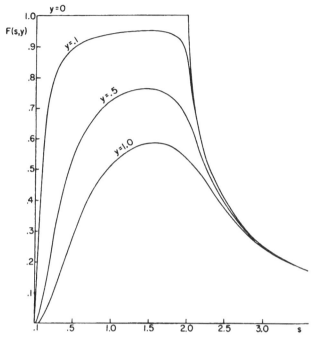

Fig. A.10. Eigenvalue function $F(s, y)$ for two dimensions [A.8] [Isihara and Toyoda: Z. Phys. B *23*, 389 (1976)]

In (A2.7), spin has been neglected. If we distinguish spin parallel and antiparallel pairs from each other, the corresponding pair distribution functions are given by

$$\rho_2^{\uparrow\uparrow}(r) = \frac{n^2}{2} - 2I_2(r) - \frac{1}{2\beta(2\pi)^3}\sum_j \int \frac{u(q)[\lambda_j^2 - \lambda_j^2(q, r)]e^{i\mathbf{q}\cdot\mathbf{r}}}{1 + \lambda_j u(q)} d\mathbf{q}, \quad (A.2.17)$$

$$\rho_2^{\uparrow\downarrow}(r) = \frac{n^2}{2} - \frac{1}{2\beta(2\pi)^3}\sum_j \int \frac{u(q)\lambda_j^2 e^{i\mathbf{q}\cdot\mathbf{r}}}{1 + \lambda_j u(q)} d\mathbf{q}. \quad (A.2.18)$$

In this case a spin factor 2 should be introduced into the eigenvalues given by (A2.15). Note that, in these expressions, only the spin parallel pair distribution function vanishes at the origin.

A.3 Ring Diagram Formulas

The pair distribution function can be used to derive the grand partition function. For this purpose, it is convenient to make use of the formula

$$\ln \Xi/\Xi_0 = -\frac{\beta V}{2} \int_0^1 \frac{d\xi}{\xi} \int \phi(r,\xi)[\rho_2(r,\xi) - n^2]dr, \qquad (A.3.1)$$

where Ξ_0 is the grand partition function of the ideal case and the potential $\phi(r,\xi)$ is defined in proportion to a coupling parameter ξ which is varied between 0 and 1. Introducing (A2.7) into (A3.1) we obtain

$$\ln \Xi/\Xi_0 = \frac{V}{2(2\pi)^3} \sum_j \int d\mathbf{q}[u\lambda_j - \ln(1 + u\lambda_j)]. \qquad (A.3.2)$$

This formula can be derived directly by a consideration of ring diagrams. Because the interaction potential $\phi(r)$ is multiplied by $\rho_2(r)$ in (A3.1), the resulting partition function is one order higher in interaction than the pair distribution function. The process of multiplying $\phi(r,\xi)$ by $\rho_2(r)$ in the integrand of (A3.1) amounts to introducing a new bond between the pair, thus forming rings from chains.

Equation (A3.2) can be used to derive the energy. A β-differentiation yields the following internal energy formula,

$$U = U_0 - \frac{V}{2(2\pi)^3} \sum_j \int \frac{u^2 \lambda_j \lambda_j'}{1 + u\lambda_j} d\mathbf{q}, \qquad (A.3.3)$$

where U_0 is the ideal gas energy and

$$\lambda_j' = \partial \ln \lambda_j/\partial\beta.$$

For an electron gas, the integrand of the above formula shows that U is at least of order e^4. Note that λ_j is still temperature dependent so that U is also a function of temperature. For absolute zero the energy can be obtained by replacing the j-sum in (A3.2) by integration over a continuous variable $y = 2\pi j/\beta$ and by using the limiting expression $\lim_{\beta\to\infty} \lambda_j$. This process results in a formula which is similar to (A3.2).

The graphs A_s^* differ from A^* in that the effective interaction line does not connect the propagation lines of electrons 1 and 2. That is, it is within the propagation line associated with one of the electrons. Hence, in their conjugate graphs, electrons 1 and 2 are not linked. For this reason, the contribution from A_s^* is related to the singlet distribution function, which is the number density.

The one-body distribution function can be evaluated in terms of graphs which are labeled by one particle. For a homogeneous system, it is constant but can be expanded in ascending powers of interaction:

$$n = n(0) = n^{(0)}(0) + n^{(1)}(0) + n^{(2)}(0) + \cdots, \qquad (A.3.4)$$

where

$$n^{(0)}(r) = \frac{1}{(2\pi)^3} \int f(p)e^{i\mathbf{p}\cdot\mathbf{r}}d\mathbf{p}, \qquad (A.3.5)$$

$$n^{(1)}(r) = \frac{\beta}{(2\pi)^6} \int f(p)f(\mathbf{p}+\mathbf{q})[1 - f(p)]u(q)e^{i\mathbf{p}\cdot\mathbf{r}}d\mathbf{p}d\mathbf{q}, \tag{A.3.6}$$

$$n^{(2)}(r) = \sum_j \frac{1}{2(2\pi)^3} \int \frac{\partial\lambda_j}{\partial\ln z} \frac{\lambda_j u^2(q)}{1 + \lambda_j u(q)} e^{i\mathbf{p}\cdot\mathbf{r}}d\mathbf{q}. \tag{A.3.7}$$

In the integrand of (A3.5), $f(p)$ appears because at least one particle must be present. For a propagation without this requirement, a hole distribution $[1 - f(p)]$ should be used. The same rule has been used in constructing the integrand of $\lambda_j(q,r)$. However, in (A3.7) its derivative $\partial\lambda/\partial\ln z$ has been used to simplify a sum of products of such distribution functions. Note that those graphs which do not have any interaction line contribute to $n^{(0)}(r)$, which has been defined by (A2.11) or (A3.5). One can easily find that for $\beta \to \infty$.

$$n^{(0)}(0) = \left(\frac{4}{3\pi^{1/2}}\right)(\ln z)^{3/2} + \left(\frac{\pi^{3/2}}{6}\right)(\ln z)^{-1/2} + \cdots. \tag{A.3.8}$$

The use of these number density functions simplifies $A_s^*(r)$ in (A3.9). The diagrams A_s^* which are first order in $u(q)$ correspond to a self-interaction. Hence, no exchange of interaction lines entering and leaving a graph should be taken into consideration. Also, any self-interaction of the same particle must be prohibited in constructing diagrams.

After some straightforward calculations [A.7], $A_s^*(r)$ is found to be given by

$$A_s^*(r) = -2n^{(1)}(r)n^{(0)}(r) + n^{(0)}\frac{1}{(2\pi)^3}\sum_j \int \left[\frac{u^2(q)\lambda_j}{1 + u(q)\lambda_j}\right]$$

$$\times \left[\frac{\partial\lambda_j(q,r)}{\partial\ln z}\right] - \mu_j(q,r)(e^{-i\mathbf{q}\cdot\mathbf{r}} - 1)\Big]d\mathbf{q}, \tag{A.3.9}$$

where

$$\mu_j(q,r) = \frac{1}{(2\pi)^3} \int\int f(p)f(\mathbf{p}+\mathbf{q})[1 - f(\mathbf{p}+\mathbf{q})]$$

$$\times \exp\{\alpha[p^2 - (\mathbf{p}+\mathbf{q})^2] - i\mathbf{p}\cdot\mathbf{r} + 2\pi ij\alpha/\beta\}d\alpha d\mathbf{p}. \tag{A.3.10}$$

The integrand of (A3.7) or (A3.9) can be constructed based essentially on the same rule which was used for constructing that of λ_j or $\lambda_j(q,r)$, and by making use of the momentum conservation law which holds at any unlabeled junction. The derivative of $\lambda_j(q,r)$ has been used to simplify a combination of Fermi distribution functions. Note that $A_s^*(0)$ does not vanish.

In Fig. A.8, B and B^* represent the contributions to the pair distribution function from the diagrams with two effective interaction lines. The two interaction lines are represented by momenta $\mathbf{q}$ and $\mathbf{q}'$, and their interchange must be considered. It can be shown that in combination these graphs yield

$$B(r) + B^*(r) = \frac{1}{2(2\pi)^6} \sum_{j,l} \int d\mathbf{q} d\mathbf{q}' \frac{u(q)u(q')}{[1 + \lambda_j u(q)][1 + \lambda_j u(q')]} e^{i(\mathbf{q}+\mathbf{q}')\cdot\mathbf{r}}$$

$$\times [\nu_{jl}(0, q, q')\nu_{jl}^*(0, q, q') - \nu_{jl}(r, q, q')\nu_{jl}^*(r, q, q')]. \quad (A.3.11)$$

This result is characterized by new eigenvalues defined by

$$\nu_{jl}(r, q, q') = \frac{1}{\beta(2\pi)^3} \int \{[1 - f(p)][1 - f(\mathbf{p}+\mathbf{q})]$$

$$\times f(\mathbf{p}+\mathbf{q}+\mathbf{q}')E_{jl}(\beta', \beta''; q, q')$$

$$+ [1 - f(p)][1 - f(\mathbf{p}+\mathbf{q}')]f(\mathbf{p}+\mathbf{q}+\mathbf{q}')E_{lj}(\beta'', \beta'; q', q)$$

$$- [1 - f(p)]f(\mathbf{p}+\mathbf{q}+\mathbf{q}')E_{lj}(\beta'', \beta'; q', q)\}d\beta'd\beta''d\mathbf{p}.$$

$$(A.3.12)$$

Note the interchange in the indices and arguments of the function E_{jl} defined by

$$E_{jl}(\beta', \beta''; q, q') = \frac{1}{\beta} \exp[-\beta' p^2 - (\beta'' - \beta')(\mathbf{p}+\mathbf{q})^2 + \beta''(\mathbf{p}+\mathbf{q}+\mathbf{q}')^2]$$

$$\times \exp[-2\pi i(j\beta'' - l\beta')].$$

Two intermediate temperature variables β' and β'' appear to specify the locations of the interaction lines. They must be integrated over between 0 and β. Therefore, the integrand of (A3.12) has been constructed to avoid a double count of the interchange between the two interaction lines.

The contribution from A_s^{*2} can be given by repeating essentially the same process for A_s^*. It is given by

$$A_s^*(r) = - \left(\frac{1}{2(2\pi)^3} \int \gamma(q, r)d\mathbf{q}\right)^2 \quad (A.3.13)$$

in terms of the new eigenvalues defined by

$$\gamma(q, r) = \sum_j \frac{1}{(2\pi)^3} \int \frac{u^2(q)\lambda_j}{1 + u(q)\lambda_j} \{f(p)[1 - f(p)]f(\mathbf{p}+\mathbf{q})$$

$$+ f(p)f(\mathbf{p}+\mathbf{q})[1 - f(\mathbf{p}+\mathbf{q})]e^{-i\mathbf{q}\cdot\mathbf{r}}$$

$$- f(p)[1 - f(p)]\} \exp\alpha[p^2 - (\mathbf{p}+\mathbf{q})^2] \exp(-i\mathbf{p}\cdot\mathbf{r})$$

$$\times \exp(-2\pi i j\alpha/\beta)d\alpha d\mathbf{p}. \quad (A.3.14)$$

Note that

$$\gamma(q, 0) = - \sum_j \{u^2(q)\lambda_j/[1 + u(q)\lambda_j]\}(\partial\lambda_j/\partial \ln z), \quad (A.3.15)$$

and that the derivative here is related to the one in $A_s^*(r)$. As a corollary, one can show that the resulting pair distribution function satisfies the correct normalization and boundary conditions. The above formulas are explicit and can be used to evaluate $\rho_2(r)$ in the ring diagram approximation.

A.4 Density Functional Approach

In the density functional approach [A.9], the local density function $n(\mathbf{r})$ of electrons is used as a variational function. It is determined such that the free energy is minimum when the variational function agrees with the actual density of a given system. It can be applied not only to homogeneous, but also to inhomogeneous electron systems. Cases with an external field or a confining potential can also be treated.

Let us first consider the ground state of an electron system which is subject to an external potential of $v(\mathbf{r})$. In the ground state, the density $n(\mathbf{r})$ of electrons depends on $v(\mathbf{r})$. Conversely, it can be shown that $v(\mathbf{r})$ is a unique functional of $n(\mathbf{r})$ apart from a trivial constant. The ground state energy is expressed by

$$E = \int v(\mathbf{r})n(\mathbf{r})d\mathbf{r} + \frac{1}{2}\int\int \frac{n(\mathbf{r})n(\mathbf{r}')}{|\mathbf{r}-\mathbf{r}'|}d\mathbf{r}d\mathbf{r}' + G[n], \qquad (A.4.1)$$

$G[n]$ is a universal functional of the density

$$G[n] = K[n] + E_{xc}[n]. \qquad (A.4.2)$$

$K[n]$ is the kinetic energy of a system of noninteracting electrons. $E_{xc}[n]$ is the exchange-correlation energy of the interacting system.

For absolute zero, E is minimum when $n(\mathbf{r})$ coincides with the actual density.

The exchange correlation energy can be expressed by a sum of the exchange energy $E_x[n]$ in the Hartree-Fock scheme and the correlation energy

$$E_{xc}[n] = E_x[n] + \int n(\mathbf{r})\varepsilon_c(n(\mathbf{r}))d\mathbf{r}. \qquad (A.4.3)$$

In the second term on the right side, $\varepsilon_c(n)$ is taken to be the correlation energy per particle of a homogeneous electron gas, and $n(\mathbf{r})$ is assumed to vary sufficiently slowly. The correlation energy of an electron gas is discussed in Chap. 4. Here, we assume that it has been given.

We introduce a variational procedure with respect to $n(\mathbf{r})$ under the condition

$$\int \delta n(\mathbf{r})d\mathbf{r} = 0. \qquad (A.4.4)$$

The condition that E of (A4.1) must be stationary yields the following set of equations:

$$\left(-\frac{1}{2}\nabla^2 + \phi(\mathbf{r}) + \mu_c(\mathbf{r})\right)\psi_i(\mathbf{r}) - \int \frac{n_1(\mathbf{r},\mathbf{r}')}{|\mathbf{r}-\mathbf{r}'|}\psi_i(\mathbf{r}')d\mathbf{r}' = \varepsilon_i\psi_i(\mathbf{r}), \quad (A.4.5)$$

where

$$\mu_c = d(n\varepsilon_c)/dn, \tag{A.4.6}$$

$$n_1(\mathbf{r},\mathbf{r}') = \sum_{j=1}^{N} \psi_j(\mathbf{r})\psi_j^*(\mathbf{r}'), \tag{A.4.7}$$

$$\phi(\mathbf{r}) = v(\mathbf{r}) + \int \frac{n(\mathbf{r}')}{|\mathbf{r}-\mathbf{r}'|}d\mathbf{r}', \tag{A.4.8}$$

$$n(\mathbf{r}) = \sum_{i=1}^{N} |\psi_i(\mathbf{r})|^2. \tag{A.4.9}$$

We start with an assumed $n(\mathbf{r})$ and evaluate μ_c and $\phi(\mathbf{r})$. The results are introduced into (A4.5) to find a new $n(\mathbf{r})$. The energy is given by

$$E = \sum_{i=1}^{N} \varepsilon_i - \frac{1}{2}\iint \frac{n(\mathbf{r})n(\mathbf{r}')}{|\mathbf{r}-\mathbf{r}'|}d\mathbf{r}d\mathbf{r}' + \frac{1}{2}\iint \frac{n_1(\mathbf{r},\mathbf{r}')n_1(\mathbf{r}',\mathbf{r})}{|\mathbf{r}-\mathbf{r}'|}d\mathbf{r}d\mathbf{r}'$$
$$+ \int n(\mathbf{r})[\varepsilon_c(n(\mathbf{r})) - \mu_c(n(\mathbf{r}))]d\mathbf{r}. \tag{A.4.10}$$

Essentially the same procedure can be used for the grand potential and distribution functions of liquids. The point is that, for a given interaction potential and an external potential, the grand potential is minimum when the density $n(\mathbf{r})$ equals the equilibrium value. The potential can be written as

$$\Omega_v[n(r)] = \int n(\mathbf{r})v(\mathbf{r})d\mathbf{r} + \Psi[n(r)] - \mu \int n(r)d\mathbf{r}. \tag{A.4.11}$$

Here, $v(\mathbf{r})$ is an external potential, and Ψ is a unique functional of the density $n(\mathbf{r})$ and is given by

$$\Psi[(n(\mathbf{r}))] = \langle (K + \Phi + \beta^{-1}\ln(f_0)) \rangle, \tag{A.4.12}$$

where K is the total kinetic energy, Φ is the total mutual interaction, and $\langle \ldots \rangle$ is the average in terms of the equilibrium distribution

$$f_0 = \exp[-\beta(H - \mu N)]/\Xi. \tag{A.4.13}$$

Ξ is the grand partition function given by

$$\Xi = \mathrm{Tr}\{\exp[-\beta(H - \mu N)]\}.$$

For equilibrium

$$\Psi[n_o(r)] = \Omega. \tag{A.4.14}$$

The minimum condition is

$$\frac{\delta\Omega_v[n(\mathbf{r})]}{\delta n(\mathbf{r})} = 0, \tag{A.4.15}$$

and at equilibrium

$$\Omega_v[n_0(\mathbf{r})] = \Omega. \tag{A.4.16}$$

We note that the Helmholtz free energy is given by

$$F = \int d\mathbf{r} n_0(\mathbf{r}) v(\mathbf{r}) + \Psi[n_0(\mathbf{r})]. \tag{A.4.17}$$

We note that in an ideal case without interaction,

$$\Psi_0[n(\mathbf{r})] = \beta^{-1} \int d\mathbf{r} n(\mathbf{r}) [\ln[\lambda^3 n(\mathbf{r})] - 1\}, \tag{A.4.18}$$

where $\lambda = (h^2\beta/2m\pi)^{1/2}$ is the de Broglie thermal wavelength. In this case, we obtain

$$n_0(\mathbf{r}) = z \exp[-\beta v(\mathbf{r})]. \tag{A.4.19}$$

$z = \lambda^{-3} \exp(\beta\mu)$ is the fugacity.

In the presence of mutual interaction Φ we define

$$W[n] = \Psi_0[n(\mathbf{r})] - \Psi[n(\mathbf{r})]. \tag{A.4.20}$$

This function generates the distribution functions. First of all, the singlet distribution function is given by

$$n(\mathbf{r}) = z \exp\{-\beta v(\mathbf{r}) + c[n, \mathbf{r}]\}, \tag{A.4.21}$$

where

$$c[n, \mathbf{r}] = \beta \frac{\delta W[n]}{\delta n(\mathbf{r})}. \tag{A.4.22}$$

Moreover, the function $c_2(r) = c_2(|\mathbf{r}_2 - \mathbf{r}_1|)$ given by the second derivative of W,

$$c_2(n, \mathbf{r}_1, \mathbf{r}_2) = \beta \frac{\delta^2 W[n]}{\delta n(\mathbf{r}_1)\delta n(\mathbf{r}_2)}, \tag{A.4.23}$$

is the direct correlation function which satisfies the Ornstein–Zernike equation for the radial distribution function $g(r)$,

$$g(r) - 1 = c(r) + n_0 \int d\mathbf{r}'[g(r') - 1]c(|\mathbf{r} - \mathbf{r}'|), \tag{A.4.24}$$

where n_0 is constant. The same differentiation process can be repeated for higher-order distribution functions.

References

Chapter 1

1.1   J. E. Lilienfeld: US Patent 1,745,175 (1930);
      O. Heil: British Patent 439,457 (1935);
      W. Shockley, G. L. Pearson: Phys. Rev. **74**, 232 (1948)
1.2   D. Kahng, M. M. Atalla: IFEF Trans. FD-**23**, 655 (1976)
1.3   L. Esaki, R. Tsu: IBM J. Res. Dev. **24**, 61 (1970)
1.4   R. de L. Kronig, W. O. Penney: Proc. R. Soc. London A **130**, 499 (1931)
1.5   See, e.g., O. Bauer, F. Kuchar, H. Heinrich (eds.): *Two-Dimensional Systems,
      Heterostructures, and Superlattices*, Springer Ser. Solid-state Sci., Vol. 53
      (Springer, Berlin, Heidelberg 1984)
1.6   M. W. Cole, M. H. Cohen: Phys. Rev. Lett. **23**, 1238 (1966)
1.7   H. Fröhlich: Proc. R. Soc. London A **233**, 296 (1954);
      R. E. Peierls: *Quantum Theory of Solids* (Oxford University Press, London
      1955)
1.8   L. B. Coleman, M. J. Cohen, D. J. Sandman, F. O. Yamagishi, A. F. Garito,
      A. J. Heeger: Solid State Commun. **12**, 1125 (1973)
1.9   See, e.g., articles in J. de Phys. Suppl. Colloq. **C3**, 543–573 (1983)
1.10 See, e.g., D. Jerome, H. J. Schultz: Adv. Phys. **31**, 299 (1982)
1.11 A. B. Fowler, A. Hartstein, R. A. Webb: Phys. Rev. Lett. **48**, 196 (1982);
      W. J. Skocpol, L. D. Jackel, E. L. Hu, L. A. Fetter: Phys. Rev. Lett. **49**, 951
      (1982);
      R. G. Wheeler, K. K. Choi, A. Goel, R. Wisnieff, D. E. Pepper: Phys. Rev.
      Lett. **49**, 1674 (1982)
1.12 See, e.g., A. J. Leggett: Rev. Mod. Phys. **47**, 331 (1975)
1.13 J. Hubbard: Proc. R. Soc. London A **276**, 238 (1963);
      M. C. Gutzwiller: Phys. Rev. Lett. **10**, 159 (1963)
1.14 J. Kondo: Prog. Theor. Phys. **32**, 37 (1964)
1.15 P. W. Anderson: Phys. Rev. **124**, 41 (1961)

Chapter 2

2.1   C. Kittel: *Quantum Theory of Solids* (Wiley, New York 1963);
      D. Pines P. Nozières: *The Theory of Quantum Liquids* (W. A. Benjamin,
      New York 1966);
      S. Doniach, F. H. Sondheimer: *Green's Functions for Solid State Physicists*
      (W. A. Benjamin, New York 1974)
2.2   J. Lindhard: Kgl. Dan. Mat. Fys. Medd. **28**, 8 (1954)

2.3   S. G. Brush, H. L. Sahlin, E. Teller: J. Chem. Phys. **45**, 2102 (1966);
       J. P. Hansen: Phys. Rev. A **8**, 3096 (1973);
       W. L. Slattery, G. D. Doolen, H. DeWitt: Phys. Rev. A **21**, 2087 (1980);
       M. Baus, J. Hansen: Phys. Rep. **59**, 1 (1980);
       S. Galam, J. P. Hansen: Phys. Rev. A **14**, 816 (1976);
       S. Ichimaru: Rev. Mod. Phys. **54**, 1017 (1982)
2.4   J. Hubbard: Proc. R. Soc. London A **243**, 336 (1957)
2.5   F. Brosens, L. F. Lemmens, J. T. Devreese: Phys. Status Solidi (a) **59**, 447
       (1980)
2.6   L. E. Ballentine: Phys. Rev. **158**, 670 (1967);
       L. Kleinman: Phys. Rev. **160**, 585 (1967);
       M. P. Tosi: Riv. Nuovo Cimento, Ser. I, **1**, 160 (1969);
       F. Brosens, J. T. Devreese, L. F. Lemmens: Phys. Rev. B **21**, 1363 (1980)
2.7   K. S. Singwi, M. P. Tosi, A. Sjölander: Nuovo Cimento, Ser. X, **54**, 160
       (1968);
       K. S. Singwi, M. P. Tosi, R. H. Land, A. Sjölander: Phys. Rev. **176**, 589
       (1968);
       K. S. Singwi, A. Sjölander, M. P. Tosi, R. H. Land: Phys. Rev. B **1**, 1044
       (1970);
       P. Vashishta, K. S. Singwi: Phys. Rev B **6**, 875 (1972);
       G. Niklasson: Phys. Rev. B **10**, 3052 (1974);
       H. Yasuhara, J. Phys. Soc. Jpn. **36**, 361 (1974)
2.8   A. Isihara: *Statistical Physics* (Academic, New York 1971) p. 175, Eq. (10.3);
       K. Tago, K. Utsumi, S. Ichimaru: Prog. Theor. Phys. **65**, 54 (1981)
2.9   J. Chihara: Prog. Theor. Phys. **50**, 409 (1973)
2.10   S. Totsuji, S. Ichimaru: Prog. Theor. Phys. **50**, 753 (1973); ibid. **52**, 42 (1974)
2.11   K. F. Berggren: Phys. Rev. A **1**, 1783 (1970)
2.12   Y. Rosenfeld, N. W. Ashcroft: Phys. Rev. A **20**, 1208 (1979);
       F. J. Rogers, A. Young: Phys. Rev. A **30**, 999 (1984);
       H. Iyetomi, S. Ichimaru: Phys. Rev. A **34**, 433 (1986); P. Nozières, D. Pines:
       Phys. Rev. **111**, 442 (1958)
2.13   K. S. Singwi, M. P. Tosi, R. H. Land, A. Sjölander: Phys. Rev. **176**, 589
       (1968)
2.14   H. Raether: *Excitation of Plasmons and Interband Transitions by Electrons*
       (Springer, Berlin, Heidelberg 1980);
       G. D. Priftis, J. Boviatsis: Phys. Status Solidi (b) **104**, 673 (1981)
2.15   A. V. Chaplik: Zh. Eksp. Teor. Fiz. **62**, 746 (1972) [Sov. Phys.–JFTP **35**,
       395 (1972)];
       M. Nakayama: J. Phys. Soc. Jpn. **36**, 393 (1974)
2.16   T. N. Theis, J. P. Kotthaus, P. J. Stiles: Solid State Commun. **26**, 803 (1978)
2.17   E. Batke, D. Heitmann, C. W. Tu: Phys., Rev. B **34**, 6951 (1986)
2.18   A. Pinczuk, M. G. Lamont, A. C. Gossard: Phys. Rev. Lett. **36**, 2092 (1986);
       T. Zettler, C. Peters, J. P. Kotthaus, K. Ploog: Phys. Rev. B **39**, 3931
       (1989-II);
       G. Abstreiter, M. Cardona, A. Pinczuk: In *Light Scattering in Solids IV*,
       ed. by M. Cardona, G. Güntherodt, Topics Appl. Phys., Vol. 54 (Springer,
       Berlin, Heidelberg 1984);
       G. Abstreiter, A. Pinczuk: In *Light Scattering in Solids V*, ed. M. Cardona,
       G. Güntherodt, Topics Appl. Phys., Vol. 66 (Springer, Berlin, Heidelberg
       1989)
2.19   A. L. Fetter: Ann. Phys. (N.Y.) **88**, 1 (1974);
       D. Grecu: J. Phys. C **8**, 2627 (1975);

S. Das Sharma, J. J. Quinn: Phys. Rev. B **25**, 7603 (1982);
W. L. Bloss, F. M. Brody: Solid State Commun. **43**, 523 (1982);
J. K. Jain, P. B. Allen: Phys. Rev. B **32**, 997 (1985)
2.20   D. B. Mast, A. J. Dahm, A. L. Fetter: Phys. Rev. Lett. **54**, 1710 (1985);
D. C. Glattli, E. Y. Andrei, G. Deville, J. Poitrenaud, F. I. B. Willams: Phys. Rev. Lett. **54**, 1714 (1985)
2.21   E. Batke, C. W. Tu: Phys. Rev. Lett. **58**, 2474 (1987)

Chapter 3

3.1   P. Debye, F. Hückel: Phys. Z. **24**, 185 (1923)
3.2   J. Mayer: J. Chem. Phys. **18**, 1426 (1950)
3.3   A. Isihara: Phys. Rev. **178**, 412 (1969);
A. Isihara, M. Wadati: Phys. Rev. **183**, 312 (1969)
3.4   R. Abe: Prog. Theor. Phys. **22**, 213 (1959)
3.5   W. L. Slattery, G. D. Doolen, H. E. DeWitt: Phys. Rev. A **21**, 2087 (1980); ibid. **26**, 2255 (1982);
H. E. DeWitt, Y. Rosenfield: Phys. Lett. A **75**, 79 (1979);
J. P. Hansen: Phys. Rev. A 8, 3096 (1973);
H. E. DeWitt, Y. Rosenfield: Phys. Lett. 75A, 79 (1979)
3.6   H. J. Hoffmann, W. Ebeling: Physica **39**, 593 (1968);
H. E. De Witt: J. Math. Phys. **3**, 1210 (1962)
3.7   M. Wadati, A. Isihara: Phys. Rev. A **4**, 344 (1971); Physica **57**, 237 (1972)

Chapter 4

4.1   A. Isihara: Physica **71**, 83 (1974);
A. Isihara, C. A. Ten Seldam: Physica **76**, 153 (1974)
4.2   J. S. Langer, S. H. Vosko: J. Phys. Chem. Solids **12**, 196 (1960)
4.3   M. Wadati, A. Isihara: Phys. Rev. A **4**, 344 (1971);
A. Isihara, M. Wadah: Physica **57**, 237 (1972)
4.4   M. Gell-Mann, K. A. Brueckner: Phys. Rev. **106**, 442 (1957)
4.5   A. Isihara, D. Y. Kojima: Z. Phys. **21**, 33 (1975); ibid. B **24**, 237 (1976)
D. Y. Kojima, A. Isihara: Z. Phys. B **25**, 167 (1976)
4.6   D. Y. Kojima, A. Ishihara: Phys. Cond. Matter **18**, 121 (1974)
4.7   L. Onsager, L. Mittag, M. J. Stephen: Ann. Phys. **18**, 71 (1966)
4.8   F. A. Stevens Jr., M. A. Pokrant: Phys. Rev. A **8**, 990 (1973)
4.9   A. Isihara, E. W. Montroll: Proc. Nat. Acad. Sci. **68**, 3111(1971)
4.10   A. Isihara, L. C. Ioriatti, Jr.: Physica **103A**, 621 (1980);
L. C. Ioriatti, Jr., A. Isihara: Z. Phys. B **44**, 1 (1981);
L. C. Ioriatti, Jr., A. Isihara: Phys. Status Solidi (b) **97**, K65 (1980)
4.11   A. Isihara, L. C. Ioriatti, Jr.: Phys. Rev. B **22**, 214 (1980)
4.12   C. I. Um, A. Isihara: J. Kor. Phys. Soc., **11**, 60 (1978)
4.13   A. Isihara, T. Toyoda: Phys. Rev. B **21**, 3358 (1980)
4.14   A. Isihara, L. C. Ioriatti, Jr.: Phys. Rev. B **25**, 5534 (1982)

Chapter 5

5.1   For a review, see, e.g.: C. D. Jefferies, L. V. Keldysh (eds.): *Electron-Hole Droplets in Semiconductors* (North-Holland, New York 1983); H. Ehrenreich, F. Seitz, D. Turnbull (eds.): *Solid State Physics*, Vol. 32 Academic, New York 1977); M. Ueta, H. Kanzaki, K. Kobayashi, Y. Toyozawa, F. Hanamura: *Excitonic Processes in Solids*, Springer Ser. Solid-State Sci., Vol. 60 (Springer, Berlin, Heideberg 1986)

5.2   S. Nikitine: In *Exeitons at High Density*, ed. by H. Haken, S. Nikitine (Springer, New York 1975)

5.3   M. A. Tamor, J. P. Wolfe: Phys. Rev. Lett. **44**, 1703 (1980)

5.4   D. Hulin, A. Mysyrowicz, M. Greenstein, J. P. Wolfe: Solid State Commun. **35**, 339 (1980); D. Hulin, A. Mysyrowicz, C. Benoit a la Guillaume: Phys. Rev. Lett. **45**, 1970 (1980); D. W. Snoke, J. P. Wolfe, A. Mysrowicz: Phys. Rev. Lett. **59**, 827 (1987); For properties of Cu_2O see: V. T. Agekyan: Phys. Status Solidi A **43**, 11 (1977); D. W. Snoke, J. P. Wolfe, A. Mysrowitz: Phys. Rev. Lett. **64**, 2543 (1990); J. Ling, P. W. Wolfe: Phys. Rev. Lett. **71**, 1222 (1993)

5.5   G. A. Thomas, V. B. Timofeev: In *Optical Properties of Solids*, Handbook on Semiconductors, Vol. 2, ed. by M. Balkanski (North-Holland, Amsterdam 1980)

5.6   L. V. Keldysh: Proc. 9th Int'l. Conf. Phys. Semicond. (Nauka, Moscow 1968) p. 1303

5.7   V. M. Asnin, A. A. Rogachov: Pis'ma Zh. Fksp. Teor. Fiz. **9**, 415 (1969) [Engl. Transl.: Sov. Phys.–JFTP Lett. **9**, 248 (1969)]; J. C. Hensel, T. G. Phillips, G. A. Thomas: In [5.1b]; P. L. Gourley, J. P. Wolf: Phys. Rev. B **20**, 3319 (1979); ibid. B **24**, 5970 (1981)

5.8   M. Greenstein, J. P. Wolfe: Phys. Rev. B **24**, 3318 (1981)

5.9   P. L. Gourley, J. P. Wolfe: Phys. Rev. Lett. **40**, 526 (1978); Phys. Rev. B **24**, 5970 (1981); ibid. B **25**, 6338 (1982); J. P. Wolfe: Phys. Today **35**, 46 (1982)

5.10 M. Combescot, P. Nozières: J. Phys. C **5**, 2369 (1972); M. Combescot: Phys. Rev. Lett. **32**, 15 (1974); J. C. Hensel, K. Suzuki: In Proc. 10th Int'l Conf. on Physics of Semiconductors, Cambridge, MA, ed. by S. P. Keller, J. Hensel, F. Stern, CONF-700501 (US AFC Div. of Technical Information, Springfield, VA 1970) p. 541

5.11 G. A. Thomas, T. G. Phillips, T. M. Rice, J. C. Hensel: Phys. Rev. Lett. **31**, 386 (1973); T. K. Lo: Solid State. Commun. **15**, 1231 (1974); A. Forchel, B. Laurich, J. Wagner, W. Schmid, T. L. Reinecke: Phys. Rev. B **25**, 2730 (1982); M. A. Vouk, E. C. Lightowlers: J. Phys. C **8**, 3695 (1975); R. B. Hammond, T. C. McGill, J. W. Mayer: Phys. Rev. B **13**, 3566 (1976)

5.12 E. Hanamura: In Proc. 10th Intl. Conf. on Physics of Semiconductors, Cambridge, MA, ed. by S. P. Keller, J. Hensel, F. Stern, CONF-700501 (US AFC DIV. of Technical Information, Springfield, VA 1970) p. 504

5.13 W. F. Brinkman, T. M. Rice, P. W. Anderson, S. T. Chui: Phys. Rev. Lett. **28**, 961 (1972);

W. F. Brinkman, T. M. Rice: Phys. Rev. B **7**, 1508 (1973)

5.14   P. Vashishta, S. G. Das, K. S. Singwi: Phys. Rev. Lett. **30**, 1248 (1973); ibid.
       33, 911 (1974);
       P. Vashishta, P. Bhattacharyya, K. S. Singwi: Phys. Rev. B **10**, 5108 (1974);
       P. Bhattacharyya, V. Massida, K. S. Singwi, P. Vashishta: Phys. Rev. B **10**,
       5127 (1974)

5.15   T. L. Reinecke, M. C. Lega, S. C. Ying: Phys. Rev. B **20**, 1562 (1979)

5.16   J. Shah, M. Combescot, A. H. Dayem: Phys. Rev. Lett. **38**, 1497 (1977);
       B. Hammond, T. G. McGill, J. W. Mayer: Phys. Rev. B **13**, 3566 (1976)

5.17   G. A. Thomas, T. M. Rice, J. C. Hensel: Phys. Rev. Lett. **31**, 219 (1974)

5.18   Y. Kuramoto, H. Kamimura: J. Phys. Soc. Jpn. **37**, 716 (1974)

5.19   P. Isihara, Y. Nakane, A Isihara: J. Phys. C **15**, 2929 (1982): ibid. **17**, 2255
       (1984)

5.20   P. Vashishta, R. K. Kalia, K. S. Singwi: In *Electron-Hole Droplets in Semi-
       conductors*, ed. by C. D. Jeffries, L. V. Keldysh (North-Holland, Amsterdam
       1983) Fig. 16

5.21   A. N. Lobaev, A. P. Silin: Fiz. Tverd. Tela **29**, 1416 (1987) [Engl. transl.:
       Sov. Phys.–Solid State **29**, 809 (1987)];
       M. Combescot: Solid State Commun. **42**, 61 (1982)

5.22   P. D. Altukhov, A. V. Ivanov, Yu. N. Lomasov, A. A. Rogachev: Pis'ma Zh.
       Fksp. Teor. Fiz. **38**, 4 (1983); ibid. **39**, 523 (1984) [Engl. transl.: JFTP Lett.
       38, 4 (1983); ibid. **39**, 523 (1984)];
       F. Martelli: Solid State Commun. **55**, 905 (1985);
       A. Pinczuk, H. L. Shah, H. L. Störmer, R. C. Miller, A. C. Gossard,
       W. Wiegmann: Solid State Commun. **50**, 735 (1984)

5.23   E. A. Andryushyn, A. P. Silin: Solid State Commun. **20**, 453 (1976);
       Y. Kuramoto, C. Horie: Solid State Commun. **32**, 713 (1978);
       I. V. Lerner, Yu. E. Lozovik: J. Low Temp. Phys. **38**, 333 (1980) Zh. Fksp.
       Teor. Fiz. **80**, 1488 (1981) [Engl. transl.: Sov. Phys.–JFJP **53**, 763 (1981)];
       Yu. A. Bychov, E. I. Rashba: Solid State Commun. **48**, 399 (1983);
       Yu. A. Bychov, S. V. Iordanskii: Fix. Tverd. Tela **29**, 2442 (1987) [Engl.
       transl.: Sov. Phys.–Solid State. **29**, 1405 (1987);
       D. Paquet, T. M. Rice, K. Ueda: Phys. Rev. B **32**, 5208 (1985)

Chapter 6

6.1   A. Isihara, J. Tsai, M. Wadati: Phys. Rev. A **3**, 990 (1971);
       J. T. Tsai, M. Wadati, A. Isihara: Phys. Rev. A **4**, 1219 (1971);
       A. Isihara, M. Wadati: Phys. Rev. A **1**, 318 (1970);
       M. J. Stephen: Proc. R. Soc. London A **265**, 215 (1962)

6.2   G. E. Uhlenbeck, L. Gropper: Phys. Rev. **41**, 79 (1932)

6.3   I. M. Lifshitz, A. M. Kosevich: Zh. Fksp. Teor. Fiz. **29**, 730 (1955) [Engl.
       transl.: Sov. Phys.–JFTP **2**, 636 (1956)]

6.4   P. C. Alkaraz, Y. Kojima, A. Isihara: J. Phys. Chem. Solids **40**, 293 (1979)

6.5   R. B. Dingle: Proc. R. Soc. London A **211**, 517 (1952)

6.6   R. A. Phillips, G. V. Gold: Phys. Rev. **178**, 932 (1969);
       D. D. Shoenberg, J. Vanderkooy; J. Low Temp. Phys. **2**, 483 (1970)

6.7   P. J. Coleridge, I. M. Templeton: J. Phys. F **2**, 643 (1972)

6.8   D. Y. Kojima, A. Isihara: Phys. Rev. B **20**, 489 (1979);
       A. Isihara, D. Y. Kojima: Phys. Rev. B **10**, 4925 (1974);

See also Ref [6.1]

6.9   D. L. Martin: Phys. Rev. A **139**, 150 (1965);
       F. S. Ham: Phys. Rev. **128**, 2524 (1962);
       A. H. Wilson: *The Theory of Metals* (Cambridge University Press, Cambridge 1953)

6.10   E. W. Collings: J. Phys. Chem. Solids **26**, 949 (1965)

6.11   J. B. Sampson, F. Seitz: Phys. Rev. **58**, 633 (1940);
       D. Pines: In *Solid State Physics*, Vols. 1, 2, ed. F. Seitz, D. Turnbull (Academic New York 1955);
       M. Shimizu: J. Phys. Soc. Jpn. **15**, 2220 (1960);
       S. D. Silverstein: Phys. Rev. **130**, 1703 (1963);
       K. A. Brueckner, K. Sawada: Phys. Rev. **112**, 328 (1958)

6.12   E. H. Sondheimer, A. H. Wilson: Proc. R. Soc. London A **210**, 173 (1951);
       H. Ichimura, S. Tanaka: Prog. Theor. Phys. **25**, 315 (1961);
       H. Kanazawa, N. Matsudaira: Prog. Theor. Phys. **23**, 426 (1960);
       A. K. Rajagopal, K. P. Jain: Phys. Rev. A **5**, 1475 (1972)

6.13   T. Kushida, J. C. Murphy, M. Hanabusa: Phys. Rev. B **13**, 5136 (1976); ibid. B **20**, 4953 (1979);
       T. Kushida, J. C. Murphy: Phys. Rev. B **21**, 4247 (1980);
       J. P. Perdew, S. B. Nickerson, S. H. Vosko, R. A. Moore: Can. J. Phys. **53**, 648 (1975)

6.14   Xiaodong Zhu, A. W. Overhauser: Phys. Rev. B **30**, 3158 (1984)

6.15   F. F. Fang, P. J. Stiles: Phys. Rev. **174**, 823 (1968);
       J. L. Smith, P. J. Stiles: Phys. Rev. L **1**, 102 (1973)

6.16   A. Isihara, L. C. Ioriatti, Jr.: Physica B **113**, 42 (1982)

6.17   A. Isihara, L. Ioriatti, Jr., D. Y. Kojima: Z. Phys. B **44**, 91(1981)

6.18   S. Yarlangadda, G. F. Giuliani: Phys. Rev. B **38**, 10966 (1988–II)

6.19   A. Isihara, D. Y. Kojima: Phys. Rev. B **19**, 816 (1979)

6.20   Y. Shiwa, A. Isihara: Phys. Rev. B **27**, 4743 (1983)

6.21   J. P. Eisenstein, H. L. Störmer, V. Narayanamurti, A. Y. Cho, A. C. Gossard: Surf Sci. **170**, 271 (1986)

6.22   J. F. Kwak, J. E. Schirber, R. L. Greene, E. M. Engler: Phys. Rev. Lett. **46**, 1296 (1981);
       P. M. Chaikin, Mu-Yong Choi, J. F. Kwak, J. S. Brooks, K. P. Martin, M. J. Naughton, E. M. Engler, R. J. Greene: Phys. Rev. Lett. **51**, 2333 (1983);
       M. Ribault, D. Jerome, J. Tuchendler, C. Weyl, K. Bechgaard: J. Phys. Lett. **44**, L-953 (1983);
       T. Takahashi, D. Jerome, K. Bechgaard: J. Phys. Colloq. **44**, C3-805 (1983)

6.23   V. M. Yakovenko: Zh. Eksp. Teor. Fiz. **93**, 627(1987) [Engl. transl.: Sov. Phys. JETP 66, 355 (1987)]

6.24   K. Yamaji: Physica B **143**, 439 (1986);
       L. Chen, K. Maki, A. Virosztek: Physica B **143**, 444 (1986)

6.25   M. J. Naughton, R. V. Chamberlin, X. Yan, S. Y. Hsu, L. Y. Chang, M. Ya. Azbel, P. M. Chaikin: Phys. Rev. Lett. **61**, 621 (1988)
       A. G. Lebed, J. Phys. I (France) **4**, 351 (1994);
       P. M. Chaikin, Phys. Rev. Lett. **69**, 2831 (1992)

6.26   L. P. Gor'kov, A. G. Lebed: J. de Phys. Lett. **45**, L440 (1984);
       L. P. Gor'kov: Sov. Phys. Usp, **27**, 803 (1984);
       M. Héritier, G. Montambaux, P. Lederer: J. de Phys. **45**, L943 (1984);
       P. M. Chaikin: Phys. Rev. B **31**, 4770 (1985);
       K. Yamaji: J. Phys. Soc. Jpn. **55**, 1424 (1986)

6.27   X. Yan, M. J. Naughton, O. S. Cheema, R. V. Chamberlin, S. Y. Hsu, L. Y. Chang, P. M. Chaikin: Solid State Commun. **66**, 905 (1988)

6.28   A G. Lebed, P. Bak: Phys. Rev. Lett. **63**, 1315 (1989);
       A. G. Lebed, P. Bak: Phys. Rev. B **40**, 11433 (1989);
       A. G. Lebed: J. Phys. I (France) **4**, 351 (1994);
       P. M. Chaikin: Phys. Rev. Lett. **69**, 2831 (1992);
       S. P. Strong, David G. Clark, P. W. Anderson: Phys Rev. Lett. **73**, 1007 (1994)

Chapter 7

7.1   A. B. Lidiard: Philos. Mag. **42**, 1325 (1951)

7.2   M. Gell-Mann: Phys. Rev. **106**, 369 (1957);
       D. Pines: Phys. Rev. **92**, 620 (1953);
       D. F. DuBois: Ann. Phys. **8**, 24 (1959);
       B. Ninham: Ann. Phys. **8**, 220 (1964)

7.3   A. Isihara, Y. Kojima: Physica **77**, 469 (1974); Z. Phys. B **21**, 33 (1975)

7.4   Y. Kojima, A. Isihara: Phys. Cond. Matter **17**, 179 (1974)

7.5   L. Ioriatti, A. Isihara: Phys. Status Solidi (b) **97**, K65 (1980)

7.6   A. Isihara, L. Ioriatti: Physica **103A**, 621 (1983)

7.7   A. Isihara, T. Toyoda: Phys. Rev. B **21**, 3358 (1980)

7.8   A. Isihara, Y. Shiwa: J. Phys. C **17**, 5705 (1984); ibid., **18**, 4703 (1985);
       Y. Shiwa, A. Isihara; Solid State Commun. **48**, 1081 (1983); ibid **49**, 497 (1984); ibid., **50**, 35, 433 (1984)

7.9   G. R. Stewart: Rev. Mod. Phys. **56**, 755 (1984)

7.10   J. Kondo: Prog. Theor. Phys. **32**, 37 (1964)

7.11   Y. Onuki, T. Komatsubara: J. Magn. Magn. Mater. **63, 64**, 281 (1987)

7.12   N. Read, M. Newns: J. Phys. C **16**, 3273, L1055 (1983); Solid State Commun. 52, 993 (1984)

7.13   P. Coleman: Phys. Rev. B **29**, 3035 (1984); J. Magn. Magn. Mater. **63, 64**, 245 (1987)

7.14   P. Fulde, J. Keller, G. Zwicknagl: In *Solid State Physics*, Vol. 41, ed. by H. Ehrenreich, D. Turnbull (Academic, New York 1988) p. 1;
       T. M. Rice: Proc. 18th Int. Conf. on Low Temperature Physics (Kyoto, 1987) Jpn. J. Appl. Phys. **26**, Suppl. 26–3, 1865 (1987);
       B. Coqblin, J. R. Schrieffer: Phys. Rev. **185**, 847 (1969)

Chapter 8

8.1   A. Isihara, L. Smrčka: J. Phys. C **19**, 6777 (1986); L. Smrčka, A. Isihara: Solid State Commun. **57**, 259 (1986)

8.2   B. Beličky, S. Kirkpatrick, H. Ehrenreich: Phys. Rev. **175**, 747 (1968)

8.3   T. Ando, Y. Matsumoto, Y. Uemura, M. Kobayashi, K. F. Komatsubara: J. Phys. Soc. Jpn. **32**, 859 (1972)

8.4   W. A. Harrison: *Solid State Theory* (McGraw-Hill, New York 1971) Chap. 2

8.5   F. F. Fang, A. B. Fowler, A. Hartstein: Phys. Rev. B **16**, 4446 (1977);

S. J. Allen, Jr., D. C. Tsui, F. DeRosa: Phys. Rev. Lett. **35**, 1359 (1975)
8.6   W. Götze, P. Wölfle: Phys. Rev. B **6**, 1226 (1972): J. Phys. C **11**, 3993 (1978)
8.7   C. S. Ting, S. Ying, J. J. Quinn: Phys. Rev. B **16**, 5394 (1977)
8.8   Y. Shiwa, A. Isihara: J. Phys. C **16**, 4853 (1983)
8.9   A. Kawabata: Surf Sci. **98**, 276 (1980)
8.10   W. Götze, J. Hajdu: J. Phys. C **11**, 3993 (1978); Solid State Commun. **29**, 89 (1979)
8.11   A. Bastin, C. Lewiner, O. Betbeder-Matibet, P. Nozières: J. Phys. Chem. Solids **32**, 1811 (1971);
       K. Ohta: Jpn. J. Appl. Phys. **10**, 850 (1971); J. Phys. Soc. Jpn. **31**, 1627 (1971);
       P. Streda, L. Smrčka: Phys. Status Solidi (b) **70**, 537 (1975)
8.12   T. Ando, Y. Uemura: J. Phys. Soc. Jpn **36**, 959 (1974)
8.13   Y. Shiwa, A. Isihara: Solid State Commun. **53**, 519 (1985)
8.14   W. Götze: Philos. Mag. B **43**, 219 (1981);
       A. Gold, W. Götze: J. Phys. C **12**, 4049 (1981)
8.15   A. Isihara, D. Y. Kojima: Phys. Rev. B **19**, 846 (1976);
       A. Isihara, L. Ioriatti, Jr., D. Y. Kojima: Z. Phys. B **44**, 91 (1981);
       A. Isihara, M. Mukai, S. J. Lee: Phys. Rev. B **26**, 3279 (1982)
8.16   S. J. Allen, Jr., A. Wilson, D. C. Tsui: Phys. Rev. B **26**, 5590 (1982);
       B. A. Wilson, S. J. Allen, Jr., D. C. Tsui: Phys. Rev. B **24**, 5887 (1981)
8.17   M. J. Chou, D. C. Tsui, G. Weimann: Phys. Rev. B **37**, 848 (1981-I)

Chapter 9

9.1   P. W. Anderson: Phys. Rev. **109**, 1492 (1958)
9.2   N. F. Mott: Rev. Mod. Phys. **50**, 203 (1978)
9.3   E. Abrahams, P. W. Anderson, D. C. Licciardello, J. V. Ramakrishnan: Phys. Rev. Lett. **42**, 673 (1979)
9.4   A. MacKinnon: In *Localization, Interaction, and Transport Phenomena*, ed. by B. Kramer, G. Bergmann, Y. Bruynseraede, Springer Ser. Solid-State Sci., Vol. 61 (Springer, Berlin, Heidelberg 1985) p. 90
9.5   D. Vollhard, P. Wölfle: In *Anderson Localization*, ed. by Y. Nagaoka, Springer Ser. Solid-State Sci., Vol 39 (Springer, Berlin, Heidelberg 1976) p. 125:
       A. Kawabata: Solid State Commun. **38**, 823 (1981)
9.6   P. A. Lee: Phys. Rev. Lett. **42**, 1492 (1979)
9.7   L. P. Gor'kov, A. I. Larkin, D. F. Khmel'nitskii: Zh. Eksp. Teor. Phys. **30**, 248 (1979);
       S. Hikami: Prog. Theor. Phys. **64**, 1466 (1980)
9.8   P. A. Lee, J. V. Ramakrishnan: Rev. Mod. Phys. **57**, 287 (1985);
       A. Kawabata: Prog. Theor. Phys. Suppl. **84**, 16 (1985)
9.9   B. L. Al'tshuler, A. G. Aronov, B. Z. Spivak: Pis'ma Zh. Eksp. Teor. Fiz. **33**, 101(1981) [Engl. transl.: Sov. Phys.–JETP Lett. **33**, 94 (1981)];
       D. Yu. Sharvin, Yu. V. Sharvin: Pis'ma Zh. Eksp. Teor. Fiz. **34**, 285 (1980) [Engl. transl. Sov. Phys.–JETP Lett. **34**, 272 (1981)]
9.10   S. Hikami, A. I. Larkin, Y. Nagaoka: Prog. Theor. Phys. **63**, 607 (1980);
       B. L. Al'tshuler, D. F. Khmelnitzkii, A. I. Larkin, P. A. Lee: Phys. Rev. B **22**, 5142 (1980)
9.11   Y. Kawaguchi, S. Kawaji: J. Phys. Soc. Jpn. **49**, Suppl A, 983 (1980);
       S. Kawaji, Y. Kawaguchi: J. Phys. Soc. Jpn. **53**, 2868 (1984)

9.12   A. Kawabata: J. Phys. Soc. Jpn. **53**, 3540 (1984)
9.13   H. Fritzsche, K. Lark-Horovitz: Phys. Rev. **99**, 400 (1955);
      W. Sasaki, Y. Kanai: J. Phys. Soc. Jpn. **11**, 894 (1956);
      W. Sasaki: J. Phys. Soc. Jpn. **20**, 825 (1965)
9.14   A. Kawabata: Solid State Commun. **34**, 432 (1980);
      Y. Ootsuka, A. Kawabata: Prog. Theor. Phys. Suppl. No. **84**, 249 (1985)
9.15   S. Morita, N. Mikoshiba, Y. Koike, T. Fukase, M. Kitagawa, S. Ishida:
      J. Phys. Soc. Jpn. **53**, 2185 (1984)
9.16   Y. Kawaguchi, S. Kawaji: Surf. Sci. **113**, 505 (1982)
9.17   B. L. Al'tshuler, A. G. Aronov: Solid State Commun. **30**, 115 (1979); Zh.
      Eksp. Teor. Fiz. **77**, 2028 (1979) [Engl. transl.: Sov. Phys.–JFTP **50**, 968
      (1979)]; B. L. Al'tshuler, A. G. Aronov, P. A. Lee: Phys. Rev. Lett. **44**, 1288
      (1980);
      H. Fukuyama: J. Phys. Soc. Jpn. **48**, 2169 (1980)
9.18   J. Kokanović, B. Leontić, J. Lukatela: Solid State Commun. **69**, 447 (1989)
9.19   K. K. Choi, D. C. Tsui, S. C. Palmateer: Phys. Rev. B **33**, 8216 (1986)
9.20   M. S. Burdis, C. C. Dean: Phys. Rev. B **38**, 3269 (1988-I)

Chapter 10

10.1   N. F. Mott: J. Noncryst. Solids **1**, 1 (1968)
10.2   S. Kirkpatrick: Rev. Mod. Phys. **45**, 574 (1973)
10.3   A. B. Fowler, A. Hartstein, R. A. Webb: Physica **117** & **118B**, 661 (1983);
      Phys.Rev. B **16**, 196 (1982)
10.4   G. Deutscher, M. Rapport: J. Phys. C **6**, 581 (1978);
      G. Deutscher: In *Disordered Systems and Localization*, ed. by C. Castellani,
      C. Di Castro, L. Peliti, Lect. Notes Phys., Vol. **149** (Springer, Berlin, Hei-
      delberg 1981) p. 26
10.5   I. Webman, J. Jortner, M. H. Cohen: Phys. Rev. B **16**, 2593 (1977)
10.6   M. E. Fisher, J. W. Essam: J. Math. Phys. **2**, 609 (1961)
10.7   A. P. Young, R. B. Stinchcombe: J. Phys. C **8**, L535 (1975); R. B. Stinch-
      combe, B. P. Watson: J. Phys. C **9**, 3221 (1976)
10.8   V. Ambegaokar, B. J. Halperin, J. S. Langer: Phys. Rev. B **4**, 2612 (1971)
10.9   P. N. Butcher, K. J. Hayden, J. A. McInnes: Philos. Mag. **36**, 19 (1977)
10.10   A. Miller, E. Abrahams: Phys. Rev. **120**, 745 (1960)
10.11   S. Kivelson: Phys. Rev. Lett. **46**, 1344 (1981)
10.12   A. J. Heeger, A. G. Macdiarmid: Mol. Cryst. Liq. Cryst. **77**, 1 (1981)
10.13   C. P. Umbach, S. Washburn, R. B. Laibowitz, R. A. Webb: Phys. Rev. B **30**,
      4048 (1984);
      R. A. Webb, S. Washburn, P. Umbach, R. B. Laibowitz: Phys. Rev. Lett.
      54, 2696 (1985);
      G. E. Blonder: Bull. Am. Phys. Soc. **28**, 535 (1984)
10.14   P. A. Lee: Phys. Rev. Lett. **53**, 2042 (1984)
10.15   A. D. Stone: Phys. Rev. Lett. **54**, 2692 (1985);
      P. A. Lee, A. D. Stone: Phys. Rev. Lett. **55**, 1622 (1985)
10.16   R. Landauer: IBM J. Res. Dev. **1**, 233 (1957)
10.17   M. Ya. Azbel: J. Phys. C **14**, L225 (1981);
      M. Büttiker, Y. Imry, R. Landauer, S. Pinhas: Phys. Rev. B **31**, 6207 (1985);
      D. C. Langreth, E. Abrahams: Phys. Rev. B **24**, 2978 (1981)
10.18   S. Maekawa, Y. Isawa, H. Ebisawa: J. Phys. Soc. Jpn. **56**, 25 (1987);

Y. Isawa, H. Ebisawa, S. Maekawa: Jpn. J. Appl. Phys. **26**, 25 (1987)

10.19 M. Büttiker: Phys. Rev. Lett. **57**, 1761(1986); IBM J. Res. Dev. **32**, 317 (1988);
A. D. Stone, A. Szafer: IBM J. Res. Dev. **32**, 384 (1988)

10.20 R.J. Haug, A. H. MacDonald, P. Streed, K. von Klitzing: Phys. Rev. Lett. **61**, 2797 (1988);
S. Washburn, A. B. Fowler, H. Schmid, D. Kern: Phys. Rev. Lett. **61**, 2801 (1988)

10.21 H. van Houton, C. Beenakker, P. van Lossdrecht, T. Thornton, H. Ahmed, M. Pepper, C. Foxon, J. Harris: Phys. Rev. B **37**, 8534 (1988);
M. Büttiker: Phys. Rev. B **38**, 9375 (1988)

10.22 B. L. Al'tshuler: Pis'ma Zh. Fksp. Teor. Fiz. **41**, 530 (1985) [Engl. transl.: Sov. Phys.–JFTP Lett. **41**, 648 (1985)]

10.23 J. C. Licini, D. J. Bishop, M. A. Kastner, J. Melngailis: Phys. Rev. Lett. **55**, 2987 (1985)

10.24 S. B. Kaplan, A. Hartstein: Phys. Rev. Lett. **56**, 2403 (1986); IBM J. Res. Dev. **32**, 347 (1988)

10.25 B. L. Al'tshuler, D. E. Khmel'nitskii: Pis'ma Zh. Fksp. Teor. Fiz. **42**, 291(1985) [Engl. transl.: Sov. Phys.–JEJP Lett. **42**, 359 (1985)];
A. I. Larkin, D. E. Khmel'nitskii: Zh. Eksp. Teor. Fiz. **91**, 1815 (1986) [Sov. Phys.–JFJP **64**, 1075 (1986)]

10.26 R. A. Webb, S. Washburn, C. P. Umbach: Phys. Rev. B **37**, 8455 (1988-I).

10.27 M. J. Cohen, L. B. Coleman, A. F. Garito, A. J. Heeger: Phys. Rev. B **10**, 1298 (1974);
M. J. Cohen, A. J. Heeger: Phys. Rev. B **16**, 688 (1977)

10.28 E. M. Conwell, N. Banik: J. de Phys **42**, C7-315 (1981)

10.29 W. P. Su, J. R. Schrieffer, A. J. Heeger: Phys. Rev. B **22**, 2099 (1980)

10.30 S. Kivelson: In "Solitons", ed. by S. E. Trullinger, V. E. Zakharov, L. Pokrovsky (Elsevier, Amsterdam 1986) p. 30;
D. Baeriswyl, K. Maki: Phys. Rev. B **31**, 6633 (1985);
S. Kivelson, D. E. Heim: Phys. Rev. B **26**, 4278 (1982);
S. Kivelson, W.-P. Su, J. R. Schrieffer, A. J. Heeger: Phys. Rev. Lett. **58**, 1899 (1987);
C.-Q. Wu, X. Sun, K. Nasu: Phys. Rev. Lett. **59**, 831 (1987)

10.31 J. Ashkenazi, W. E. Pickett, H. Krakauer, C. S. Wang, B. M. Klein, S. R. Chubb: Phys. Rev. Lett. **62**, 2016 (1989)

Chapter 11

11.1 J. G. Bednorz, K. A. Müller: Z. Phys. B **64**, 189 (1986)

11.2 L. N. Cooper: Phys. Rev. **104**, 1189 (1956)

11.3 J. Bardeen, L. N. Cooper, J. R. Schrieffer: Phys. Rev. **108**, 1175 (1957)

11.4 J. R. Schrieffer: *Theory of Superconductivity* (W. A. Benjamin, New York 1964);
G. Rickayzen: *Theory of Superconductivity* (Interscience, New York 1965);
R. D. Parks: *Superconductivity I, II* (Marcel Dekker, New York 1969)

11.5 L. F. Schneemeyer, J. K. Thomas, J. Siegrist, B. Batlogg, L. W. Rupp, R. L. Opila, R. J. Cava, D. W. Murphy: Nature **335**, 421 (1988)

11.6 D. R. Harshman et al.: Phys. Rev. B **36**, 2386 (1987)
More recent data with angle dependences can be found in:

D. R. Harshman, L. F. Schneemeyer, J. V. Waszczak, O. Aeppli, R. J. Cava, B. Batlogg, L. W. Rupp, E. J. Ansaldo, D. L. Williams: Phys. Rev. B **39**, 851 (1989)

11.7   H. C. Yang, B. D. Yao, H. E. Horng: Supercond. Sci. Tech. **1**, 160 (1988); S. Ahmed, S. Bungre, B. Cyca, R. Meisels, A. Tacconelli, A. D. Caplin: Supercond. Sci. Tech. **1**, 281 (1989)

11.8   K. Kumagai, Y. Nakamichi, I. Watanabe, Y. Nakamura, H. Nakajima, N. Wada, P. Ledder: Phys. Rev. Lett. **60**, 724 (1988); S. E. Inderhees, M. B. Salamon, J. P. Rice, D. M. Ginsberg,: Phys. Rev. Lett. **66**, 232 (1991)

11.9   S. E. Stupp, D. M. Ginsberg: Physica C **158**, 299 (1989); N. E. Phillips, R. A. Fisher, J. E. Gordon, A. Kim, A. M. Stacey, M. K. Crawford, E. M. McCarron III: Phys. Rev. Lett. **65**, 357 (1990)

11.10   J. S. Tsai et al.: Physica C **163**, 1385 (1988); M. Gurvitch, J. M. Valles, Jr., A. M. Cucolo, R. C. Wynes, J. P. Garno, L. F. Schneemeyer, J. V. Waszczak: Phys. Rev. Lett. **63**, 1008 (1989)

11.11   B. Friedl, C. Thomsen, M. Cardona: Phys. Rev. Lett. **65**, 915 (1990)

11.12   D. E. Morris, R. M. Kuroda, A. G. Markelz, J. H. Nickel, J. Y. T. Wei: Phys. Rev. B **37**, 5936 (1988)

11.13   C. E. Gough et al.: Nature **326**, 855 (1987)

11.14   J. T. Market, T. W. Noh, S. E. Russek, R. M. Cotts: Solid State Commun. **63**, 847 (1987); G. J. Kramer, H. B. Brom, J. van den Berg, P. H. Kes, D. J. W. Ijdo: Solid State Commun. **6** 4, 705 (1987); H. Alloul, P. Mendels, G. Collin, P. Monod: Phys. Rev. Lett. **61**, 746 (1988)

11.15   Y. Kitaoka et al.: IBM J. Res. Dev. **33**, 277 (1989); M. Mehring: IBM J. Res. Dev. **33**, 342 (1989)

11.16   R. J. Birgeneau et al.: Phys. Rev. B **38**, 6614 (1988); G. Shirane et al: Phys. Rev. Lett. **63**, 330 (1989); N. Bulut, D. J. Scalapino: Phys. Rev. Lett. **68**, 706 (1992)

11.17   J. Schümann, W. Ose, J. Keller, K. F. Renk, B. Roas, L. Shultz, L. Saemann-Ischenko: Europhys. Lett. **8**, 679 (1989); J. Orenstein, G. A. Thomas, A. J. Millis, S L. Cooper, D. H. Rapkine, T. Timusk, L. F. Schneemeyer, J. V. Waszczak: Phys. Rev. B **42**, 6342 (1990); K. Kamaras et al.: Phys. Rev. Lett. **64**, 84 (1990); Z. Schlesinger, R. T. Collins, F. Holtzberg, C. Feild, S. H. Blanton, V. Welp, G. W. Crabtree, V. Fang, J. Z. Liu: Phys. Rev. Lett. **65**, 679 (1990); F. Slakey, M. V. Klein, J. P. Rice, D. M. Ginsberg: Phys. Rev. B **43**, 3764 (1991)

11.18   J. M. Tranquada, S. M. Heald, A. R. Moodenbaugh: Phys. Rev. B **36**, 5263 (1987); Z.-X. Shen et al.: Phys. Rev. B **36**, 8414 (1987)

11.19   P. W. Anderson: Science **235**, 1196 (1987); P. W. Anderson, G. Baskaran, Z. Zou, T. Hsu: Phys. Rev. Lett. **58**, 2790 (1987); G. Baskaran, P. W. Anderson: Phys. Rev. B **37**, 580 (1988); J. M. Wheatley, T. C. Hsu, P. W. Anderson: Phys. Rev. B **37**, 5897 (1988)

11.20   V. J. Emery: Phys. Rev. Lett. **58**, 2794 (1987); V. J. Emery, G. Reiter: Phys. Rev. B **38**, 4547 (1988)

11.21   J. Hirsch: Phys. Rev. B **31**, 4403 (1985)

11.22   R. J. Birgeneau, M. A. Kastner, A. Ahrony: Z. Phys. B **71**, 57 (1988);

A. Aharony, R. J. Birgeneau, A. Coniglio, M. A. Kartner, H. E. Stanley: Phys. Rev. Lett. **60**, 1330 (1988)

11.23 J. R. Schrieffer, X.-G. Wen, S.-C. Zhang: Phys. Rev. Lett., **60**, 944 (1988)

11.24 C. M. Varma, S. Schmitt-Rink, E. Abrahams: Solid State Commun. **62**, 681(1987)

11.25 N. E. Bickers, D. J. Scalapino, S. White: Phys. Rev. Lett. **62**, 961 (1991);
T. Moriya, Y. Takahashi, K. Ueda: Physica C **185**, 114 (1991);
N. Bulut, D. J. Scalapino: Phys. Rev. Lett. **68**, 706 (1992);
D. Thelen, D. Pines, J. P. Lu: Phys. Rev. B **47**, 9151 (1993);
P. Monthoux, D. Pines: Phys. Rev. B **47**, 6069 (1993);
D. J. Scalapino: Phys. Rep. **250**, 329 (1995)

11.26 J. A. Martindale, S. E. Barrett, K. E. O'Hara, C. P. Slichter, W. C. Lee, D. M. Ginsberg: Phys. Rev. **47**, 9155 (1993)

11.27 B. J. Sternlieb, G. Shirana, J. M. Tranquada, M. Sato, S. Shamoto: Phys. Rev. B **47**, 5320 (1993)

11.28 W. H. Hardy, D. A. Bonn, D. Morgan, R. Liang, K. Zhang: Phys. Rev. Lett. **70**, 3999 (1993);
D. J. Van Harlington: Rev. Mod. Phys. **67**, 515 (1995)

11.29 P. W. Anderson, Science **256**, 1526 (1992);
S. Chakravarty, A. Sudobø, P. W. Anderson, S. Strong: Science **261**, 337 (1993);
A. Sudobø, S. Chakravarty, S. Strong, P. W. Anderson: Phys. Rev. B **49**, 12245 (1994)
For experiments, see: A. G. Sun, D. A. Gajewski, M. B. Maple, R. Dynes: Phys. Rev. Lett. **72**, 2267 (1994);
P. Chaudri, S. Y. Lin: Phys. Rev. Lett. **72**, 1084 (1994)

11.30 J. M. Luttinger: J. Math. Phys. **4**, 1154 (1963)

Chapter 12

12.1 *The Quantum Hall Effect*, ed. by R. E. Prange, S. M. Girvin (Springer, New York 1990)
K. V. Klitzing, G. Dorda, M. Pepper: Phys. Rev. Lett. **45**, 494 (1980)

12.2 D. C. Tsui, A. C. Gossard, B. F. Field, M. E. Cage, R. F. Dziuba: Phys. Rev Lett **48**, 3 (1982)

12.3 B. Tausendfreund, K. von Klitzing: Surf Sci. **142**, 220 (1984);
J. V. Branch et al.: See [13.26 d]

12.4 K. Yoshihiro, J. Kinoshita, K. Inagaki, C. Yamanouchi, J. Moriyama, S. Kawaji: Physica B **117**, 706 (1983);
M. E. Cage, B. F. Field, R. F. Dziuba, S. M. Girvin, A. C. Gossard, D. C. Tsui: Phys. Rev. B **30**, 2286 (1984)

12.5 A. Isihara: In *Solid State Physics*, Vol 42, ed. by H. Ehrenreich, D. Turnbull (Academic, New York 1989) p. 271

12.6 J. Kinoshita, K. Inagaki, C. Yamanouchi, K. Yoshihiro, J. Wakabayashi, S. Kawaji: Proc. 2nd Int'l Symp. on the Foundations of Quantum Mechanics, Tokyo, 1986 (Phys. Soc. Jpn., Tokyo 1986) p. 150;
T. Okuno et al.: J. Phys. Soc. Jpn **64**, 1881 (1995)

12.7 G. S. Boebinger, H. W. Jiang, L. N. Pfeiffer, K. W. West: Phys. Rev. Lett. **64**, 1793 (1990)

12.8 H. A. Fertig: Phys. Rev. B **40**, 1087 (1989-I);

A. H. MacDonald, P. M. Platzman, G. S. Boebinger: Phys. Rev. Lett. **65**, 775 (1990);

V. Nikos Nicopoulos, S. A. Trugman: Phys. Rev. Lett. **65**, 779 (1990)

12.9   F. Kuchar, R. Meisels, G. Weimann, W. Schlapp: Phys. Rev. B **33**, 2965 (1986)

12.10   S. Washburn, A. B. Fowler, H. Schmid, D. Kern: Phys. Rev. Lett. **61**, 2801(1988)

12.11   H. L. Störmer, J. P. Eisenstein, A. C. Gossard, W. Wiegmenn, K. Baldwin: Phys. Rev. Lett., **56**, 85 (1986)

12.12   D. F. Khmel'nitskii: Pis'ma Zh. Eksp. Teor. Fiz. **38**, 454 (1983) [Engl. transl.: Sov. Phys.–JFTP **38**, 552 (1983)];

A. M. M. Pruisken: Phys. Rev. B **32**, 2636 (1985); Phys. Rev. Lett. **61**,1297 (1988);

H. P. Wei, D. C. Tsui, M. A. Paalanen, A. M. M. Pruisken: In *High Magnetic Fields in Semiconductor Physics*, ed. by G. Landwehr, Springer, Ser. Solid-State Sci., Vol. 71 (Springer, Berlin, Heidelberg 1987) p. 11; Phys. Rev. Lett. **61**, 1294 (1988)

12.13   J. J. Chalker, G. J. Daniell: Phys. Rev. Lett. **61**, 593 (1988)

12.14   R. B. Laughlin: Phys. Rev. B **23**, 5632 (1981);

B. I. Halperin: Phys. Rev. B **25**, 2185 (1982)

12.15   D. J. Thouless: Phys. Rev. **31**, 8305 (1985);

Q. Niu, J. Thouless, Y. Wu: Phys. Rev. B **31**, 3372 (1985)

12.16   O. Heinonen, P. L. Taylor: Phys. Rev. B **28**, 6119 (1983)

12.17   A. Isihara: Surf. Sci. **170**, 267 (1986)

12.18   F. Wegner: Z. Phys. **B 51**, 279 (1983);

H. Levine, S. B. Libby, A. M. M. Pruisken: Phys. Rev. Lett. **51**, 1915 (1983);

H. Aoki, J. Ando: Surf. Sci. **170**, 249 (1986)

12.19   V. J. Goldman, J. K. Jain, M. Shayegan: Phys. Rev. Lett. **65**, 907 (1990)

12.20   S. E. Barrett, G. Dabbagh, L. N. Pfeiffer, K. W. West, R. Tycko: Phys. Rev. Lett. **74**, 5112 (1995);

R. Tycko, S. E. Barrett, G. Dabbagh, L. N. Pfeiffer, K. W. West: Science **268**, 1460 (1995);

A. Schmeller, J. P. Eisenstein, L. N. Pfeiffer, K. W. West: Phys. Rev. Lett. **75**, 4290 (1995)

12.21   X. G. Wu, J. K. Jain: Phys. Rev. B **19** 7515 (1994)

Chapter 13

13.1   T. Chakraborty, P.Pietiläinen: *The Fractional Quantum Hall Effect.* (Springer, Berlin 1995);

D. C. Tsui, H. L. Stormer, A. C. Gossard: Phys. Rev. Lett. **48**, 1559 (1982)

13.2   I. V. Kukushkin, V. B. Timofeev: In *High Magnetic Fields in Semiconductor Physics*, ed. by O. Landwehr, Springer Ser. Solid-State Sci., Vol. 71 (Springer, Berlin, Heidelberg 1987) p. 136

13.3   G.S. Boebinger, A. M. Chang, H. L. Stormer, D. C. Tsui: Phys. Rev. Lett. **55**, 1606 (1985);

See also: J. Wakabayashi, S. Kawaji, J. Yoshino, H. Sakaki: J. Phys. Soc. Jpn. **55**, 1319 (1986);

L. Engel, H. P. Wei, D. C. Tsui, Reported in the Int'l Conf. on Electronic Properties in Two-Dimensional Systems, Grenoble, 1989

13.4   R. J. Haug, K. von Klitzing, R. J. Nicholas, J. C. Maan, O. Weimann: Phys. Rev. **B 36**, 4528 (1987-I)

13.5   J. F. Furneaux, D. A. Syphers, A. O. Swanson: Phys. Rev. Lett. **63**, 1098 (1989)

13.6   J. P. Eisenstein, H. L. Stormer, L. Pfeiffer, K. W. West: Phys. Rev. Lett. **62**, 1540 (1989)

13.7   B. I. Halperin: Helv. Phys. Acta **56**, 75 (1983)

13.8   R. J. Willett, J. P. Eisenstein, H. L. Stormer, D. C. Tsui, A. C. Gossard, J. H. English: Phys. Rev. Lett., **59**, 1776 (1987);
R. G. Clark, J. R. Mallett, S. R. Haynes, P. A. Maksym, J. Harris, C. J. Foxon: Surf Sci. **170**, 141 (1986);
R. L. Willett, R. Ruel, M. Paalanen, K. W. West, L. Pfeiffer: Phys. Rev. B **47**, 7344 (1993);
R. R. Du, A. S. Yeh, H. L. Stormer, D. C. Tsui, L. N. Pfeiffer, K. W. West: Phys. Lett. **75**, 3926 (1995)

13.9   G. Timp, R. Behringer, J. E. Cunningham, R. E. Howard: Phys. Rev. Lett. **63**, 2268 (1989)

13.10   R. B. Laughlin: Phys. Rev. Lett., **50**, 1395 (1983)

13.11   J. M. Caillol, D. Levesque, J. J. Weis, J. P. Hansen: J. Stat. Phys. **28**, 325 (1982);
R. Morf: Phys. Rev. Lett. **43**, 931 (1979)

13.12   K. Takano, A. Isihara: Phys. Rev. B **34**, 1399 (1986)

13.13   D. Yoshioka, B. I. Halperin, P. A. Lee: Phys. Rev. Lett. **50**, 1219 (1983)

13.14   D. Levesque, J. J. Weis, A. H. MacDonald: Phys. Rev. B **30**, 1056 (1984)

13.15   K. Takano, A. Isihara: J. Phys. C **20**, L281 (1987); In *Anderson Localization*, ed. by T. Ando, H. Fukuyama, Springer Proc. Phys., Vol. 28 (Springer, Berlin, Heidelberg (1988) p. 268

13.16   F. D. M. Haldane, F. H. Rezayi: Phys. Rev. Lett. **54**, 237 (1985)

13.17   R. Morf, B. I. Halperin: Phys. Rev. B **33**, 2221 (1986)

13.18   W. P. Su: Phys. Rev. B **30**, 1069 (1984);
Q. Niu, D. J. Thouless, Y. Wu: Phys. Rev. B **31**, 3372 (1985); See also [13.12]

13.19   R. B. Laughlin: Surf. Sci. **142**, 163 (1984)

13.20   S. M. Girvin, A. H. MacDonald, P. M. Platzman: Phys. Rev. Lett. **54**, 581 (1985)

13.21   V. B. Bargmann: Rev. Mod. Phys. **34**, 829 (1962)

13.22   A. H. MacDonald, S. M. Girvin: Phys. Rev. B **33**, 4009 (1986)

13.23   A. Pinczuk, J. P. Valladares, D. Heiman, A. C. Gossard, J. H. English, C. W. Tu, L. Pfeiffer, K. West: Phys. Rev. Lett. **61**, 2701 (1988)

13.24   S. M. Girvin, A. H. MacDonald: Phys. Rev. Lett. **58**, 1252 (1987)

13.25   D. Yoshioka: J. Phys. Soc. Jpn. **55**, 885 (1986);
A. Gold: Phys. Rev. B **40**, 3375 (1989-I)

13.26   J. A. Simmons, H.P. Wei, L. W. Engel, D. C. Tsui, M. Shayegan: Phys. Rev. **63**, 1731 (1989);
R. G. Clark, J. R. Mallett, S. R. Haynes, J. J. Harris, C. J. Foxon: Phys. Rev. Lett. **60**, 1747 (1988);
R. G. Clark, S. R. Haynes, J. V. Branch, A. M. Suckling, P. A. Wright, P. M. Oswald: Surf. Sci. **229**, 25 (1990);
J. V. Branch, R. G. Clark, C. Andrikidis, S. J.Collocott, G. J. Griffiths, J. J. Harris, T. Foxon, J.T. Chalker: Int'l Conf. Phys. Semicond. Thessaloniki **2**, 841 (1990);
A. Sachrajda, R. Boulet, Z. Wasilewski, P. Coleridge: Solid State Commun. **74**, 1021 (1990);

S. I. Dorozhkin, R. J. Haug, K. von Klitzing, K. Ploog: Phys. Rev. B **51**, 14729 (1995)

13.27 J. K. Jain: Phys. Rev. B. **41**, 7653 (1990);
J. K. Jain, S. A. Kivelson, N. Trivedi: Phys. Lett. **64**, 1297 (1990);
J. K. Jain, X. G. Wu: Phys. Rev. B **49**, 5085 (1995)

13.28 B. I. Halperin, P. A: Lee, N. Read: Phys. Rev. B **47**, 7312 (1993)

Chapter 14

14.1 E. P. Wigner: Phys. Rev. **46**, 1002 (1934);
E. P. Wigner: Trans Faraday Soc. **34**, 678 (1938)

14.2 C. C. Grimes, G. Adams: Phys Rev. Lett. **42**, 795 (1979)

14.3 J. M. Kosterlitz, D. J. Thouless: J. Phys. C**6**, 1181 (1973);
D. J. Thouless: J. Phys. C **11**, L189 (1978)

14.4 R. Morf: Phys. Rev. Lett. **43**, 931 (1979);
D. S. Fisher, B. I. Halperin, R. Morf: Phys. Rev. B **20**, 4692 (1979);
D. S. Fisher, B. I. Halperin, P. M. Platzman: Phy. Rev. Lett. **42**, 798 (1979)
D. S. Fisher: Phys. Rev. B **26**, 5009 (1982)

14.5 D. C. Glattli, E. Y. Andrei, F. I. B. Williams: Phys. Lett. **60**, 420 (1988)

14.6 M. Imada, M. Takahashi: J. Phys. Soc. Jpn **52**, 3770 (1984)

14.7 P. Leiderer, W. Ebner, V. B. Shikin: Physica **107**, 217 (1981)

14.8 W. J. Carr, Jr.: Phys. Rev. **122**, 1437 (1961)

14.9 A. Isihara, E. W. Montroll: Proc. Natl. Acad. Sci. USA. **68**, 3111 (1971)
A. Isihara: Phys. Konden. Mater. **15**, 225 (1972)

14.10 Lynn Bonsall, A. A. Maradudin: Phys. Rev. **15**, 1959 (1979)

14.11 A. Isihara: In *Solis State Physics* Vol. 42, ed. by H. Ehrenreich, D. Turnbull (Academic, New York 1989)

14.12 R. L. Dobrushin: Th. Prob. Appl. (USSR) **9**, 646 (1964);
M. E. Fisher, D. Ruelle: J. Math. Phys. **7**, 260 (1966);
D. Ruelle: *Statistical Mechanics* (Benjamin, New York 1969);
J. L. Lebowitz, E. H. Lieb: Phys. Rev. Lett. **22**, 631 (1969);
A. Isihara: Physica **71**, 75 (1974)

14.13 A. Isihara: J. Phys. A **1**, 539 (1968)

14.14 L. L. Foldy: Phys. Rev. **124**, 649 (1961);
J. M. Stephen, Proc. Phys. Soc. (London) **79**, 994 (1962);
T. Samulski, A. Isihara: Physica **82A**, 294 (1979)

14.15 A. Lenard, F. J. Dyson: J. Math. Phys. **9**, 698 (1968);
F. J. Dyson: In *Brandeis University Summer Institute in Theo. Phys.* Vol. 1 ed. by M. Chretien, E. P. Gross, S. Deser (Gordon and Breach, New York 1969) p. 179;
A. Lenard: In *Lecture Notes in Physics*, Vol. 20, ed. by J. Ehlers, K. Hepp, H. A. Weidenmüller (Springer, Berli, Heidelberg 1973) p. 114;
E. Lieb: Rev. Mod. Phys. **48**, 553 (1976)

14.16 *The Stability of Matter, from Atoms to Stars*, ed. by W. Thirring (Springer, Berlin, Heidelberg 1991)

Appendix

A.1   S. Fujita, A. Isihara, F. W. Montroll: Bull. Cl. Sci., Acad. R. Belg. **44**, 1018
      (1958);
      A. Isihara: *Statistical Physics*, (Academic, New York 1971)
A.2   A. Isihara: Phys. Rev. **172**, 166 (1968); ibid. **178**, 412 (1969)
A.3   F. W. Montroll, J. C. Ward: Phys. Fluids **44**, 1(1958)
A.4   F. Lado: Phys. Rev. B **17**, 2827 (1978)
A.5   J. M. Caillol, D. Levesque, J. J. Weis, J. P. Hansen: J. Stat. Phys. **28**, 325
      (1982)
A.6   J. F. Springer, M. A. Pokrant, F. A. Stevens, Jr.: J. Chem. Phys. **58**, 4863
      (1973);
      K. Ng: J. Chem. Phys. **61**, 2680 (1974)
A.7   A. Isihara: Prog. Theor. Phys. Suppl. **44**, 1 (1969);
      M. Wadati, A. Isihara: Phys. Rev. B **1**, 2156 (1970)
A.8   A. Isihara, T. Toyoda: Z. Phys. B **23**, 389 (1976)
A.9   P. Hohenberg, W. Kohn: Phys. Rev. **136**, B864 (1964);
      W. Kohn, L. J. Sham: Phys. Rev. **140**, A 1133 (1965);
      N. D. Mermin: Phys. Rev. **137**, A 1441 (1965);
      R. Evans: Adv. Phys. **28**, 143 (1979)

Subject Index

Springer Series in Solid-State Sciences

Editors: M. Cardona P. Fulde K. von Klitzing H.-J. Queisser

Springer Series in Solid-State Sciences

Editors: M. Cardona   P. Fulde   K. von Klitzing   H.-J. Queisser

Springer
and the
environment

At Springer we firmly believe that an international science publisher has a special obligation to the environment, and our corporate policies consistently reflect this conviction.
We also expect our business partners – paper mills, printers, packaging manufacturers, etc. – to commit themselves to using materials and production processes that do not harm the environment. The paper in this book is made from low- or no-chlorine pulp and is acid free, in conformance with international standards for paper permanency.

Springer